Creating Wealth from Knowledge

Creating Wealth from Knowledge

Knowledge

Meeting the Innovation Challenge

Edited by

John Bessant

Tanaka Business School, Imperial College London, UK

Tim Venables

Tanaka Business School, Imperial College London, UK

Edward Elgar

Cheltenham, UK • Northampton, MA, USA

Published by
Edward Elgar Publishing Limited
Glensanda House
Montpellier Parade
Cheltenham
Glos GL50 1UA
UK

Edward Elgar Publishing, Inc.
William Pratt House
9 Dewey Court
Northampton
Massachusetts 01060
USA

A catalogue record for this book
is available from the British Library

ISBN 978 1 84720 348 9 (cased)

Printed and bound in Great Britain by MPG Books Ltd, Bodmin, Cornwall

Contents

Contributors

Richard Adams is AIM Research Fellow in Business Engagement at Cranfield School of Management, Cranfield University, UK.

John Bessant is AIM Senior Fellow and Professor of Innovation and Technology Management at Tanaka Business School, Imperial College London, UK.

Kate Bishop is an AIM Research Fellow at the Innovation and Entrepreneurship Group, Tanaka Business School, Imperial College London, UK.

Neil Burns is Professor of Manufacturing Systems in the Wolfson School of Mechanical and Manufacturing Engineering at Loughborough University, UK.

Linus Dahlander is an AIM Research Fellow at the Innovation Studies Centre, Tanaka Business School, Imperial College London, UK.

Mark Dodgson is Director of the Technology and Innovation Management Centre at the University of Queensland Business School, Australia.

Pablo D'Este is Research Fellow at SPRU – Science and Technology Policy Research Unit, University of Sussex, and AIM Research Fellow at Cranfield School of Management, Cranfield University, UK.

Simone Ferriani is Associate Professor of Management at the Management Department of the University of Bologna, Italy.

David Gann is Professor of Innovation and Technology Management and Head of Innovation and Entrepreneurship, Tanaka Business School, Imperial College London, UK.

Elizabeth Garnsey is Reader in Innovation Studies at the Centre for Technology Management, University of Cambridge, UK.

Alan Hughes is Margaret Thatcher Professor of Enterprise Studies, Judge Business School and Director, Centre for Business Research, University of Cambridge, UK.

Dennis Kehoe is Saxby Professor of e-Business at the University of Liverpool Management School, UK.

Weisheng Liu is a doctoral research student at the University of Liverpool Management School, UK.

Brian McCaul is Director of Business Development at AiMES Centre, the University of Liverpool, UK.

Roula Michaelides is AIM Research Fellow in the e-Business Division, The Management School, University of Liverpool, UK.

Tim Minshall is a lecturer in Technology Management at the University of Cambridge Institute for Manufacturing, UK.

Sue Morton is AIM Research Fellow at the Wolfson School of Mechanical and Manufacturing Engineering, Loughborough University, UK.

Andy Neely is Professor of Operations Strategy and Performance at Cranfield School of Management, UK.

Markus Perkmann is AIM Research Fellow at the Wolfson School at Loughborough University, UK.

David Probert is Reader in Technology Management at the University of Cambridge Institute for Manufacturing, UK.

Toke Reichstein is a lecturer in the Department of Industrial Economics and Strategy, Copenhagen Business School, Denmark.

Ammon Salter is a reader in Technology and Innovation Management, Tanaka Business School, Imperial College London, UK.

Hossein Sharifi is a lecturer on Operations Management and e-Business at the University of Liverpool Management School, UK.

Erik Stam is AIM Research Fellow at the Centre for Technology Management, University of Cambridge, UK and Research Fellow at the Entrepreneurship, Growth and Public Policy Group of the Max Planck Institute of Economics, Jena, Germany.

Bettina von Stamm runs the Innovation Leadership Forum and is a Visiting Professor at Middlesex University Business School, UK.

John Steen is a lecturer at the University of Queensland Business School, Australia.

Tim Venables is Strategic Research Manager, Tanaka Business School, Imperial College London, UK.

Kathryn Walsh is Director of the Electronics-enabled Products Knowledge Transfer Network and Senior Research Fellow at the Wolfson School at Loughborough University, UK.

Acknowledgements

Putting this book together involved extensive work on the part of all the contributors – but also a great deal of 'backstage' support, and we'd like to thank all those friends and colleagues who have helped bring this project to fruition. We have been fortunate to receive very active support and guidance from members of the Innovation and Productivity Grand Challenge Advisory Board under the able chairmanship of Richard Lambert, for which we are very grateful. We'd also like to thank Matthew Pitman and the team at Edward Elgar Publishing for their support and patience in helping make the book happen, and Edward Elgar himself, who made the original suggestion.

A key theme in the book is that of networking – and we have tried to practise what we preach and build an active, knowledge-sharing network among the Fellows and associated colleagues. But networks don't happen by accident, and we'd also like to express our thanks to friends and colleagues at our various institutions who have helped coordinate the many meetings and other activities that underpin the book. In particular we'd like to thank Shelley Meehan at Imperial and Esme Foster, Jacqueline Brown and Claire Fitzpatrick at the Advanced Institute for Management Research (AIM) for their continuing and patient support.

Much of the work reported here was supported by research grants under the Innovation and Productivity Grand Challenge programme, an initiative of the Engineering and Physical Sciences Research Council (EPSRC) and the Economic and Social Science Research Council (ESRC). The programme is being delivered as one of the activities within the Advanced Institute for Management Research (AIM) which is also EPSRC/ESRC supported. We would like to record our thanks to these organizations for their generous assistance.

1. Introduction

John Bessant and Tim Venables

AN OLD QUESTION . . .

Innovation is vital to meeting the national challenges of the twenty-first century. As the economist William Baumol put it: 'Under capitalism, innovative activity – which in other types of economy is fortuitous and optional – becomes mandatory, a life-and-death matter' (Baumol 2002: 1). It lies at the heart of discussion of the concept of the 'knowledge economy', that is, the theory that advanced economies are increasingly based on the production, distribution and use of knowledge, and that their future competitive advantage lies in how efficiently and effectively they are able to engage in these activities (Baumol 2002; DTI 2003; NESTA 2006).

Every year the UK spends around £21 billion of public and private sector money creating knowledge – an impressive figure but only a drop in the global ocean of R&D spending, which the OECD estimates at around £500 billion ($720 bn in 2004). Even this is likely to be an underestimate because we don't really know how much some of the major new players like India and China are spending. As we move from the eras of labour and of capital intensity, so knowledge is increasingly recognized as the next major source of international competitiveness. No enterprise – and at aggregate level, no nation – wants to see itself left behind in this race, and so we see the rise of the 'knowledge economy' as the next step in economic evolution.

Small trading nations like Denmark and Ireland see their future as depending not on owning or running large factories but in contributing the design and service elements of a globally distributed manufacturing value network. And larger players like Germany and France wrestle with the difficult choices of supporting domestic industries or following the trend towards 'outsourcing' much of the downstream manufacturing activity and concentrating on the high-value aspects. Even giant economies like the USA are facing the challenge of lower-cost manufacturing and looking to regroup around their core strengths in knowledge-based work. The result is a race to invest more and more in creating new knowledge to put the clear blue water between the different national ships in the competitive race.

Faced with such widening gaps, most advanced economies have adopted some version of the 'knowledge economy' as an 'escape route' – climbing up the value-added ladder for goods and services. For example, in 2003 the UK Department of Trade and Industry stated that the government wanted 'the UK to be a key knowledge hub in the global economy . . . In terms of business R&D and patenting we will aim to be the leading major country in Europe within ten years' (DTI 2003: 1). It went on to make a number of general recommendations, including the development of a national Technology Strategy. Across the European Union (EU) since the mid-1990s a succession of measures to increase R&D spend have been introduced, including a series of Research and Technological Development (RTD) framework programmes. In March 2000, the Lisbon Council committed the EU to the objective of becoming the 'most competitive and dynamic knowledge-based economy in the world' by 2010. To achieve this it established (in 2002) the goal of increasing its R&D expenditure to 3 per cent of GDP by 2010.

Services account for 75 per cent or more of these economies, but here too the chilly winds of international competition are being felt. With infrastructures such as the Internet, the traditional problems of local production and distribution have given way to many services being globally traded – and to the same kinds of shift towards lower-cost providers. To counter this, service industry firms are increasingly committing to knowledge-based competition, investing in R&D and putting in place structures to support a much more innovation-led kind of business.

The result is a huge push down on the accelerator pedal of knowledge creation. Any way you look at it, there's a great deal of knowledge being generated. But investing on that scale inevitably prompts the question – not least from the taxpayers and shareholders whose money is being spent – is it worth it? Do we get the maximum return for our dollar, pound or yen – does the R&D goose really lay golden eggs? We can certainly find examples of spectacular gains which follow good research – it's the rationale, for example, behind the pharmaceutical industry's continuing to spend 15–20 per cent of its turnover on fuelling the discovery cycle through frontier research. When they get it right – as with blockbuster drugs such as Zantac or Viagra – the returns run into millions of pounds per day. The relentless pursuit of progress along the pathways laid down by Moore's law in electronics (that performance doubles every 18 months) is driven by a huge research investment on the part of firms such as Intel and AMD and their extended equipment supply chains.

And public sector projects do pay off – not just via the occasional spin-off like Teflon coatings for non-stick pans, but in the deep and wide-ranging creation of opportunities – for example, much of the information and communications technology (ICT) revolution has its roots in public sector programmes, primarily around defence activities.

But although there are returns on R&D investment, the question remains whether these are sufficient to justify the huge input. Does new knowledge create wealth or social value – and does the system do it as 'efficiently' as it might? The UK, for example, is often typecast as being good at inventing and weak at capitalizing on those initial inventions. The usual suspects are wheeled out in evidence – the first computer, whose roots go back to wartime experiences at the code-breaking centre in Bletchley Park. Or the body scanner (which has become 'the photocopier of the medical sector' but whose later (and profitable) exploitation was not carried through by EMI but by GE, Siemens and Philips). But this negative view ignores innovation success stories in areas such as pharmaceuticals and aerospace – and it underlines the need for patience. Lead times between initial knowledge production and later commercialization in sectors such as these routinely exceed 10–15 years but their impact is none the less significant when they do finally come through as widely diffused innovations.

INVENTION ISN'T ENOUGH

The reality is that investing in creating knowledge *does* create value – through new products and services which can satisfy existing markets and create new ones, and through improvements in productivity brought about by improved or radically new and more effective processes. But we also need to recognize that innovation is a *process* – an extended set of activities that translate new knowledge into something of value. It isn't – despite the persistence of images such as Archimedes in his bath or cartoon characters with light bulbs flashing above their heads – simply a matter of a 'Eureka!' moment, but rather a long and painstaking process of translating the initial idea into something useful – and used.

The UK DTI's definition of innovation as 'the successful application of new ideas' sums this up quite well. Most people would accept that there's more to innovation than the invention stage – for example: 'Industrial innovation includes the technical, design, manufacturing, management and commercial activities involved in the marketing of a new (or improved) product or the first commercial use of a new (or improved) process or equipment' (Freeman and Soete 1997: 3), or 'Innovation does not necessarily imply the commercialization of only a major advance in the technological state of the art (a radical innovation) but it includes also the utilization of even small-scale changes in technological know-how (an improvement or incremental innovation) . . .' (Rothwell and Gardiner 1985: 168).

Moreover, innovation involves two different strands being woven together – a 'supply' strand of knowledge about possible means and another 'demand' strand of knowledge about needs. Innovation results from the intertwining of these two – as Chris Freeman, one of the pioneer economists who worked in industrial innovation put it, 'necessity may be the Mother of invention but procreation still needs a partner!'

That takes us to the core of the problem – getting value from knowledge depends on understanding and organizing this innovation *process*. And whether at the firm or the national level, we need to recognize that this is a complex *system* of interacting players – innovation is emphatically not a solo act. If we wish to increase the effectiveness of that system, we should do more than simply call for more money to be thrown in at the start, and instead try to understand and improve the workings of the system itself. What are the mental models that underpin our thinking about how innovation works? Do we know who the relevant actors are and what 'good practice' might be in terms of getting them to work better together? Are some innovation systems – whether across a particular sector, in a region or around a major transnational firm – more effective than others – and if so, why?

These are not trivial questions – take the idea of mental models as an example. If we believe innovation works in a particular way, then we will organize and structure the system around that model and exclude alternative views. But this carries the risk that we work with simplistic or inappropriate models and create less than effective systems as a result. If we believe that innovation is, like the cartoon characters with the light bulb flashing on at the 'Eureka!' moment, then we'll build innovation systems that are great at invention. But coming up with new ideas is no guarantee of successful innovation – as many originators of good ideas know to their cost. (For example, Elias Howe invented the sewing machine, but it was Singer who actually made the money from promoting the diffusion of the idea.) If we believe innovation is a linear process, a logical sequence of activity in which the R&D front end creates a knowledge push and innovations happen as a result of this momentum, then we'll create structures and mechanisms that assume ideas will be taken up – despite the clear evidence that the world does *not* beat a path to the doors of every mousetrap salesman with a novel idea.

In similar vein, if we believe that the main contribution of universities to innovation is to incubate entrepreneurial ideas which then spin off, then we'll invest in structures and mechanisms to support this activity. Yet the evidence is that – while the model works spectacularly well for a few high-tech firms in sectors such as electronics and biotechnology – the reality is that very few firms are born out of universities, even in the USA, where this model is widely thought to work best (Hughes 2007).

The issue is not about debunking models that are too simplistic, but rather to develop richer, more accurate understanding of where and how innovation works – and then to use that as a template for policy and practice. We need to experiment around emergent good practices and to evolve more appropriate local, regional and national innovation *systems*. Indeed, in such a globalized and technically interconnected world we may need to look at developing an international or globally networked innovation system.

So the challenge in getting value from knowledge involves rethinking our innovation systems and making sure that their design is fit for purpose and their operation is effective. The trouble is that we are not dealing with simple cuckoo-clock mechanisms but with complex, dynamic and multi-player systems. As if that weren't hard enough, we also need to recognize that the innovation game itself is changing and being played by a different set of emergent rules in the twenty-first century.

NEW RULES OF THE GAME

Whilst it is clear that the 'exploitation/exploration' question is not new and has led to the development of firm-specific approaches to resolving the tension between them, we would argue that this old question is now challenged by a significantly different context within which firms operate. Changes along several core environmental dimensions mean that the incidence of discontinuities is likely to rise – for example in response to a massive increase in the rate of knowledge production and the consequent increase in the potential for technology-linked instabilities. But there is also a higher level of interactivity among these environmental elements – a complexity that leads to unpredictable emergence. For example, the rapidly growing field of VoIP (voice over Internet protocol) communications is not developing along established trajectories towards a well-defined end-point. Instead it is a process of *emergence*. The broad parameters are visible – the rise of demand for global communication, increasing availability of broadband, multiple peer-to-peer networking models, growing technological literacy among users – and the stakes are high, both for established fixed-line players (who have much to lose) and new entrants (such as Skype, recently bought by eBay for $2.6 bn). The dominant design isn't visible yet – instead there is a rich fermenting soup of technological possibilities, business models and potential players from which it will gradually emerge.

Table 1.1 summarizes some of the key changes in the context within which search behaviour is located. Arguably these require firms to pay more

Table 1.1   Changing context for search behaviour

Context change	Indicative examples
Acceleration of knowledge production	OECD estimates that close to $1 trillion is spent each year (public and private sector) in creating new knowledge – and hence extending the frontier along which 'breakthrough' technological developments may happen
Global distribution of knowledge production	Knowledge production is increasingly involving new players, especially in emerging market fields such as the BRIC nations (Brazil, Russia, India and China) – so the need for search routines to cover a much wider search space increases
Market fragmentation	Globalization has massively increased the range of markets and segments so that these are now widely dispersed and locally varied – putting pressure on search routines to cover much more territory, often far from 'traditional' experiences – such as the 'bottom of the pyramid' conditions in many emerging markets (Prahalad 2006)
Market virtualization	Increasing use of the Internet as marketing channel means different approaches need to be developed. At the same time emergence of large-scale social networks in cyberspace poses challenges in market research approaches – for example, My Space currently has over 80 million subscribers. Further challenges arise in the emergence of parallel world communities as a research opportunity – for example, Second Life now has over 6 million 'residents'
Rise of active users	Although von Hippel long ago identified the active role that users can pay in innovation, there has been an acceleration in the ways in which this is now taking place – for example, the growth of Linux has been a user-led open community Development (Von Hippel 2005). In sectors such as media the line between consumers and creators is increasingly blurred – for example, You Tube has around 100 million videos viewed each day but also has over 70 000 new videos uploaded every day from its user base

Table 1.1   (continued)

Context change	Indicative examples
Development of technological and social infrastructure	Increasing linkages enabled by ICTs around the Internet and broadband have enabled and reinforced alternative social networking possibilities. At the same time the increasing availability of simulation and prototyping tools has reduced the separation between users and producers (Schrage 2000; Gann 2004)

attention to the limits of their current models and add an element of urgency to the need to extend and develop new routines.

TOWARDS A 'FIFTH-GENERATION' MODEL OF INNOVATION

The question of mental models of innovation is a powerful one, since what we think about shapes what we pay attention to and what we create as an innovation system. One of the key thinkers on innovation for many years was Roy Rothwell, professor at SPRU who worked on Project SAPPHO, an influential piece of research on how organizations manage the innovation process (Rothwell 1992). In this important paper he mapped out the changing landscape of our thinking around innovation models, grouping them into five generations – essentially dominant modes of thinking about the process. Over time we move from simplistic understanding to a richer and more complex one – reflected in the innovation systems we built and managed. Table 1.2 illustrates this evolution, and Figure 1.1 shows the process.

Arguably Rothwell's predicted 'fifth-generation' model reflects the emergent picture in the twenty-first century and gives us some clues as to the design rules for our complex innovation system. Elements of this will include:

From	**Towards**
Knowledge creation emphasis	Knowledge flows and their management
Knowledge ownership and use	Knowledge trading and use
Closed innovation models	Open innovation models
Linear models of knowledge flow within innovation	'Spaghetti' models with inside and outside connections
Passive users as consumers	Active users, co-creators

Table 1.2   Rothwell's five generations of innovation models

Generation	Key features
First and second	Simple linear models – need pull, technology push
Third	Coupling model, recognizing interaction between different elements and feedback loops between them
Fourth	Parallel model, integration within the firm, upstream with key suppliers and downstream with demanding and active customers, emphasis on linkages and alliances
Fifth	Systems integration and extensive networking, flexible and customized response, continuous innovation

Figure 1.1   Fifth-generation innovation process

WEALTH FROM KNOWLEDGE

An early and influential innovation study back in the 1970s took this heading as its title – and it remains an important challenge (Langrish et al. 1972). Innovation can create value – not just financial wealth but also social value – but only if the system is well tuned and appropriate. So what is an appropriate model, deploying the principles outlined in the preceding section? And how far away from this model are we? For that matter, who are the 'we' who might share a concern for exploring these issues? There are certainly a number of interested players:

- Entrepreneurs trying to create new start-up businesses that deploy knowledge in new ways

- Managers in established businesses trying to improve productivity through the effective application of new knowledge, and who are also concerned with the 're-invention' of the business through a form of corporate entrepreneurship
- R&D managers in the public and private sectors trying to improve the take-up of knowledge created in their groups
- Knowledge-intensive business service providers trying to build a business through enabling connections and flows within the knowledge system
- Government policy agents at local, regional and national level charged with improving the efficiency and effectiveness of the innovation system
- Supply chain 'owners' concerned to upgrade the system-level efficiency and effectiveness of their networks through innovation.

This book tries to bring together some of the thinking and research around the questions we have outlined above. It draws particularly on a large-scale research programme funded by the UK's Engineering and Physical Sciences Research Council (EPSRC) and involving five universities (Cambridge, Cranfield, Liverpool, Loughborough and Imperial College London), together with the Advanced Institute for Management Research (AIM). The 'Grand Challenge' this group is addressing is that of innovation and productivity in the UK context, and we are trying to understand the system-level questions and how they might be addressed by focusing on four core themes:

- What's going on at the moment? What is the context within which innovation happens in the UK system – who are the actors, what are the linkages, where are the strengths and weaknesses?
- How do new firms form on the basis of knowledge and its deployment? How can such entrepreneurial behaviour be understood and best enabled?
- How do established firms access and use knowledge to improve their current activities and generate new directions?
- What is the enabling infrastructure, both technological and organizational? What opportunities and challenges exist in mobilizing this infrastructure to underpin an effective innovation system?

Within each of these areas we have a number of direct research activities going on and we are – in keeping with a fifth-generation approach – trying to establish links and connections with other researchers and their knowledge to help populate the emerging map.

Although we focus primarily on the UK experience, we would argue that · this is a problem of concern to a much wider audience and that the challenges posed are broadly similar. The book represents some of the early map-making and also highlights key research questions that need further exploration. It is organized as follows.

Part I looks at the context question. What is the current pattern in terms of knowledge production and does this translate into an enhanced competitive position? In their chapter on the UK Science Census Pablo D'Este and Andy Neely look at some of the statistical evidence and draw out some comforting but also some disturbing trends. Perhaps the most significant theme that emerges is that the game itself may be shifting away from concerns about relative amounts of expenditure and success in different R&D domains towards a global game where the ability to trade with other knowledge-producing partners becomes critical.

Linus Dahlander and David Gann look at the key question of 'open innovation' and explore the extent to which this is a new phenomenon as distinct from a rediscovery of an old and important innovation principle. They look at some of the forces shaping firms' search behaviour and the growing need for taking a more 'open' and proactive perspective on exploration activity.

Alan Hughes looks again at the evidence for the effectiveness of one of the persuasive models for capitalizing on our knowledge production investment – the university spin-off. Although this is a popular view, he suggests that the reality is that relatively little direct productivity growth arises from such activities and that simply pumping in more money to either knowledge creation or to fund the spin-off infrastructure may be misdirected. He shows, for example, that US productivity growth owes much more to Wal-Mart than to high-tech firms and their growth (although there have been some spectacular exceptions to this). Instead he argues that we might need to look at alternative views of universities as knowledge centres that create 'public space' within which a diverse set of network-based knowledge flows can happen.

Although much of the book focuses on UK experience, the chapter by Mark Dodgson and John Steen offers a different perspective – that of Australia. Significantly, although the economy has a very different shape, dominated by primary industries and operating in a different geographical zone, the challenges have a marked similarity. Future economic growth is likely to require a much higher level of knowledge intensity, based on increasing the rate of innovation. Despite its huge physical size, the economy is relatively small and cannot sustain a huge R&D investment; instead future growth will need to depend on a policy much more aligned with the principles of 'open innovation', trading knowledge in and out and using it to enhance added value of primary resource exports. The chapter looks particularly at *how* open innovation can be configured, focusing on the issue of broking and intermediary services which can facilitate connections across a national and out into an international knowledge market.

Whilst innovation happens at the level of the individual enterprise, the rapidly rising stakes in the international competitiveness game make a persuasive case for looking at what governments might do to help. Whilst heavy hands-on intervention is often clumsy and ineffective – and increasingly outlawed under the emerging world trade regime – there are many ways in which state and regional development agencies can help improve the conditions within which innovation can thrive. But what are the effective modes of intervention and how can they be fine-tuned to support a vibrant open innovation system? In his chapter on UK government policy Tim Minshall looks at the historical evolution of innovation support and explores new directions for the future.

In Part II we shift our attention to the enterprise level and how knowledge can create and grow new business opportunities, for both established firms and new start-ups. Here the question of entrepreneurial firms and how they are formed and grow assumes particular relevance – and one of the places where this has happened in the context of a region famous for knowledge production *and* exploitation is Cambridge. In their chapter, Erik Stam and Elizabeth Garnsey look in detail at this experience, reviewing the literature on knowledge as a source of entrepreneurial opportunities, with evidence at both the regional and organizational levels. In addition they explore the causal mechanisms of new firm growth, discussing longitudinal case study research on problem solving and competence creation in such firms.

The other side of this coin concerns the question of how established firms can continue to innovate through the use of new knowledge. What stops them from doing so and what might facilitate greater use of such knowledge? What channels are more or less effective in enabling knowledge flow? When should they use established networks and when does it make sense to look for new and different connections to enable wider exploration of new options as well as more efficient exploitation of established fields?

In Part III these themes are explored in the chapters by Simone Ferriani, Elizabeth Garnsey and David Probert, and by John Bessant and Bettina von Stamm, which look at the phenomenon of radical or 'discontinuous' innovation. Under certain circumstances firms are forced to extend their repertoire of search strategies in order to deal with significant shifts in their operating environments – emergence of new market constituencies, of radically different technologies, of shifts in the regulatory context, etc. How do they do this – and what implications does this shift in search behaviour have for the design and operation of effective innovation models that emphasize knowledge flows?

At the other end of the spectrum Richard Adams and John Bessant explore the question of slow take-up of well-established knowledge. Productivity gains can emerge as much from the faster adoption of proven knowledge among the mainstream population of firms as from deploying breakthrough innovations at the frontier. Using the example of advanced manufacturing techniques, they use the lens of diffusion theory to try and throw some light on why firms are slow and often resistant adopters – and what might be done to reverse this trend. Resistance to change is also a theme take up by Sue Morton and Neil Burns in their chapter on the sources of resistance at a psychological and organizational level.

Part IV looks at the ways in which connections between knowledge-based institutions like universities and research institutes can be made. What types of infrastructures can help – or hinder – the knowledge flows that underpin effective innovation? What role do factors such as geographical proximity or technological channels (including the rapidly increasing Web-enabled options) play in facilitating rich and multidirectional flows? How can social systems and networks play a role in the innovation process? And what is – or could be – the role of intermediaries and brokers (individuals and agencies) in a high-performing innovation system?

Much depends on the nature of the relationship established between different actors in the knowledge flow process. Part of the difficulty with earlier models of innovation has been the use of terms such as 'technology transfer', which imply a somewhat serial and static process whereas – as Markus Perkmann and Kathryn Walsh argue in their chapter – the reality is that successful innovation results form strong *relationships* rather than short-term contracting behaviours.

At the heart of the knowledge flow question is, of course, the individual motivation to engage in this kind of activity, and in their chapter Pablo D'Este and Andy Neely look more closely at this. In particular they examine two key issues – *who* in the academic world interacts with industry and *why* do they do so? This assumes considerable relevance when we consider that, over the past two decades, there has been a significant increase in the perception that knowledge transfer/interaction is an important mission of universities alongside their more traditional roles in teaching and fundamental research.

In their chapter on geographical proximity Kate Bishop, Toke Reichstein and Ammon Salter look at the evidence for the argument that being close to universities helps facilitate innovation. Their conclusions are that, despite the significant advances in communications technology, innovation still depends heavily on proximity factors. But they also suggest that the impact of geographic proximity on university–industry links is not always positive, stressing the importance of agency in making the most of physical location.

This theme is picked up in the chapter by Hossein Sharifi, Weisheng Liu, Brian McCaul and Dennis Kehoe, who look at the role and experience of technology transfer offices (TTOs) in the UK. Recent years have seen significant expansion in this kind of agency – in Britain alone 71 new agencies were established during the 1990s according to the Lambert Review. And these are not small operations – the average staff level in Europe is eight people in a TTO. But throwing resources at the problem does not necessarily solve it, and the chapter looks at where, when and how effective TTOs work – and the messages for developing agencies of this kind to contribute to improved knowledge flow in the wider innovation system.

One of the key developments in the twenty-first-century innovation model is, of course, the increasingly significant role that technology can play as an enabler of knowledge flow. Developments, particularly in information and communications technology (ICT) allow for massively greater interaction on a global scale – indeed the Internet itself was originally developed by Tim Berners-Lee and colleagues as an aid to their collaboration and knowledge sharing. In their chapter on the enabling technological infrastructures Roula Michaelides and Dennis Kehoe look in depth at the opportunities but also the challenges posed by such developments and their implications for improving the innovation system.

A GRAND CHALLENGE?

There is little argument about the importance of innovation in the early twenty-first century – whether at the level of the firm, in universities and research institutes or among policy actors at regional and national level. The challenge is to make it happen more effectively – getting more leverage from existing investments as well as increasing the overall rate and scale of knowledge production.

Our argument in this book is for a rethink by all the players around their mental models – and operational structures resulting from those models – of innovation. Are they really using a 'fifth-generation' model – highly networked, technologically and socially enabled and predominantly 'open' in character, or are they still trying to force-fit the 'spaghetti' reality of knowledge flows into outdated and limited linear models from an earlier generation?

This isn't an easy transition, but we hope that some of the ideas presented in this book give some indications of the new directions that such fifth-generation thinking might take.

REFERENCES

Baumol, W. (2002), *The Free-Market Innovation Machine: Analyzing the Growth Miracle of Capitalism*, Princeton, NJ: Princeton University Press.

DTI (2003), *Competing in the Global Economy: The Innovation Challenge*, London: Department of Trade and Industry.

Freeman, C. and Soete, L. (1997), *The Economics of Industrial Innovation*, Cambridge, MA: MIT Press.

Gann, D. (2004), 'Think, play, do: the business of innovation', Inaugural Lecture, Imperial College London.

Hughes, A. (2007), 'University–industry links and U.K. science and innovation policy', in Y. Shahid and K. Nabeshima (eds), *How Universities Promote Economic Growth*, Washington, DC: The World Bank.

Langrish, J., Gibbons, M., Evans, W. and Jevons, F. (1972), *Wealth from Knowledge*, London: Macmillan.

NESTA (2006), *The Innovation Gap*, London: National Endowment for Science, Technology and the Arts (NESTA).

Prahalad, C.K. (2006), *The Fortune at the Bottom of the Pyramid: Eradicating Poverty through Profits*, Upper Saddle River, NJ: Wharton School Publishing.

Rothwell, R. (1992), 'Successful industrial innovation: critical success factors for the 1990s', *R&D Management*, **22**(3): 221–39.

Rothwell, R. and Gardiner, P. (1985), 'Invention, innovation, re-innovation and the role of the user', *Technovation*, **3**: 167–86.

Schrage, M. (2000), *Serious Play: How the World's Best Companies Simulate to Innovate*, Boston, MA: Harvard Business School Press.

Von Hippel, E. (2005), *The Democratization of Innovation*, Cambridge, MA: MIT Press.

PART I

Context

2. Science and technology in the UK[1]

Pablo D'Este and Andy Neely

1. INTRODUCTION

Innovation is increasingly about managing *flows* of knowledge across complex – and often global – networks.[2] The aim of this chapter is to provide a census of where the UK's 'innovation from knowledge' (IfK) system stands in light of publically available data. Ongoing research will supplement these data with new data that will be incorporated into subsequent science base census publications to take account of global developments – for example the rise of China and India as economic powerhouses. At this stage, however, we are concerned with mapping out the UK's current position and considering the implications of this for the policy and practitioner communities.

2. THE SCIENCE AND TECHNOLOGY PERFORMANCE FRAMEWORK

This section focuses on the description of the indicators used to conduct the cross-country comparison, complemented with a brief discussion of the strengths and weaknesses of the indicators used, and a description of the data sources employed.

Before addressing the main focus of this section, however, it is important to examine three issues underlying the indicators used in this chapter: (1) why a country focus?; (2) why yet another science and technology study?; and (3) does the chapter implicitly support the linear model of innovation?

In an Increasingly Globalized Scenario, Why a Country-level Focus?

While open research and innovation has led to an increasingly internationalized scientific and innovation system, there is no such thing as a 'free lunch' for innovative firms or countries. Countries need to sustain and nurture their science and technology bases, partly to enable the country to connect to the scientific and technological advances occurring elsewhere, and partly because a healthy science base makes the country an attractive place for international players to be located.

In brief, the advances in scientific understanding and technological know-how are unforeseeable, and knowledge spillovers are unlikely to occur in a context of underinvestment, since knowledge can only be absorbed, applied and transmitted by educated minds. For these reasons, countries seek to retain and nurture their science and technology bases in order to encourage competitiveness and increase wealth.

Therefore we think it justified to take the country level as a unit of analysis, and to examine what are the strengths and weaknesses of the UK's science and technology system compared to other countries.

Do We Need yet another Science and Technology Study?

All major international organizations, such as the EU and the OECD, periodically produce an exhaustive account on how countries stand in terms of their science and technology indicators.[3] In addition to this, most countries produce their own periodical reviews on this matter (e.g. Office of Science and Technology (OST) in the UK; the National Science Foundation (NSF) in the USA). The intention of this chapter is neither to duplicate the information provided elsewhere nor to introduce novel indicators that have yet to be contemplated in the existing literature. Rather, the objective of this chapter is to use the extensive data collected mainly by the EU, NSF, OECD and OST to provide a distinct picture on how the UK stands internationally. Therefore, while the present chapter necessary overlaps with existing studies in the use of available data, it goes beyond existing reports in a number of respects. First, by taking a UK focus, the chapter looks specifically at how the UK stands when compared to three different blocks of countries: major industrialized countries – France, Germany, the USA and Japan; small, high-tech countries – Denmark, Finland, Sweden and Switzerland; and newcomer countries – Brazil, China, India, Singapore, South Korea and Taiwan. Second, the chapter examines the long-term trends in the UK by exploring a range of indicators. We argue that an accurate assessment about how well the UK is doing in terms of science and technology performance, and how well it can be expected to perform in the future, can only be achieved by identifying the long-term trends for the country (relative to comparator countries) along a range of indicators, rather than via snapshots of recent performance. Finally, in addition to the prior two points, the chapter examines the UK's performance across a range of scientific disciplines, in order to get a balanced picture regarding the country's areas of strength and weakness.

Does this Chapter Implicitly Support the Linear Model?

In developing this chapter we have received valuable comments and questions from many colleagues. One recurring theme is the question of whether this chapter assumes a linear model of innovation. For years the dominant paradigm was that innovation flowed from scientific progress. Indeed many policy documents still seem to cling to this notion. However, in the latter decades of the twentieth century, the linear model was challenged by both the science and the policy communities. In essence their argument was that the pathways from scientific to technological advance are not uniform, that advances in technology often inspire science, and that many technological developments and innovations have very little, if any, direct reliance on science (Rosenberg 1991; Stokes 1997).

Whilst we would concur with these arguments, for the sake of analytical clarity we have chosen to present the data in this chapter using a framework that distinguishes between investment in knowledge; scientific performance; and technological performance. Although analytically cleaner, this framework clearly oversimplifies reality and hence at various points in the text we will refer to the more complex issues of innovation and knowledge flows.

Indicators Used, Cautions Required and Sources of Data

In line with the comments made above, this subsection describes the indicators used in the chapter, the sources from which the data have been drawn, and the strengths and weaknesses of these indicators.

The indicators used in this chapter address the three main areas of a science and technology system: investment in knowledge; scientific performance; and technological performance. Section 3 examines the UK's investment in knowledge (relative to comparator countries), making use of the standard statistics of research and development (R&D) expenditures, both overall and broken down by sector of performance of R&D (i.e. business, government and higher education). Section 4 examines the UK's performance in producing knowledge by using indicators based on scientific publications and citations. Finally, Section 5 examines the UK's innovative capacity by looking at patent records.

Section 3 examines and compares the expenditures on R&D in the UK and the three blocks of comparator countries, focusing on trends in R&D expenditures since the early 1980s. As an overall measure of R&D investment, we use the 'gross domestic expenditure on R&D' (GERD), which is the total intramural[4] expenditure on R&D performed on the national territory during a given period. GERD includes R&D performed within a country, including R&D funded from abroad, but excludes payments for R&D performed abroad.[5] The main source for data on R&D expenditures is OECD – *Main Science and Technology Indicators*.[6]

GERD is constructed by adding together the intramural expenditures of four sectors: business enterprise, government, private non-profit and higher education. Since the private non-profit sector contributes only marginally, as an R&D-performing sector, to GERD (less than 5 per cent of total GERD in most countries), we are not explicitly considering this sector in this study. Therefore GERD is mainly composed of BERD (business expenditure on R&D), GOVERD (government expenditure on R&D) and HERD (higher education expenditure on R&D).

Business expenditure on R&D (BERD) refers to the total amount of R&D performed in the business enterprise sector, which includes all firms, organizations and institutions whose primary activity is the market production of goods and services for sale to the general public. This includes public and private enterprises, and non-profit institutions such as research institutes, hospitals or chambers of commerce (see OECD 2002).[7]

Government expenditure on R&D (GOVERD) refers to the total amount of R&D performed by all bodies, departments and establishments of government that engage in a wide range of activities such as: administration, defence and regulation of public order, health, education (with the exception of those administered by the higher education sector), cultural and recreational services, etc.

Higher education expenditure on R&D (HERD) refers to the total amount of R&D performed by all universities, colleges of technology and other institutions of post-secondary education, whatever their source of finance or legal status.

Although OECD member countries commit important resources to setting up surveys based on the *Frascati Manual* (OECD 2002) for collecting data on R&D expenditures, a number of pitfalls remain in regard to internationally comparable R&D input data. Two limitations are of particular importance. First, the lack of systematic data as a consequence of countries not reporting timely data (or reporting data that are not comparable), which is particularly noticeable in the case of R&D personnel data. Second, the OECD's methodological framework for classifying R&D activities in scientific sub-disciplines is not sufficiently detailed, which makes combining input data on R&D from OECD and publication data at disciplinary levels extremely problematic, and the outcomes at best only indicative (Luwel 2004).

Finally, in Section 3 we use both R&D intensity (this is R&D expenditure as a proportion of GDP) and R&D expenditure (at constant prices) per capita. This is largely done to take advantage of the complementary information that each indicator provides. While R&D intensity is a useful measure to indicate how much countries invest in R&D in relation to the value of their total production, and facilitates the comparison of countries of different sizes, this indicator may lead to a distorted picture caused by the different rate of growth of GDP across countries. For this reason, we complement that indicator with R&D expenditures per capita.

It is also important to acknowledge that both indicators may differ across countries as a consequence of a different industrial structure, for instance as a consequence of a different composition of R&D-intensive industries in the business sector (NESTA 2006). However, Griffith and Harrison (2003) found that, when comparing manufacturing sectors, most of the difference in R&D intensity between UK and major industrialized countries is accounted for by differences within industries (as opposed to shifts in the composition of industries).

Section 4 examines and compares the production of scientific publications as an indicator of a country's scientific activity (i.e. the country's share of world publications and the growth of that share). Since the number of publications *per se* does not reveal much about quality or usefulness, citations are used as an indicator of the usefulness and importance of publications.

The data on publications and citations used in this research come from NSF (2006).[8] The NSF counts all science and engineering articles, notes and reviews (excluding letters to the editor, news pieces and editorials) published in the scientific and technical journals tracked by Thompson ISI in the Science Citation Index (SCI) and the Social Science Citation Index (SSCI). Articles are attributed to countries by the author's institutional affiliation at the time of the publication. The counting of articles and citations is based on the fractional count method, where the credit for an article with authors from more than one institution is divided among the collaborating institutions based on the proportion of their participating institutions.

Section 5 examines and compares the UK's innovative capacity by looking at patent records in the UK and the three blocks of comparator countries, focusing on trends in patenting since the early 1980s.

The creation and use of technological knowledge are key drivers of an economy's international competitiveness due to their impact on improving product quality and process efficiency. Patent records are a codified measure of a country's capacity to generate technological knowledge since they capture an outcome of technologically oriented inventive activity, and therefore patents have been extensively used to measure technological performance.

However, some cautions with the use of patents are in order. On the one hand, not all innovative activities result in a patent, since in many sectors patenting is not the preferred mechanism to protect inventions from being copied by competitors. For instance, most innovation and inventions in the service sector are not captured by patents. On the other hand, not all patents are exploited economically – patents do not measure innovation *per se*, but rather the existence of knowledge which has the potential for innovation – and even when they are economically exploited, only a very small proportion of patents leads to a successful innovation in the marketplace.

Notwithstanding all these cautions, patents have traditionally represented an important source of data to measure technological performance, and allow us to trace technology dynamics in detail and over long time periods (EC-RSTI 2003).

Following OECD (2003) methodological notes, patent counts are presented here according to the country(ies) of residence of the inventor(s), thus giving a measure of technological innovativeness of researchers and laboratories located in a country (as opposed to assigning the patent to the country of residence of the applicant, which could be a holding organization whose location might differ from that where the research was actually conducted).

Finally, three sets of patent indicators are analysed here: patents granted by the US Patent and Trademark Office (USPTO), triadic patents (i.e. the set of patented inventions for which protection has been sought at all three major patent offices: the European Patent Office, the USPTO and the Japanese Patent Office), and patents applied for at the European Patent Office (EPO). The advantage of triadic patents is that they can eliminate the 'home bias' effect (i.e. US inventors may have a dominance in the US patent system because it is their home market). Triadic patents may also be associated with patents of a higher expected commercial value, since it is costly to file through three patent systems (however, it is also likely that they tend to reflect the patenting activity of larger companies who seek, and can afford, broader international protection) (EC-RSTI 2003).

3. INVESTMENT IN RESEARCH AND DEVELOPMENT

Overall Picture

We first examine the extent to which there are significant differences in terms of the levels of investment in research and development (R&D) across countries. As an overall measure of R&D investment, we use the 'gross domestic expenditure on R&D' (GERD).[9]

As Figure 2.1 shows, we observe that there are significant differences in terms of inputs invested in R&D between the UK and its major industrial competitors. In the 20 years to 2003, the UK has invested, on average, 2 per

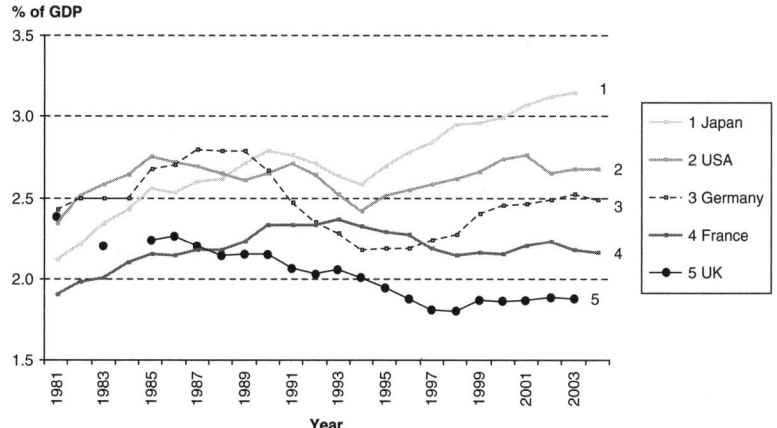

Note: Gaps in lines reflect periods for which no data was available.

Source: OECD-MSTI.

Figure 2.1   GERD as a percentage of GDP: UK versus major comparators

cent of annual GDP on R&D. This compares with 2.6 per cent for the USA
and 2.8 per cent for Japan. Even more striking, the UK investment as a proportion of GDP has fallen over this 20-year period, from 2.2 per cent in
1983 to 1.9 per cent in 2003, while the USA has remained stable in its commitments to R&D (from 2.6 per cent in 1983 to 2.7 per cent in 2003) and
Japan has increased its investment (from 2.3 per cent to 3.2 per cent over
the same period). Given these trends, it is difficult to see how the UK can
conform either to the UK's government target of 2.5 per cent of GDP by
2014,[10] or the EU target of 3 per cent by 2010.

In short, the gap between the UK and its four major competitors has
grown in terms of R&D investment to GDP. That picture does not change
much when we compare the UK with a group of high-tech, small countries – Denmark, Finland, Sweden and Switzerland. As shown in Figure
2.2, while the four countries considered had similar or lower levels of
GERD per GDP to the UK in the early 1980s, all four countries have experienced a significant growth in their investments in GERD. By 2003 they
were between 0.7 (compared to Denmark) and 2.1 (compared to Sweden)
percentage points higher than in the UK.

In the case of the newcomer countries, GERD to GDP ratios are only
available for China, Singapore and South Korea. The investments made by
both Singapore and South Korea have been steadily increasing. In the case
of South Korea, GERD to GDP was similar to the UK's in the early 1990s,

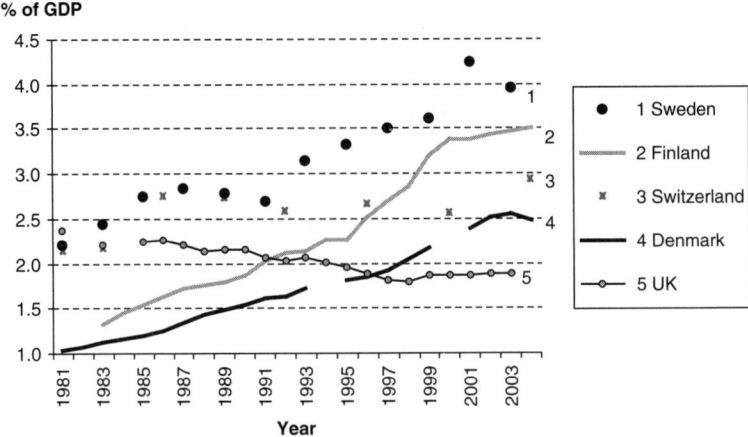

Note: Gaps in lines reflect periods for which no data was available.

Source: As for Figure 2.1.

Figure 2.2   GERD as a percentage of GDP: UK versus small, high-tech countries

but the ratio for South Korea was 2.6 per cent by 2003. Singapore caught up with the UK by 1998 and jumped ahead of the UK in 2000, a position that it has maintained. China's investment is still below the UK's, but the gap between the two countries has been progressively reduced. China's GERD per GDP was 29 per cent of the UK figure in 1995, and 60 per cent of the UK level by 2003 (see Figure 2.3).

Breakdown by R&D performing sector
The data already presented illustrate that the UK's GERD as a percentage of GDP is not only lower than most of its comparator countries, but has also declined relative to them. Why is this the case? By examining in more detail the three main constituent elements to the gross expenditures on R&D, we gain a better understanding on the relative efforts on R&D expenditures.

Figures 2.4, 2.5 and 2.6 show the UK's investment in BERD, GOVERD and HERD respectively, both in terms of percentage to GDP and in terms of the absolute level of investment relative to country population. From these figures it can be deduced that the primary areas of concern underlying the UK's underinvestment in R&D are the BERD and GOVERD components.

As shown in Figure 2.4(a), UK business expenditure on R&D relative to GDP has been decreasing (reaching a level of 1.2 per cent by 2003), while

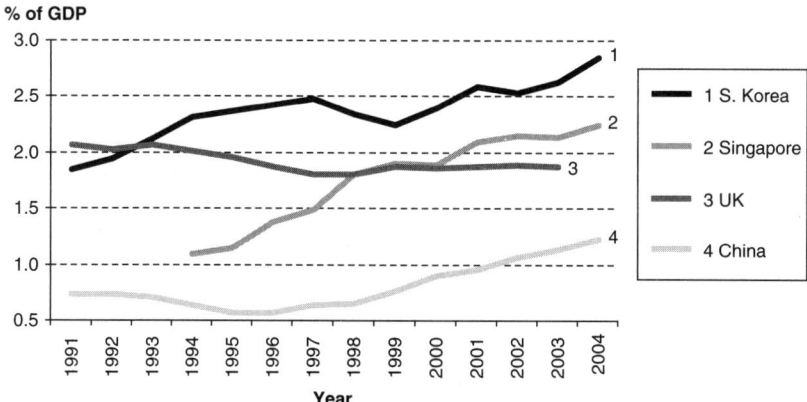

% of GDP

Note: Gaps in lines reflect periods for which no data was available.

Source: As for Figure 2.1.

Figure 2.3   *GERD as a percentage of GDP: UK versus newcomer countries*

the investments made by its major comparator countries have been increasing (in 2003 Germany was 0.5 percentage points above the UK, while the USA was 0.6 percentage points above the UK, and Japan 1.12). These differences are not a result of higher GDP growth rates in the UK relative to comparator countries, but a consequence of a lower volume of investment in R&D (relative to country size). When we look at the absolute levels of investment per capita (at constant prices) (Figure 2.4(b)), the annual growth rate in BERD per capita between 1981 and 2003 was 1.5 per cent for the UK, while the USA had an annual growth rate of 2.7 per cent and Japan had an annual growth rate of 4.5 per cent.

A similar story unfolds when looking at the small, high-tech countries. Since 1997, BERD as a percentage of GDP has been increasingly higher for the four high-tech countries compared to the UK figures. By 2003, the R&D intensity of the business sector ranged from 0.5 percentage points higher (in the case of Denmark), to 1.7 percentage points higher (in the case of Sweden). Moreover, the annual growth rate of BERD per capita (in constant prices) has been higher than 5 per cent for almost all these countries (with the exception of Switzerland) for the period 1991–2003.

When examining the figures of BERD as a percentage of GDP in emerging economies, the picture is mixed. South Korea is above the UK's levels (having reached a 2 per cent of GDP by 2003, 0.8 percentage points above the UK); Singapore has overtaken the UK's levels (since 2001 Singapore's

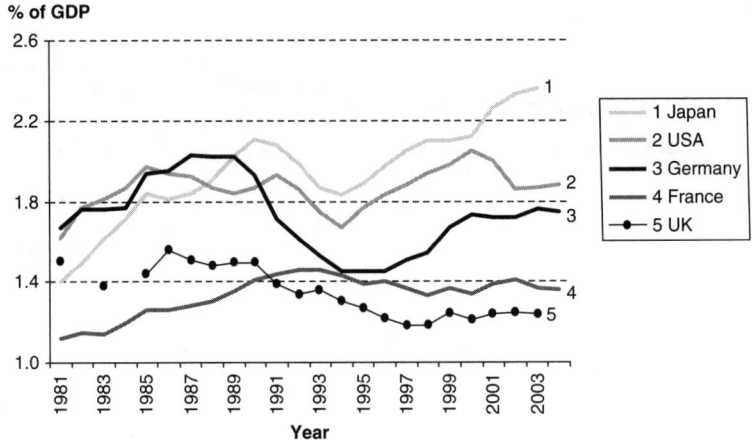

Note: Gaps in lines reflect periods for which no data was available.

Source: As for Figure 2.1.

Figure 2.4(a)   Business expenditure on R&D to GDP: UK versus major comparators

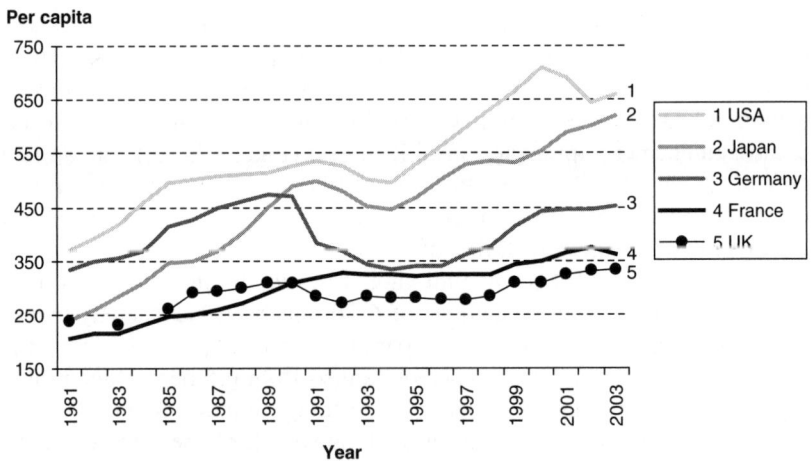

Note: Gaps in lines reflect periods for which no data was available.

Source: As for Figure 2.1.

Figure 2.4(b)   Business expenditure on R&D per capita (in US$ 2000, PPP adjusted): UK versus major comparators

business has invested 1.3 per cent of GDP in R&D, and is moving ahead of the UK); while China has been steadily catching up. China has increased its BERD to GDP from a 0.3 per cent in 1991 (1.1 percentage points below the UK) to a 0.7 per cent in 2003 (0.5 percentage points below the UK).

UK GOVERD also displays a decreasing trend over the same time period. By 2003 the UK was the country with the lowest percentage of GOVERD to GDP (0.18 per cent). This compares to 0.33 per cent for the USA and 0.43 per cent for France. Moreover, in contrast with BERD, in the case of GOVERD, there is also a decreasing trend in terms of volume of investment per capita over the whole period 1981–2003 (see Figures 2.5(a) and 2.5(b)).

In higher education expenditure, the UK has followed a pattern of investment similar to that of major competitors, both in terms of investment to GDP and investment per capita. It is worth noting that, regarding HERD per capita, the UK has the second highest annual growth rate over the period 1981–2003, only behind the USA (3.5 per cent and 4.3 per cent, respectively) (Figures 2.6(a) and 2.6(b)).

However, the UK figures on R&D performed in UK higher education are less in line with those of the small, high-tech countries. Since the early 1990s, all these four countries are well above UK figures on R&D intensity in the higher education sector (see Figure 2.6(c)). By 2003, Denmark was 0.2 percentage points above the UK, Finland and Switzerland 0.3 percentage points above, and Sweden 0.5 percentage points above.

Regarding the emerging economies, only Singapore displays HERD to GDP figures above the UK levels (i.e. 0.56 per cent in 2003), while both China and South Korea are below the UK (with figures of HERD to GDP of 0.12 per cent and 0.27 per cent respectively). However, over the period 1995–2003, these three emerging economies have observed an average annual growth rate in their HERD to GDP of 6.5 per cent, compared to the 0.7 per cent of the UK.

Internationalization of business R&D
In a context of increasing internationalization of R&D investment flows by multinational enterprises, a country's competitiveness largely relies on the capacity to retain access to centres of science and technological excellence and to become an attractive place to perform R&D activities. Several reports have argued that the UK is well placed to gain from an increase in international flows of R&D, since the UK is an attractive place for foreign-owned firms to conduct R&D activities and it is a major investor in R&D overseas (see DTI 2005).

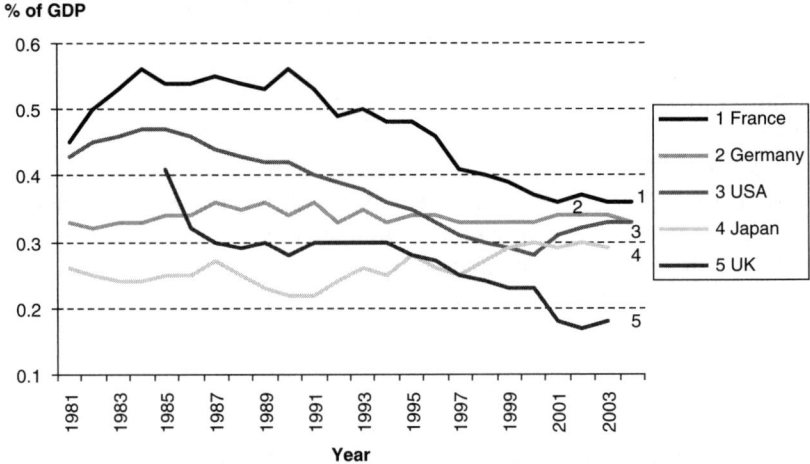

Note: Gaps in lines reflect periods for which no data was available.

Source: As for Figure 2.1.

Figure 2.5(a)   Government expenditure on R&D to GDP: UK versus major comparators

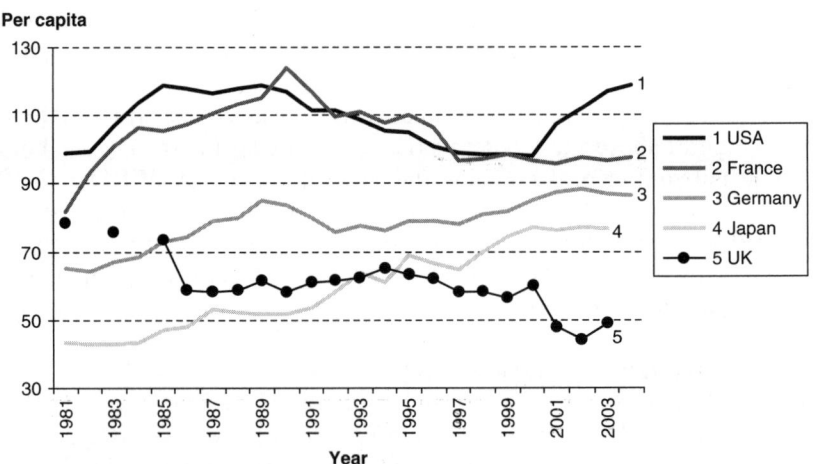

Note: Gaps in lines reflect periods for which no data was available.

Source: As for Figure 2.1.

Figure 2.5(b)   Government expenditure on R&D per capita (in US$ 2000, PPP adjusted): UK versus major comparators

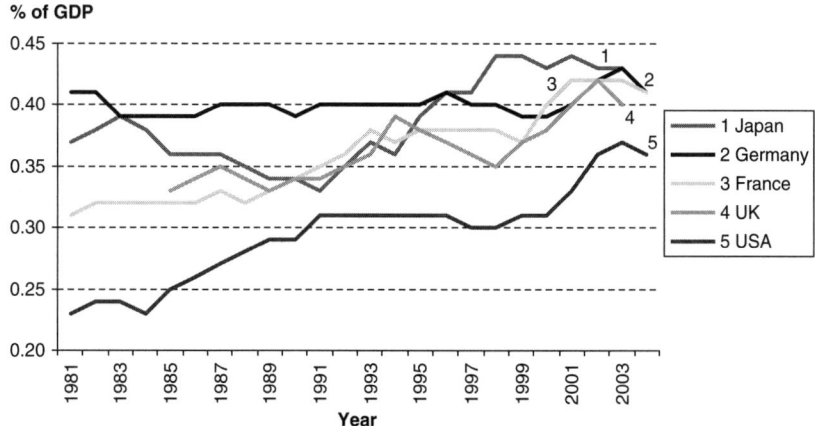

Note: Gaps in lines reflect periods for which no data was available.

Source: As for Figure 2.1.

Figure 2.6(a)   *Higher education expenditure on R&D as a percentage of GDP: UK versus major comparators*

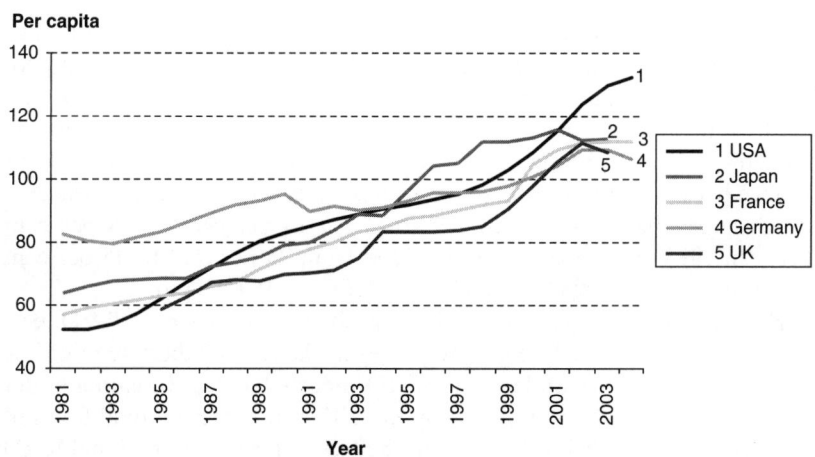

Note: Gaps in lines reflect periods for which no data was available.

Source: As for Figure 2.1.

Figure 2.6(b)   *Higher education expenditure on R&D per capita (in US$ 2000, PPP adjusted): UK versus major comparators*

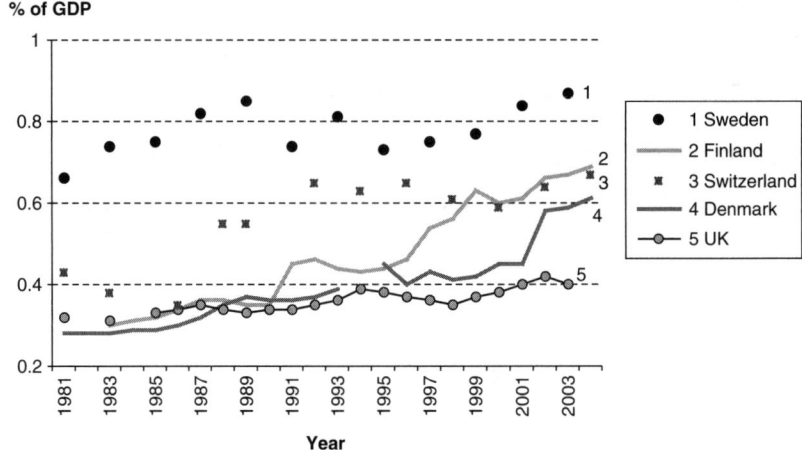

Note: Gaps in lines reflect years for which no data was available.

Source: As for Figure 2.1.

Figure 2.6(c)   *Higher education expenditure on R&D per capita: UK*
versus small, high-tech countries

Indeed, most industrialized countries have experienced a continuous increase in the proportion of business R&D performed by affiliates of foreign companies, and this trend has been particularly strong in the case of the UK. For instance, the proportion of business R&D performed by affiliates of foreign companies increased from 14 per cent to 23 per cent for France between 1994 and 2003, from 16 per cent to 25 per cent in the case of Germany between 1993 and 2001, from 14 per cent to 34 per cent between 1993 and 2002 for Sweden, and from 28 per cent to 45 per cent between 1994 and 2003 for the UK (see OECD-MSTI 2005).

However, the comparatively high proportion of business R&D performed by affiliates of foreign companies in the case of the UK might be driven by the low overall level of BERD reported above. To gain a better understanding of the role played by the UK in the international flows of business R&D, it is helpful to look at the US profile of international R&D flows, examining the foreign-owned R&D in the USA and the US-owned R&D overseas. We observe that the position of the UK is very strong, although Germany stands in a very similar position to the UK.

Figure 2.7(a) shows the R&D performed abroad by majority-owned foreign affiliates of US parent companies over the period 1994–2002 by country (in percentages) (NSF 2006), while Figure 2.7(b) shows the R&D

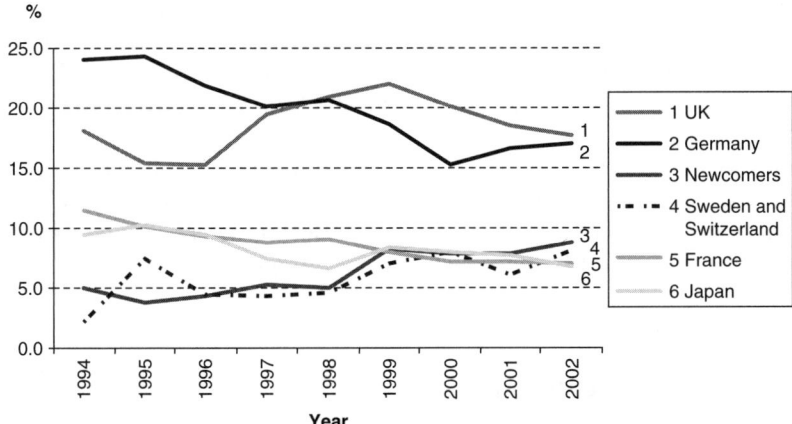

Note: Newcomers include Brazil, China (with Hong Kong), India, S. Korea, Singapore and Taiwan.

Source: NSF (2006).

Figure 2.7(a)   *R&D performed abroad by majority-owned foreign affiliates of US parent companies, by country: 1994–2002*

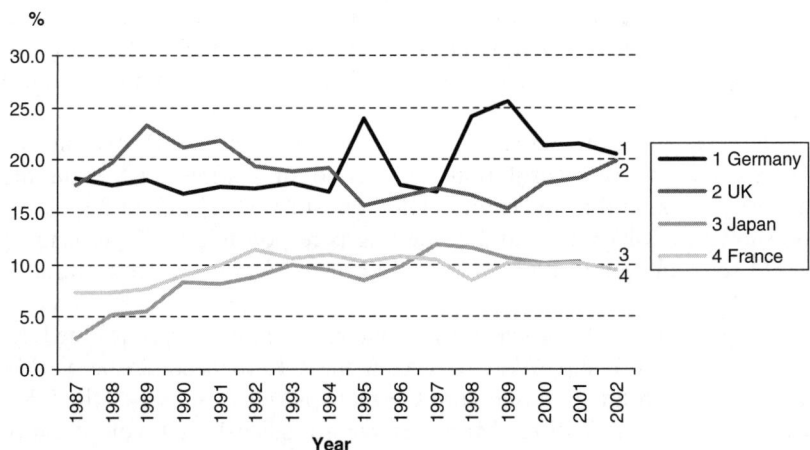

Note: Figures for the emerging economies were not available.

Source: As for Figure 2.7(a).

Figure 2.7(b)   *R&D expenditure by majority-owned affiliates of foreign companies in the USA, by country: 1987–2002*

expenditure by majority-owned affiliates of foreign companies in the USA (period 1987–2002) by country (in percentages) (ibid.). Both figures show that the UK and Germany are the major investors of R&D in the USA and the major recipients of R&D investment from the USA.

In short, while the UK has a comparative low level of domestic business R&D relative to major competitors, it has a much more active role in terms of international flows of business R&D, both as an investor overseas and in its capacity to attract business R&D from abroad.

4.   SCIENTIFIC PERFORMANCE[11]

The Overall Picture

While the previous section focused on how much money countries are investing in R&D (i.e. the input data), this section addresses one of the performance indicators of a country's R&D system: scientific performance. This will be examined mainly through the patterns of scientific publications displayed by the UK and our comparator countries.

As mentioned in Section 2, this section considers the production of scientific publications as an indicator of a country's scientific activity (i.e. the country's share of world publications and the growth of that share), and the number of citations as an indicator of the usefulness and importance of publications.

The UK scientific research base has long been considered at the highest level in the world ranking, second only to the USA on the majority of scientific indicators (HM Treasury, DTI and DfES 2005). As shown in Tables 2.1 and 2.2, the UK is responsible for 7 per cent of world scientific publications and 8 per cent of world scientific citations (NSF 2006).

However, in recent years significant challenges have emerged for the UK. On the one hand, Japan has overtaken the UK to become the world's second most prolific publisher and Germany is getting closer to the UK's world share of publications. Moreover, regarding world share of citations, UK has maintained second position, but Japan and Germany have significantly reduced the gap with the UK.

On the other hand, the emergence of newcomer countries has had an impact on the scientific production, causing a shift in the composition of major players in the world landscape of scientific producers (Zhou and Leydesdorff 2006). As Table 2.1 shows, a number of newcomer countries

Table 2.1   Ranking of countries by publication shares

Countries	World share 1988	Countries	World share 2003
1 USA	38.09	1 USA	30.23
2 UK	7.83	2 Japan	8.60
3 Japan	7.38	3 UK	6.91
4 USSR	6.78	4 Germany	6.34
5 Germany	6.28	5 France	4.58
6 France	4.59	6 China	4.18
7 Canada	4.59	7 Canada	3.55
8 Italy	2.41	8 Italy	3.53
9 Australia	2.12	9 Spain	2.41
10 India	1.90	10 Australia	2.26
11 Netherlands	1.84	11 Russia	2.26
12 Sweden	1.62	12 South Korea	1.97
13 Spain	1.16	13 Netherlands	1.93
14 Switzerland	1.14	14 India	1.83
15 Israel	1.05	15 Sweden	1.47
16 China	0.99	16 Taiwan	1.33
17 Poland	0.86	17 Brazil	1.24
18 Belgium	0.77	18 Switzerland	1.22
19 Denmark	0.74	19 Israel	0.99
20 Finland	0.60	20 Poland	0.97

Source: NSF (2006).

have entered the list of the 20 most prolific publishers, for example China, South Korea, Taiwan and Brazil. Moreover, increasing numbers of scientific producers may be one of the driving forces underlying the sharp fall of US world shares in terms of both volume of publications and citations.

Finally, the UK is the leading country among the G8 in terms of number of publications per capita (HM Treasury, DTI and DfES 2005). As shown in Figure 2.8, the UK is well ahead in terms of publications per million capita relative to Germany, France and Japan (Figure 2.8 shows the publications per million capita of each comparator country relative to UK figures over two periods of time). Moreover, none of the major competitor countries is reducing the gap with the UK on this science productivity indicator.

However, it is important to note that when considering a wider range of countries, the small, high-tech countries have higher productivity ratios than the UK, as shown in Table 2.3.

Table 2.2   Ranking of countries by world citations

	Countries	World share 1992		Countries	World share 2003
1	USA	51.75	1	USA	42.39
2	UK	8.27	2	UK	8.10
3	Japan	6.50	3	Japan	7.34
4	Germany	5.86	4	Germany	7.04
5	France	4.34	5	France	4.65
6	Australia	4.16	6	Australia	3.72
7	Netherlands	2.14	7	Italy	3.01
8	Italy	2.04	8	Netherlands	2.29
9	Canada	1.86	9	Canada	2.11
10	Sweden	1.82	10	Spain	1.90
11	Switzerland	1.62	11	Switzerland	1.71
12	USSR	1.16	12	Sweden	1.65
13	Spain	0.83	13	China	1.51
14	Israel	0.79	14	South Korea	0.94
15	Belgium	0.77	15	Belgium	0.90
16	Denmark	0.76	16	Israel	0.89
17	Finland	0.55	17	Denmark	0.88
18	India	0.54	18	Finland	0.80
19	Austria	0.41	19	Russia	0.74
20	Norway	0.38	20	India	0.73

Source: As for Table 2.1.

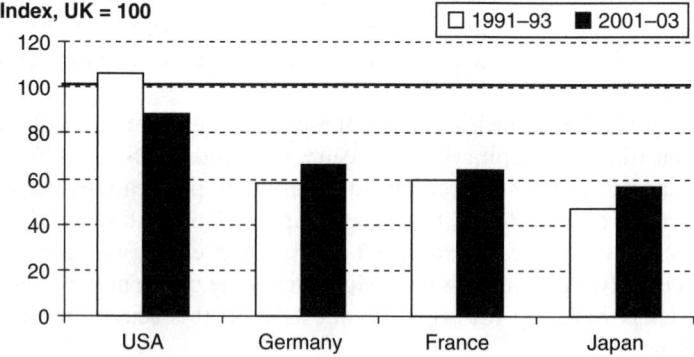

Source: As for Figure 2.7(a).

Figure 2.8   Papers per head of population

Table 2.3   Number of publications per million capita

	1990–92		2001–03
1 Sweden	951	1 Sweden	1135
2 Switzerland	904	2 Switzerland	1107
3 Canada	831	3 Finland	972
4 USA	768	4 Denmark	931
5 Denmark	752	5 Netherlands	797
6 UK	**705**	**6 UK**	**794**
7 Netherlands	693	7 Australia	765
8 Finland	646	8 Canada	744
9 Australia	630	9 New Zealand	735
10 New Zealand	624	10 Norway	713
11 Norway	585	11 USA	703
12 Belgium	423	12 Iceland	660
13 Germany	418	13 Belgium	598
14 France	414	14 Austria	571
15 Iceland	385	15 Germany	525
16 Austria	365	16 France	507
17 Japan	329	17 Japan	451
18 Czech Republic	294	18 Ireland	431
19 Ireland	261	19 Italy	398
20 Italy	248	20 Spain	397

Source: As for Table 2.1.

Discussion on scientific performance

The first thing to highlight regarding scientific performance is that most countries have been increasing the absolute number of publications over the last 15 years. Indeed, the world number of publications, as collected by the NSF (2006), has expanded from 466 419 in 1988 to almost 698 726 in 2003, at an annual growth rate of 2.8 per cent.

Such a positive growth trend has been present in our set of comparator countries. As shown in the figures below, all the countries considered in this study have experienced an expansion in their volume of scientific publications. The issue then is not a question of whether our set of countries has experienced positive growth in their scientific production but a question of how much. It is clear that the pace at which different countries have expanded their scientific production dramatically differs across groups of countries, and between the countries in each group. Indeed, while on average our set of major industrial competitors (i.e. the UK, the USA, Germany, France and Japan) has an average annual growth rate of publications of 2.5 per cent over the period 1988–2003, the growth rate for the

Index, 1988 = 100

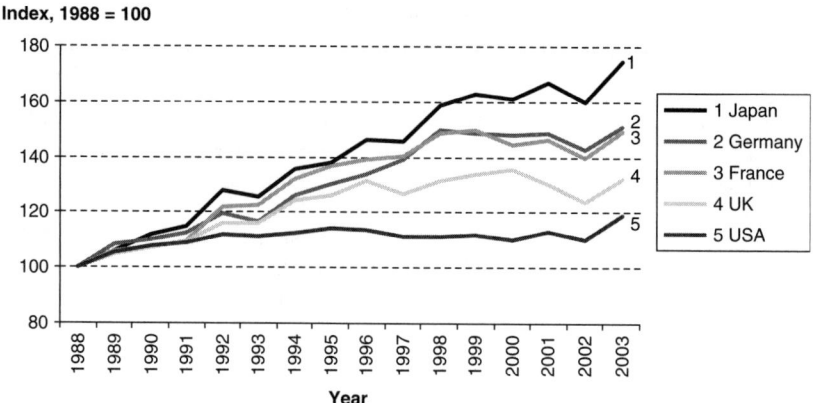

Figure 2.9   Trends in publication output: UK versus major comparators

Index, 1988 = 100

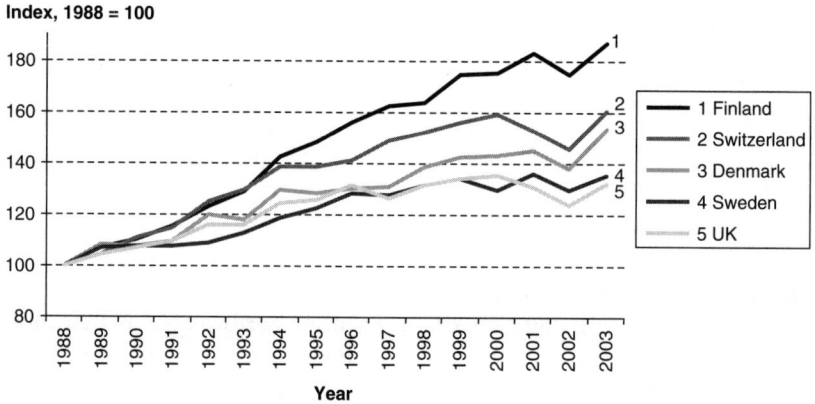

*Figure 2.10   Trends in publication output: UK versus small, high-tech
                countries*

small, high-tech countries has been 3.1 per cent, and that for the group of
newcomer countries has been 12.6 per cent (see Figures 2.9, 2.10 and 2.11).

The large increase in publication output from the emerging economies has
led, as mentioned in Table 2.1, to a situation in which several newcomers
appear listed among the 20 largest publishing countries. Looking at the
trends in more detail, we observe that not only China has raised to an outstanding sixth position in terms of world publication shares in 2003, but also
that the evolution of its publication share points to a convergence with the
UK in the medium term. Moreover, if all newcomer countries are considered

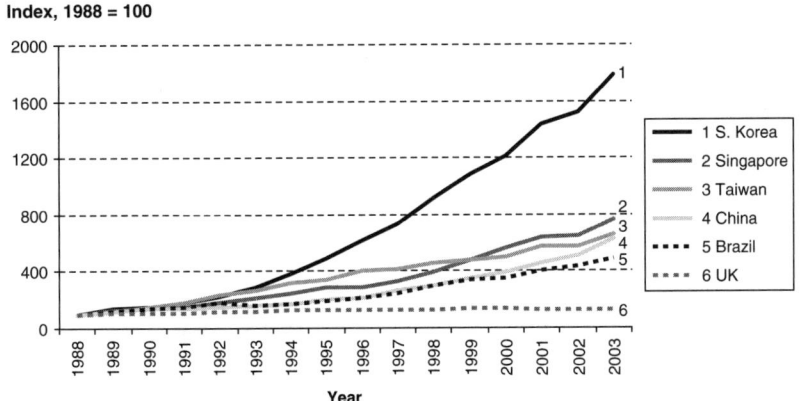

Figure 2.11   Trends in publication output: UK versus newcomer countries[12]

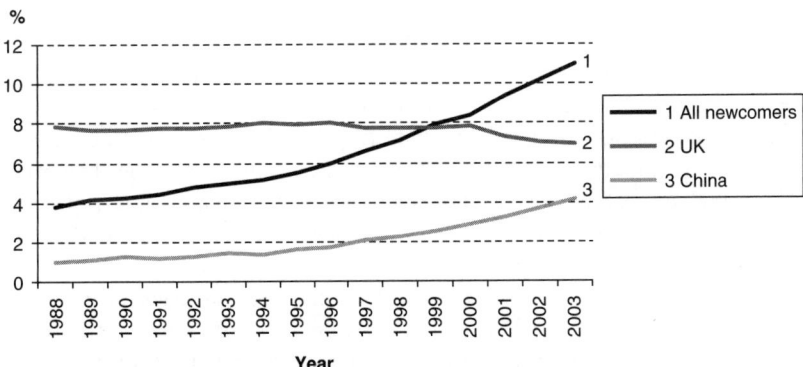

Figure 2.12   Trends in publications' world market shares: UK versus newcomer countries

together, they have already reached the second largest world share in 2003, only behind the USA, with about 11 per cent of world publications (see Figure 2.12), and have overtaken the UK's figures since 1999. In addition to this, while China plays an important role in the overall growth of the newcomer countries, its contribution to the newcomer shares represents less than 40 per cent: the remaining 60 per cent is largely evenly distributed across the other five countries that form our group of newcomer countries.

As the figures above show, the emerging economies are becoming outstanding centres of scientific production within the world scientific landscape,

Ratio

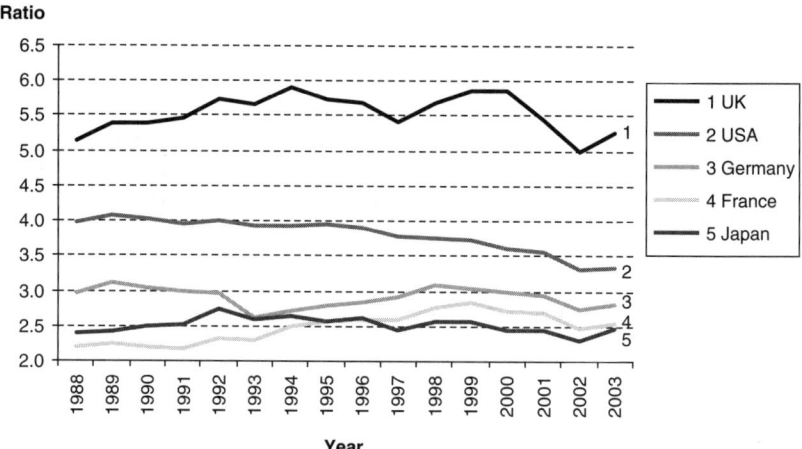

Figure 2.13   Publications per GOVERD+HERD (2000 $, PPP adjusted):
            UK versus major comparators

pointing to a shift in the centre of gravity of the world system of science. This
clearly poses a challenge to the traditional nodes of scientific production,
since the emerging countries are disputing the world publication market
shares of incumbent countries on the scientific production scene. The other
side of the coin is that the emerging economies provide a great opportunity
for incumbent countries to draw upon the new knowledge generated by the
emerging centres of knowledge production.

Since one of the consequences that the emerging countries have had for
the world system of science is to put an increasing pressure on the traditional advantages of world leading countries, it is likely that this may have
had an impact on the relationship between the number of publications
and R&D investment. That seems to be happening, in particular in the
case of the USA. The USA has experienced a substantial drop in
the number of publications per unit of investment in R&D, when measured both in terms of total R&D performed by government and higher
education together, and by higher education alone (see Figures 2.13 and
2.14). The ratio measures the number of publications at a point in time
relative to the weighted average investment in R&D over the previous
three years.

While the UK is well above other major industrial competitors in terms
of the ratio of publications per unit of investment performed in the public
sector, it seems also to have been at strain regarding this indicator, particularly from 2000 onwards. It is interesting to note that neither Germany,

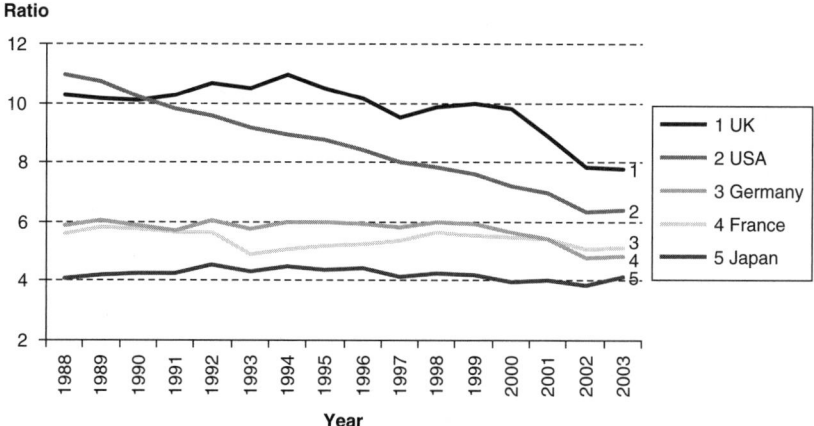

Figure 2.14   Publications per HERD (2000 $, PPP adjusted): UK versus major comparators

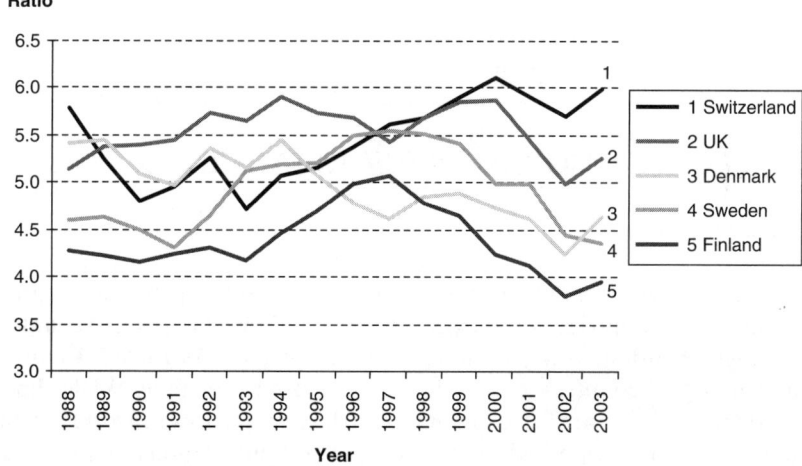

Figure 2.15   Publications to GOVERD+HERD (2000 $, PPP adjusted): UK versus small, high-tech countries

France nor Japan displays a decreasing trend over the whole period. Also, most small, high-tech countries follow a pattern similar to that of the UK, both in terms of overall values of the ratio and in terms of its trend (Figure 2.15). The only two countries that have experienced a substantial growing trend in the ratio of publications to public R&D investment are

Ratio

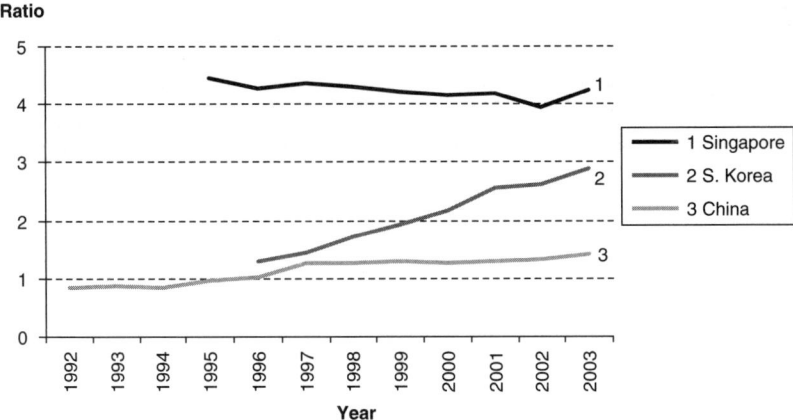

Note: Gaps in lines reflect periods for which no data was available.

Figure 2.16   Publications per GOVERD+HERD (2000 $, PPP adjusted): newcomer countries

South Korea and Switzerland – in this latter case, from 1993 onwards – and to a lesser extent China (Figure 2.16).[13]

5.   TECHNOLOGICAL PERFORMANCE

Overall Picture

In order to assess technological performance we examine patenting records for our set of countries over the period 1981–2003. We first look at country rankings according to world shares of triadic patents. As Table 2.4 shows, the ten largest countries remain largely unaltered over time. While three emerging countries are listed among the 20 largest in terms of world patent shares by 2001, as opposed to the case of world publications shares, none of them is still among the ten largest countries.

While the UK is ranked fifth in terms of its (triadic) patent world shares, when patent figures are normalized by country size, its ranking drops dramatically below the tenth position in the world ranking of triadic patents per capita (see Table 2.5).

Examining in more detail the long-term trends of triadic patents per capita, it becomes clear that the UK has been performing worse than our set of comparator countries. Compared to the major industrial competitors, since the early 1990s the UK has been growing at a lower pace than

Table 2.4   World shares of triadic patents (%)

Country rank	1995	Country rank	2001
1 USA	35.0	1 USA	34.9
2 Japan	27.0	2 Japan	24.9
3 Germany	13.7	3 Germany	15.8
4 France	5.4	4 France	5.2
5 UK	4.3	5 UK	4.6
6 Switzerland	2.1	6 Netherlands	2.1
7 Netherlands	2.1	7 Switzerland	1.8
8 Sweden	2.0	8 Italy	1.8
9 Italy	1.7	9 Sweden	1.7
10 Canada	1.1	10 Canada	1.4
11 Belgium	1.1	11 Finland	1.1
12 S. Korea	0.9	12 S. Korea	1.1
13 Finland	0.9	13 Belgium	0.9
14 Australia	0.6	14 Australia	0.8
15 Austria	0.6	15 Austria	0.6
16 Denmark	0.5	16 Denmark	0.5
17 Spain	0.3	17 China	0.3
18 Norway	0.3	18 Spain	0.3
19 Ireland	0.1	19 Norway	0.2
20 Hungary	0.1	20 Singapore	0.2

Source: MSTI-OECD (2005).

the comparator countries (with the exception of France), and thus, Germany, the USA and Japan are moving further apart from UK records of patenting per capita (see Figure 2.17).

When compared with the small, high-tech countries, the UK is even further below, and with little signs of convergence. In particularly, Sweden and Finland have experienced much higher growth rates since 1991 (annual growth rates of 6.7 per cent and 12.7 per cent for Sweden and Finland, respectively, relative to 4.2 per cent for the UK) (see Figure 2.18).

Finally, regarding the newcomer countries the picture is mixed, largely caused by the different country scales. Regarding triadic patenting records, we could get information only for China, South Korea and Singapore (OECD-MSTI various years). While China's figures of patenting records per capita are still very low relative to international standards, South Korea and Singapore have been clearly catching up with major industrialized countries, such as the UK (see Figure 2.19), although they still remain low.

Table 2.5   Triadic patents per million capita

Country rank	1995	Country rank	2001
1 Switzerland	105.5	1 Switzerland	118.2
2 Sweden	79.3	2 Finland	98.5
3 Japan	75.3	3 Japan	92.3
4 Finland	61.1	4 Sweden	91.8
5 Germany	59.0	5 Germany	90.7
6 Netherlands	46.8	6 Netherlands	61.9
7 USA	46.7	7 USA	57.7
8 Belgium	36.8	8 Belgium	42.1
9 Denmark	36.2	9 Denmark	41.4
10 France	32.8	10 France	40.3
11 Austria	27.3	**11 UK**	36.7
12 UK	25.9	12 Austria	34.9
13 Iceland	22.5	13 Norway	23.9
14 Norway	18.4	14 Iceland	21.1
15 Canada	12.8	15 Canada	20.6
16 Australia	12.5	16 Singapore	20.3
17 Italy	10.7	17 Ireland	19.2
18 Ireland	8.6	18 Australia	19.2
19 S. Korea	7.3	19 Italy	14.8
20 Singapore	6.8	20 S. Korea	10.6

Source: As for Table 2.4.

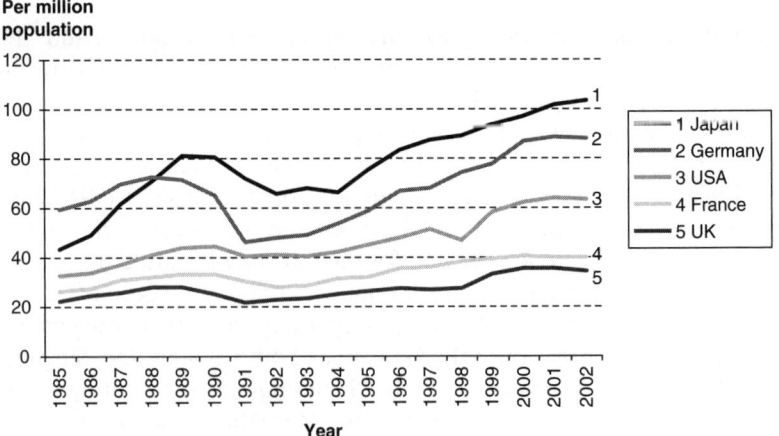

Figure 2.17   Triadic patents per million population: UK versus major comparators

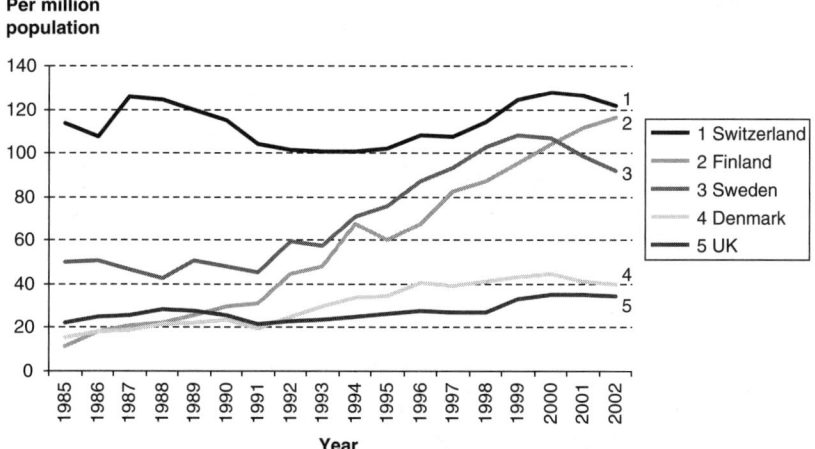

Figure 2.18   Triadic patents per capita: UK versus small, high-tech countries

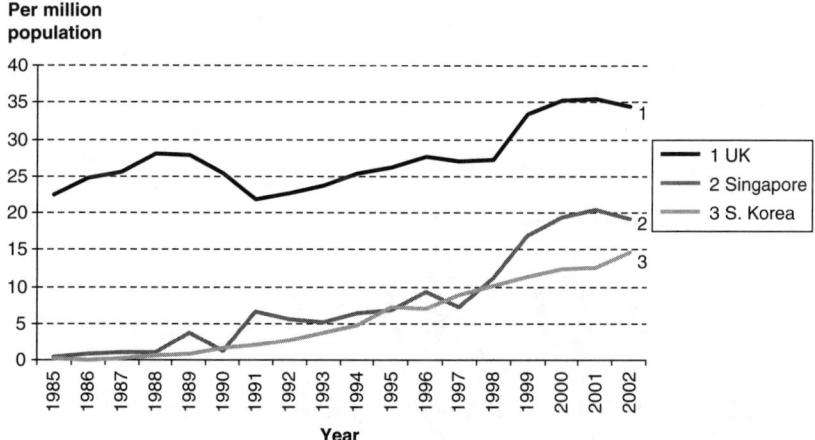

Figure 2.19   Triadic patents per capita: UK versus South Korea and Singapore

A similar picture emerges when we look at the patents granted by the USPTO. To avoid problems of home bias, we have not included data corresponding to the USA in the figures. As Figures 2.20, 2.21 and 2.22 show, the UK is well below major industrial competitors and small, high-tech countries, and both South Korea and Singapore have overtaken UK figures from 1998 onwards.

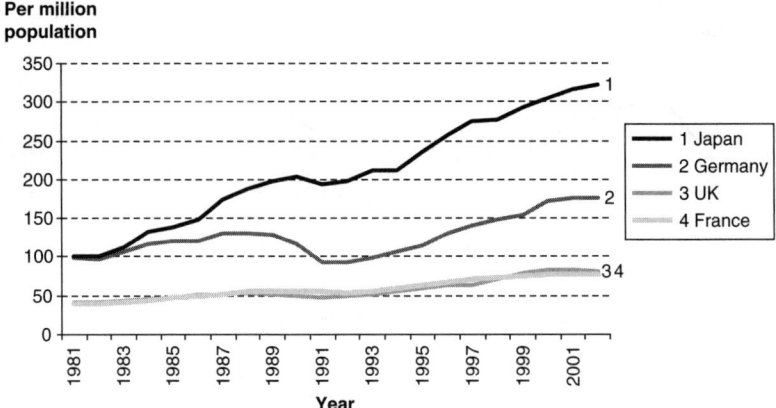

Figure 2.20   USPTO granted patents: UK versus major comparators

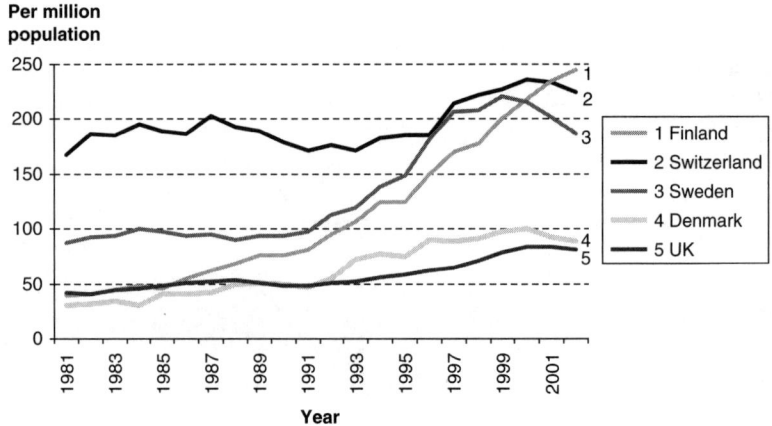

Figure 2.21   USPTO granted patents: UK versus small, high-tech
*   countries*

Finally, similarly as we did with the publications and public R&D investment, we have compared in this case the country patenting records (i.e.
triadic patents) to business-performed R&D investments.[14] As Figures 2.23
and 2.24 show, the UK is well below most of the small, high-tech countries
(with the exception of Sweden and, more recently, Denmark), and well
below Japan and Germany, and since the early 1990s the UK has followed
a very similar pattern to that of the USA and France. When compared to
the newcomer countries, the UK is still well ahead of Singapore and South

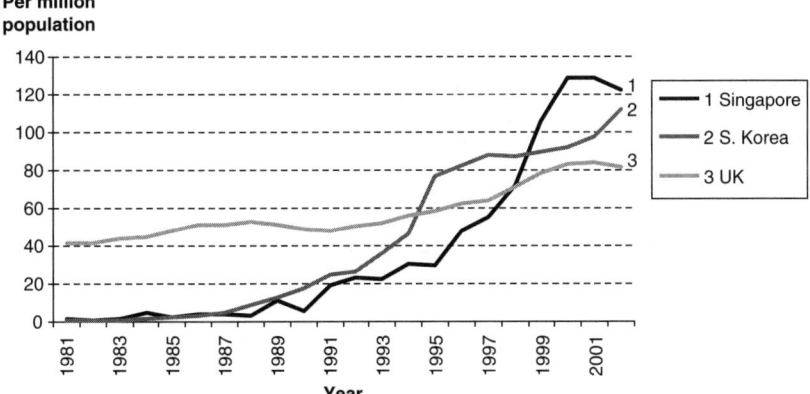

Figure 2.22   USPTO granted patents: UK versus South Korea and
Singapore

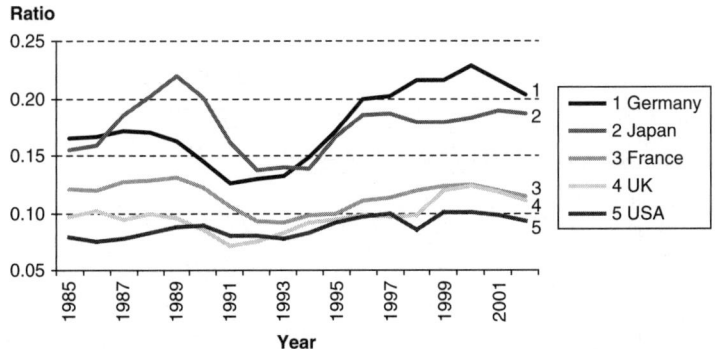

Figure 2.23   Triadic patents per unit of BERD: UK versus major
comparators

Korea in terms of patenting per unit of business-performed R&D (i.e. in
2002, South Korea had a ratio of triadic patents per BERD that was 46 per
cent of the UK's ratio, while Singapore was 63 per cent of the UK's ratio).

Finally, the UK does seem to perform much better than comparator
countries when looking at the technology produced abroad by its own
multinationals, indicating a rather strong globalization of their R&D
activities.

Figure 2.25 shows the percentages of EPO patents owned by a country
(i.e. by its companies or other major patenting institutions) that are
invented abroad by affiliates of its own multinationals (source RSTI 2003)

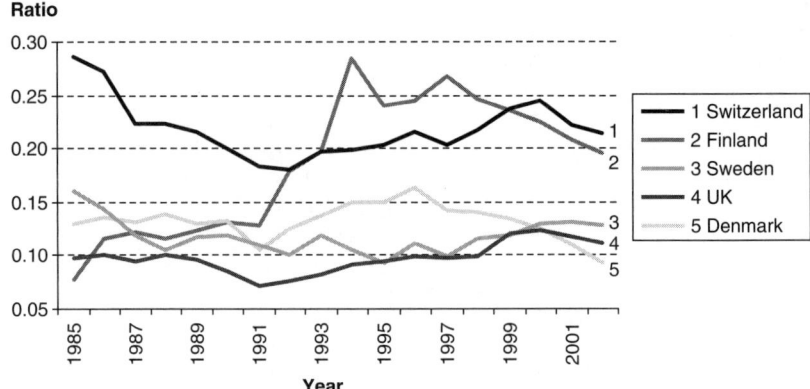

Figure 2.24   Triadic patents per unit of BERD: UK versus small, high-tech countries

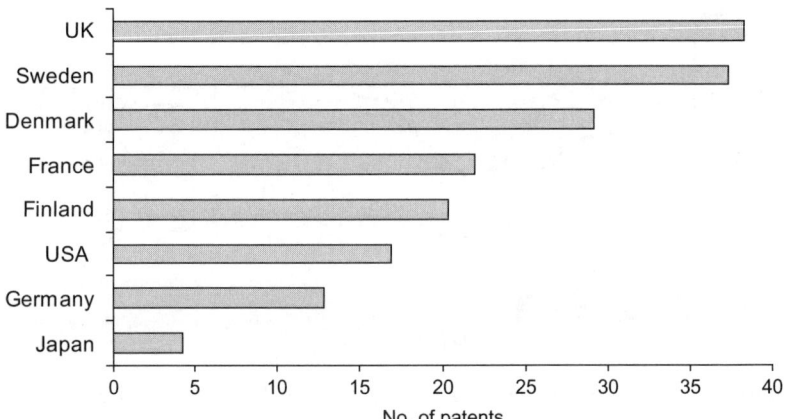

Source: EC-RSTI (2003).

Figure 2.25   Patents produced by a country's foreign-based affiliates (%): 1999

in 1999. The UK has more than one-third of its patents invented by its multinationals' affiliates abroad. Only Sweden has a similar proportion among our set of comparator countries (no data are available here for Switzerland or the newcomer countries), while the USA, Germany and Japan have comparatively lower percentages of patents produced by their overseas affiliates.

It is important to add that all these countries have more than 90 per cent of the patents produced by their overseas affiliates either in Europe or the USA, and thus the extent of the globalization of patenting remains highly concentrated mainly in these two economic blocs. However, with this caution in mind, Figure 2.25 shows that the UK's companies have a much stronger internationalization of their technological activities (as measured by overseas affiliate patenting).

6.   THE UK PERFORMANCE OVERALL

This section summarizes the various figures that we have presented throughout this chapter. To explore the aggregate trends we have reviewed these data in terms of each country's *position* and *growth* for a range of indicators. By position we refer to the country's level within a given indicator (e.g. GERD per capita) in 2003, compared to the comparator group average. By growth we refer to the country's annual growth rate within a given indicator compared to the comparator group average. Annual growth rates have been calculated over time periods that differ across indicators, as a consequence of the lack of comparable information. We have prioritized to compute growth rates for time periods longer than five years, in order to get a more reliable indicator of trends. So, for instance, while annual growth rates for GERD to GDP refer to the period 1981–2003, the annual growth rate for the relative citation index covers the period 1995–2003. While the time periods to compute growth rates differ across indicators, they are consistently the same across countries within each indicator.

Tables 2.6 and 2.7 show the relative position and growth of each country for each indicator, with the following patterns:

- ▨ indicating that the country is 'moving ahead': the country is above the comparator group average in terms of both level and growth rate.
- ▤ indicating that the country is 'catching up': the country is currently below the comparator group average but experiencing a higher than average growth rate.
- ▦ indicating that the country is 'losing momentum': the country is currently above the comparator group average but experiencing a lower than average growth rate.
- ■ indicating that the country is 'falling behind': the country is below the comparator group average in terms of both level and growth rate.

Table 2.6 Overall performance: UK versus major comparator countries

Measure	UK	USA	Germany	France	Japan
Investment in R&D					
GERD to GDP					
GERD per capita					
BERD to GDP					
BERD per capita					
GOVERD to GDP					
GOVERD per capita					
HERD to GDP					
HERD per capita					
Scientific performance					
Publications per capita					
Publ. share: engineering					
Publ. share: physics					
Publ. share: chemistry					
Publ. share: biomedical research					
Publ. share: social sciences					
Citations per paper					
R. citation index: engineering					
R. citation index: physics					
R. citation index: chemistry					
R. citation index: biomedical res.					
R. citation index: social sciences					
Publications per HERD					
Publications per PUBERD					
Technology performance					
Triadic patents per capita					
USPTO per capita					
EPO per capita					
Patents (triadic) per BERD					

Note: ▨ Moving ahead; ▦ losing momentum; ▤ catching up; ■ falling further behind.

Tables 2A.1, 2A.2 and 2A.3 in the Appendix provide the raw data on each country's position and growth across a range of indicators. This forms the basis for the symbols used in Tables 2.6 and 2.7.

As Tables 2.6 and 2.7 show, the UK is within the 'falling behind' territory regarding both investment in R&D and technology performance. With the exception of higher education expenditures in R&D, the UK has been clearly underinvesting in R&D compared to both the major comparator countries and the overall set of comparator countries. Among the major

Table 2.7 *Overall performance: UK versus all comparator countries*

Measure	UK	USA	Germany	France	Japan	Sweden	Finland	Switz.	Denmark	China	S. Korea	Singapore	Taiwan	India	Brazil
Investment in R&D															
GERD to GDP															
BERD to GDP															
GOVERD to GDP															
HERD to GDP															
Scientific performance															
Citations per paper															
R. citation index: engineering															
R. citation index: physics															
R. citation index: chemistry															
R. citation index: biomedical res.															
R. citation index: clinical medicine															
R. citation index: biology															
R. citation index: social sciences															
Outputs per unit of investment															
Patents (triadic) per BERD															
Publications per HERD															

Note: Moving ahead; losing momentum; catching up; falling further behind.

49

comparator countries, the USA remains well positioned in terms of R&D investments levels, showing comparatively higher growth rates in almost all R&D investment indicators (the only exception is GOVERD to GDP).

Regarding indicators on technology, the UK shows a poor performance not only when compared to major comparator countries, but also relative to our whole set of comparator countries. Germany and Japan are performing particularly well on technology-related indicators, while South Korea and Singapore have the largest growth rates in terms of (triadic) patents per unit of business-performed R&D.

Regarding scientific performance, while the UK does very well in terms of overall number of publications per capita, this is likely to be a consequence of its comparatively high level of specialization in biomedical-related fields, where it does particularly well in terms of world publication market shares. However, in disciplines such as chemistry, engineering and physics, the UK is performing poorly, showing a decreasing trend in its world publication shares in these three disciplines.

Regarding scientific impact, the UK performs relatively well overall. However, it is falling behind in engineering compared to the major comparator countries, and its relative citation index tends to be lower than average when compared with the small, high-tech countries across a wide range of disciplines such as chemistry, engineering, physics and social sciences.

Finally, it is noticeable not only that the newcomer countries are playing an increasingly important role in terms of the volume of publications in all disciplines, but also that they are catching up in terms of the impact of their publications, as shown by the predominance of ▭ regarding the relative citation index across disciplines.

7.   IMPLICATIONS FOR POLICY, PRACTICE AND RESEARCH

The data presented in this chapter raise some significant questions about the likelihood of the UK sustaining its traditionally held position as one of the world's leading generators of scientific knowledge. Two specific issues underpin this concern. First, the data show that the UK has systematically underinvested in science and technology relatively to its comparator countries for over 20 years. Second, the rapid emergence of new centres of knowledge production poses some potentially significant challenges for the UK, as well as for other developed economies.

UK science and technology policy is often framed in terms of UK Plc – how do we make interventions to enable the UK to become more productive and competitive as a nation? Innovation research and industrial practice suggest that the innovation process is changing. Increasingly executives are discussing open innovation, with major R&D players, such as Procter & Gamble, now talking of sourcing half or more of their innovations from outside the company. Executives in firms recognize that no single firm can hope to have a monopoly on knowledge production in any specific field. Hence the challenge for firms – in an era of open innovation – is how to sustain their own ability to innovate, while simultaneously developing their capability to access and absorb relevant innovations made by others. These twin capabilities – in-house innovation coupled with the appropriation of externally produced knowledge – raise some important challenges, not least regarding how these activities should be resourced and how firms should balance their effort between them. If the firm devotes too much resource to looking outside, searching for the innovations of others, there is a danger that the firm's own capacity to innovate is likely to atrophy and over time it will lose its ability to understand and absorb the innovations of others. Of course the counter-argument also applies. If the firm is too internally focused it will miss those important innovations made by others.

Should we apply the same thinking to science and technology policy? Instead of worrying solely about the output of the UK's science and technology base, perhaps we should also worry about how to enhance the country's capability to engage in the global knowledge production process. For it is by engaging in this process that the UK will be able to access and absorb scientific developments made elsewhere that might be of value. In policy terms, this dual focus – on the UK's scientific capability and the extent to which its scientific community engages in the global knowledge production process – requires a shift in policy emphasis. It can be argued that UK policy is currently too US-centric, with the result that developments in the small, high-tech and newcomer countries might be missed. In the research community too much emphasis is placed on the US model of research, with numerous disciplines seeking to conform to US norms, rather than arguing the case for a pluralist definition of good research. Striking the right balance between different models of research is clearly complex, but ignoring the strength of some of the European research traditions is naive. The issue for the UK and particularly for those who influence research in the country is how to strike an appropriate balance and focus in light of the country's investments in science and technology.

An Emerging Research Agenda

In addition to important policy questions the data presented in the census raise some significant topics for future research. First there is the question of why the small, high-tech countries perform so well across the board. The data suggest that Denmark, Finland, Sweden and Switzerland all achieve outstanding levels of performance in terms of publication per capita and triadic patents per capita. Why do these countries achieve such high levels of performance?

A second question relates to the UK's outstanding performance in bio-medical fields. What is it about the UK's scientific community that allows the country to perform so well in terms of biomedical research? To what extent does the UK's traditionally strong position in pharmaceuticals influence its performance? If this were the sole explanatory variable, then one could argue that the UK would also perform well in engineering. Yet the data suggest that the UK is losing its status in engineering research – which, of course, raises another question.

A third question concerns the propensity of UK scientists to engage with global knowledge production – especially centres of knowledge production outside the USA. Given the rapid emergence of the newcomer countries and the sustained strength of the small, high-tech countries, the question is to what extent the UK engages with these communities.

And this question of engagement raises a fourth issue – namely the extent of collaboration between the science base and the industrial community. The dominant model of scientific research is the linear model – namely that investment in science and technology should result in knowledge that can be commercialized and hence generate economic value. As discussed at the outset of this census, this linear model is questioned by many. The notion of knowledge flowing from the science community to the business community is regarded as outdated. Instead the assumption is that both communities inform one another. However, to what extent this is true and, if it is true, what are the processes of knowledge exchange (as opposed to knowledge flow) that have a larger impact on innovation, are issues that deserve further investigation. It is to these questions that we will turn our attention in the next chapters.

NOTES

1. This chapter is based on a report by the authors produced as part of the AIM – Advanced Institute of Management Research – initiative funded by the ESRC and EPSRC. The full chapter is available from the website www.aimresearch.org.
2. It is important to note that IfK – innovation from knowledge – does not assume a linear flow from knowledge creation to innovation. It is widely recognized that what matters is the *exchange* of knowledge – innovation influences the creation of new knowledge and vice versa.
3. OECD (2003) and EC-RSTI (2003), among others.
4. Intramural expenditures are all expenditures for R&D performed within a statistical unit (i.e. entity for which the required statistics are compiled) or sector of the economy during a specific period, whatever the sources of funds. These comprise labour costs of R&D

personnel and purchases of materials, supplies and equipment to support R&D performed by the statistical unit in a given year. In contrast, the extramural R&D expenditures are the sums a unit or sector reports having paid to another unit or sector for the performance of R&D during a specific period (see OECD 2002).

5. OECD (2002).
6. OECD-MSTI (various years), www.oecd.org.
7. Any private enterprises, public enterprises, or non-profit institutions producing higher education services should be included in the higher education sector.
8. www.nsf.gov/statistics/seind06.
9. When using R&D expenditures, absolute values are expressed in real terms from data on purchasing power parities at 2000 prices and exchange rates (see OECD-MSTI).
10. HM Treasury, DTI and DfES (2004).
11. More detail, comparing scientific performance by discipline, can be found in the original report: 'Science and Technology in the UK: 2006 Census': available at http://www. aimresearch.org/publications/scitech.pdf.
12. While the average annual growth rate of 12.6 per cent for emerging economies includes India, this country has not been included in Figure 2.11 since its growth rate, while higher than the UK's, is well below the average of the other newcomer countries, and in a figure such as 2.11 would largely overlap with the UK trend, giving a misleading signal that its growth is undifferentiated from that of the UK's.
13. We cannot assess the ratios of publications per unit of investment for the other newcomer countries because we do not have comparable data from the OECD regarding R&D investments.
14. Business R&D (BERD) measured in million PPP at 2000 prices, and calculated using a three-year time lag between year of R&D expenditures and year of patenting. In addition, we have computed a three-year weighted average for BERD, using a weighting score structure of 0.25/0.5/0.25.

REFERENCES

DTI (2005), 'PSA target metrics for the UK research base', Office of Science and Technology, December.

EC-RSTI (2003), 'Third European Chapter on Science & Technology Indicators 2003. Towards a knowledge-based economy', Brussels: EC.

Griffith, R. and Harrison, R. (2003), 'Understanding the UK's poor technological performance', IFS Briefing Note no. 37.

HM Treasury, DTI and DfES (2004), *Science & Innovation Investment Framework 2004–2014*, London, July.

HM Treasury and DTI (2005), 'R&D intensive businesses in the UK', DTI Economic Papers No. 11, March.

HM Treasury, DTI and DfES (2005), *The Ten-year Science & Innovation Investment Framework*, Annual Chapter 2005, July.

Luwel, M. (2004), 'The use of input data in the performance analysis of R&D systems', in H.F. Moed et al. (eds), *Handbook of Quantitative Science and Technology Research*, London: Kluwer Academic Publishers, pp. 315–38.

NESTA (2006), *The Innovation Gap. Why Policy Needs to Reflect the Reality of Innovation in the UK*, London, October.

NSF (2006), *Science and Engineering Indicators, 2006*, www.nsf.gov/statistics/seind06.

OECD-MSTI (various years), *Main Science and Technology Indicators*, Paris: OECD, www.oecd.org.

OECD (2002), *Frascati Manual. Proposed Standard Practice for Surveys on Research and Experimental Development*, Paris: OECD.

OECD (2003), 'OECD science, technology and industry scoreboard', OECD: Paris.

Rosenberg, N. (1991), 'Critical issues in science policy research', *Science and Public Policy*, **18**: 335–46.

Stokes, D. (1997), *Pasteur's Quadrant. Basic Science and Technological Innovation*, Washington, DC: Brookings Institution Press.

Zhou, P. and Leydesdorff, L. (2006), 'The emergence of China as a leading nation in science', *Research Policy*, **35**: 83–104.

APPENDIX

Table 2A.1   Raw data on position and growth for the UK and the group of major comparator countries

	France		Germany		Japan		UK		USA	
	Level 2003	Growth rate (%)	Level 2003	Growth rate (%)	Level 2003	Growth rate (%)	Level 2003	Growth rate (%)	Level 2003	Growth rate (%)
Investment in R&D										
GERD to GDP (%)	2.2	0.7	2.5	0.2	3.1	1.8	1.9	−0.9	2.7	0.7
GERD per capita ($ 2000, PPP)	569.6	2.4	629.9	1.5	793.5	3.8	486.6	1.4	932.3	2.7
BERD to GDP (%)	1.4	0.9	1.7	0.3	2.2	2.5	1.2	−0.8	2.0	0.7
BERD per capita ($ 2000, PPP)	358.6	2.7	439.6	1.6	579.3	4.5	321.0	1.5	674.1	2.7
GOVERD to GDP (%)	0.4	−0.9	0.3	0.1	0.3	0.7	0.2	−4.2	0.3	−1.1
GOVERD per capita ($ 2000, PPP)	96.9	0.9	85.9	1.4	76.3	2.7	51.6	−1.9	106.6	0.8
HERD to GDP (%)	0.4	1.4	0.4	0.2	0.4	0.8	0.4	1.1	0.3	2.2
HERD per capita ($ 2000, PPP)	106.1	3.1	104.4	1.3	113.2	2.7	102.9	3.5	116.1	4.3
Scientific performance										
Publications per (thousand) capita	0.5	2.3	0.5	1.0	0.4	3.6	0.8	1.6	0.7	0.0
World publication share: engineering (%)	4.4	0.2	5.4	0.2	11.6	−0.1	5.5	−3.6	23.5	−3.9

Table 2A.1   (continued)

	France		Germany		Japan		UK		USA	
	Level 2003	Growth rate (%)	Level 2003	Growth rate (%)	Level 2003	Growth rate (%)	Level 2003	Growth rate (%)	Level 2003	Growth rate (%)
World publication share: physics (%)	5.6	−0.8	7.7	−2.1	12.9	0.8	4.6	−3.0	19.1	−2.3
World publication share: chemistry (%)	5.0	−2.2	6.6	−3.7	10.6	−0.4	4.8	−3.7	19.1	−1.5
World publication share: biomedical res. (%)	4.8	−2.3	6.3	−0.7	8.3	0.6	7.2	−1.8	36.0	−0.9
World publication share: social sciences (%)	2.7	0.8	3.9	1.5	1.4	3.9	13.6	0.2	44.9	−1.5
Citations per paper	2.1	1.9	2.3	3.1	1.9	1.6	2.4	1.9	3.1	1.9
Relative citation index: engineering	0.7	−0.2	0.9	5.3	0.6	0.6	0.7	0.1	0.9	0.6
Relative citation index: physics	0.8	0.2	1.0	0.0	0.6	0.5	0.9	0.8	1.1	−0.5
Relative citation index: chemistry	0.8	1.9	0.8	2.0	0.6	0.6	0.9	1.4	1.1	−0.1
Relative citation index: biomedical research	0.7	0.7	0.8	1.0	0.6	1.0	0.9	1.0	1.1	−0.2
Relative citation index: social sciences	0.6	7.8	0.4	7.8	0.3	0.8	0.7	3.3	0.8	−0.4
Publications per (million) HERD	5.3	−1.2	5.3	−0.4	4.0	0.2	8.9	−1.7	6.9	−3.5

Publications per (million) PUBERD	2.7	1.1	2.9	−0.3	2.4	0.3	5.5	0.2	3.5	−1.2
Technology performance										
Triadic patents per (million) capita	39.7	2.6	83.1	2.9	97.0	5.6	33.2	2.8	59.0	4.3
USPTO per (million) capita	75.9	3.1	165.2	3.1	303.1	5.7	79.7	3.4	399.5	4.9
EPO per (million) capita	116.3	4.1	252.8	4.6	145.1	8.0	91.7	3.5	101.7	5.6
Patents (triadic) per (million) BERD	0.1	−0.1	0.2	1.6	0.2	1.6	0.1	1.2	0.1	1.3

Note: Levels in 2003 are the current values in 2003 (as in the case of scientific performance) or an average for the period 1999–2003 for the figures on R&D expenditures (in terms of both percentage of GDP and per capita) and patents. The period for which growth rates are calculated varies across indicators. For all R&D investment indicators and patent indicators, growth rates refer to the period 1981–2003 (except for the newcomer countries, for which R&D data are available from 1991 or later). For world publication shares, growth rates refer to the period 1996–2003. For the relative citation index, the growth rates refer to the period 1995–2003. For citations per paper, the growth rates are calculated for the period 1992–2003.

Table 2A.2   Raw data on position and growth for the group of small, high-tech countries

	Denmark		Finland		Sweden		Switzerland	
	Level 2003	Growth rate (%)	Level 2003	Growth rate (%)	Level 2003	Growth rate (%)	Level 2003	Growth rate (%)
Investment in R&D								
GERD to GDP (%)	2.4	4.4	3.4	5.2	3.9	2.7	2.6	0.9
BERD to GDP (%)	1.6	5.9	2.4	6.3	3.0	3.4	1.9	0.9
GOVERD to GDP (%)	0.2	−0.9	0.4	1.4	0.1	0.3	0.0	−6.7
HERD to GDP (%)	0.5	3.3	0.6	4.4	0.8	1.3	0.6	1.9
Scientific performance								
Citations per paper	2.6	2.9	2.3	3.1	2.4	1.3	3.0	1.2
Relative citation index: engineering	1.0	−2.7	0.6	1.2	0.8	−0.9	1.2	2.1
Relative citation index: physics	1.3	0.3	0.8	−0.2	0.9	0.5	1.3	−0.3
Relative citation index: chemistry	1.1	0.9	0.7	0.2	1.0	−2.3	1.1	0.3
Relative citation index: biomedical res.	0.8	2.7	0.7	1.3	0.7	0.4	1.2	−0.1
Relative citation index: clinical med.	0.8	2.1	0.9	1.6	0.8	0.5	1.0	0.0
Relative citation index: biology	0.9	−0.2	0.8	2.2	1.0	1.1	1.3	2.7
Relative citation index: social sciences	0.8	−2.2	0.7	1.5	0.9	1.7	1.0	7.3
Outputs per unit of investment								
Patents (triadic) per (million) BERD	0.1	−1.4	0.2	7.0	0.1	−0.6	0.2	−1.4
Publications per (million) HERD	7.7	−2.0	6.7	−1.8	5.6	−0.3	6.3	−1.7

Table 2A.3   Raw data on position and growth for the group of newcomer countries

	Brazil		China		India		Singapore		S. Korea		Taiwan	
	Level 2003	Growth rate (%)	Level 2003	Growth rate (%)	Level 2003	Growth rate (%)	Level 2003	Growth rate (%)	Level 2003	Growth rate (%)	Level 2003	Growth rate (%)
Investment in R&D												
GERD to GDP (%)	–	–	1.0	4.0	–	–	2.0	8.0	2.5	3.1	–	–
BERD to GDP (%)	–	–	0.6	8.7	–	–	1.3	7.7	1.9	1.9	–	–
GOVERD to GDP (%)	–	–	0.3	–1.2	–	–	0.3	10.9	0.3	–2.1	–	–
HERD to GDP (%)	–	–	0.1	6.4	–	–	0.5	8.3	0.3	4.8	–	–
Scientific performance												
Citations per paper	1.2	3.7	1.2	6.2	1.0	6.1	1.4	5.7	1.4	5.8	1.3	3.2
Relative citation index: engineering	0.5	2.5	0.5	3.3	0.5	3.4	0.6	8.8	0.6	2.3	0.5	1.5
Relative citation index: physics	0.4	–0.1	0.3	4.7	0.4	4.7	0.6	12.0	0.5	2.6	0.5	4.5
Relative citation index: chemistry	0.4	–0.2	0.4	1.7	0.4	5.0	0.8	2.8	0.5	4.6	0.5	1.6
Relative citation index: biomedical research	0.2	3.6	0.2	–1.4	0.2	7.8	0.7	2.8	0.4	3.6	0.4	1.5
Relative citation index: clinical med.	0.4	1.9	0.5	5.0	0.3	4.1	0.6	4.6	0.6	0.7	0.5	1.0
Relative citation index: biology	0.4	2.4	–	–	–	–	0.5	–0.9	0.6	–1.4	0.4	5.0
Relative citation index: social sciences	0.4	–0.2	0.4	1.7	0.4	5.0	0.8	2.8	0.5	4.6	0.5	1.6

59

Table 2A.3 (continued)

	Brazil		China		India		Singapore		S. Korea		Taiwan	
	Level 2003	Growth rate (%)	Level 2003	Growth rate (%)	Level 2003	Growth rate (%)	Level 2003	Growth rate (%)	Level 2003	Growth rate (%)	Level 2003	Growth rate (%)
Outputs per unit of investment												
Patents (triadic) per (million) BERD	–	–	0.002	1.7	–	–	0.1	6.4	0.05	9.7	–	–
Publications per (million) HERD	–	–	6.1	0.5	–	–	6.1	0.0	5.6	7.2	–	–

Note: – = data not available.

60

3.   How open is innovation?

Linus Dahlander and David Gann

INTRODUCTION

During the past three or four years, the concept of 'open innovation' has received wide attention by practitioners and researchers. This concept implies that firms increasingly rely on external sources of innovation by emphasizing that ideas, resources and individuals flow in and out of organizations (Chesbrough 2003a). In this chapter we assume a critical view of the concept and show that the idea of systematically using external sources of innovation is not particularly new, and that there has been a strong research tradition on this topic for decades. We undertake a critical examination of the concept, as defined by Chesbrough, and expose its weaknesses and limitations. Reviewing earlier work, we demonstrate that the dichotomy between open versus closed is not useful and that openness provides a more interesting avenue to explore. Extending this argument, we propose that openness has been conceptualized and operationalized in fundamentally different ways. In doing so, we chart an area for future studies to explore.

It is important to remember that the concept is gaining in significance in the context of research on university–industry interactions (Fontana et al. 2006; Laursen and Salter 2004), and among policy makers who appear to be promoting greater degrees of collaboration. Emphasis on collaboration between universities and industry is accentuated in the UK, where, in spite of a strong research tradition, it is considered that there is less success in harnessing research expertise than in some other countries. The *Lambert Review*, for example, underpins the importance of the role of industry in raising demand for university research to be commercialized in order for the UK to remain ahead of fierce global competition (Lambert 2003). The underlying assumption is that there is a potential for increased collaboration between industries and universities which would result in greater possibilities to stimulate commercialization and economic performance.

While not exclusively focusing on university–industry interaction, this chapter aims to shed light on open innovation. The idea has become influential in firm strategies and government policies, and it is therefore timely to examine evidence from practice in order to advance the concept further. Scholars have recognized the need to explore the idea in greater detail (Gann 2005; Helfat 2006), yet to date there are few papers systematically undertaking this endeavour.

The chapter is structured as follows. The next section explains the open innovation concept, followed by a discussion of prior literature, illustrating the importance of understanding historical antecedents. The subsequent section reviews strengths and limitations of the concept, focusing on a detailed examination of the literature on different types of openness. The conclusions draw out areas for future research.

WHAT IS OPEN INNOVATION?

The idea that innovations are brought to market through a sequential, linear process has long been superseded by a more realistic account of the innovation process as iterative and involving feedback loops (Kline and Rosenberg 1986). Innovation can stem from a firm's internal investments in R&D, but it is widely acknowledged that firms are active in acquiring important inputs from a variety of external sources including competitors, consumers, public research organizations, universities and other types of organizations (von Hippel 1988). Firms use formal relationships, such as licensing agreements, alliances, joint ventures and other types of contracts, as well as informal relationships (non-contractual personal relationships) to source expertise outside their boundaries (Powell 1990; Powell et al.1996). There has been a plethora of research resulting from these observations which examine the nature of the innovation process in itself (Freeman 1991), in addition to its effect on firm behaviour and performance (Baum et al. 2000).

The practice of searching for and using a wider range of internal and external ideas is captured in the concept of open innovation. According to Henry Chesbrough, whose book on the subject has created much recent debate, the concept challenges the notion that innovation is a single process of internal firm-specific research and development which ultimately leads to commercialization of new ideas (Chesbrough 2003a). Chesbrough (2003a: XXIV) argues that 'open innovation is a paradigm that assumes that firms can and should use external ideas as well as internal ideas, and internal and external paths to market, as firms look to advance their technology'. He argues that what was previously closed within centralized R&D facilities has since been transformed into a more distributed environment with knowledge and information transcending firm boundaries.

Within the closed model, the innovation process is characterized by firms which invest in their own R&D, employing talented researchers and engineers to outperform their competitors in new product and service development. After coming up with a stream of new cutting-edge ideas, firms defend their intellectual property rigorously to prevent competitors from using it.

Chesbrough (2003a) portrays the trend of open innovation as progressive. Firms are becoming increasingly active in acquiring inputs from competitors, universities and other types of organizations. In contrast to the closed model, the open innovation approach contends that an individual firm cannot necessarily attract the most talented employees it needs in order to innovate. Chesbrough argues that it is necessary to develop processes to ensure a flow of ideas across firm boundaries because 'not all the smart people work for you', and because there is increasing geographical dispersion of knowledge. Innovation is becoming increasingly global in its nature and distributed more widely among different types of actors (von Zedtwitz and Gassmann 2002). It is thus necessary to recognize that firms cannot innovate in isolation. They may acquire or license relevant intellectual property and integrate it into internal processes. Furthermore, even though firms do not develop exclusively all research that they use, it is still possible for them to profit from it. Within the old model of closed innovation, firms adhered to the idea that successful innovation requires control. The underlying idea was that 'if you want something done right, you've got to do it yourself'. Boundaries between the firm and the environment are much more porous in the open innovation approach, and the logic is to extract as much knowledge from the external environment as possible (Chesbrough 2006).

Several factors have been identified as driving a trend towards open innovation (Chesbrough et al. 2006; Chesbrough 2003a; Dahlander and Wallin 2006; Gassmann 2006):

1. Globalization that extends geographical markets and enables an increased division of labour and specialization.
2. Improved market institutions, e.g. intellectual property rights (IPR), venture capital (VC), and technology standards that enable resources more easily to transcend firm boundaries.
3. Technological change which reduces the minimum efficient scale of production, fuelling increased specialization.
4. Increased labour market mobility, especially for specialist personnel, in the so-called 'knowledge economy'.
5. New technologies, such as ICTs, have affected the way firms communicate, collaborate and coordinate innovation.

It is argued that with a more distributed environment and a wider variety of organizations of different sizes that can potentially develop new innovations, firms have to make extensive use of external ideas. Firms therefore have to build processes to acquire and integrate such ideas into the organization. Yet there are substantial variations in the degree to which firms use external expertise (Laursen and Salter 2006). Inert structures within organizations favour expertise that has been developed in house, known as the 'not invented here' syndrome (Katz and Allen 1982). Managers thus tend to overestimate the value of internally developed ideas, and cannot realize the full potential of external ideas. Internal ideas may also not be used in the most fruitful way. When R&D managers evaluate the performance of their organization, they often use patents and publications as a metric. R&D departments respond by generating a vast number of patents and publications, with little regard to their eventual business relevance (Chesbrough 2006). Firms therefore develop a wide array of technologies that will never be used or brought to the market and which remain on the shelf.

The open innovation approach underscores the importance of intellectual property strategies. Patents provide opportunities for firms to overcome the information problem where the innovator has to reveal the innovation to a potential licensee (Arrow 1962). Information problems cause market failures because inventors are reluctant to reveal their developments. Open innovation requires that buyer and seller reach an agreement, so appropriability regimes allow the seller to disclose information. The contention is thus that intellectual property is important in order to trade innovations (West 2006). If competitors use intellectual property, the innovating firm has to be able to profit from it.

Chesbrough's work draws heavily on case studies of large US companies in high-tech industries such as Lucent, Intel, 3Com and Millennium Pharmaceuticals (Chesbrough 2003a). Exploring other empirical settings, however, is critical in gaining external validity (Chesbrough and Crowther 2006). Indeed, there are few studies exploring open innovation using large-scale data sets over a variety of industries (with the notable exception of Laursen and Salter 2006). This suggests that we have quite limited evidence of how widespread the phenomenon is in practice.

SOME HISTORICAL ANTECEDENTS OF OPEN INNOVATION

In spite of its rapid diffusion among academics and practitioners, surprisingly few studies critically examine open innovation. In many ways, the strength of the idea is its breadth: put simply, it claims that 'valuable ideas can come from inside or outside the company and can go to market from inside or outside the company as well' (Chesbrough et al. 2006: 1). Chesbrough draws on a long legacy of research underscoring the open, distributed and interactive nature of innovation. It is, however, unclear whether the model can be substantiated against evidence of a previous era when firms operated a 'closed' innovation approach.

Chesbrough brings to the fore a number of stimulating case studies of open innovation that illustrates the research and development at AT&T and Lucent, Intel, IBM and Xerox PARC. His overall argument is that we have witnessed a shift among these organizations to a greater openness in the innovation process, with inflow and outflow of ideas, resources and people. In reality, however, firms have always sourced expertise from partners. The Edison laboratory is cited as a typical twentieth-century closed innovation behemoth. Yet, at least in its origins as the Menlo Park 'invention factory' of the late nineteenth century, it displayed characteristics remarkably similar to those of the open innovation approach. As Hargadon (2003) shows, the development of the light bulb was the product of a team of engineers, recombining ideas from previous inventions and collaborating with scientists, engineers, financiers and people in marketing outside the laboratory. This team was adept at operating through and within networks of innovators, and Menlo Park became an early example of the contract research laboratory.

Another case in point is Chandler's (1990) classical example of DuPont as a vertically integrated organization, which in Chesbrough's terminology would be a typical closed innovator. A closer investigation of this particular organization reveals that the R&D department was not closed as some might have assumed. Indeed, the R&D organization played a central role in developing cartels with competitors in order to compete with third parties (Hounshell and Smith 1988). Building on their extensive case study, Hounshell (1996: 136) asserts that 'sharing technical information across firm boundaries and international borders no doubt had major effects on firm capabilities'. The R&D boundaries of DuPont in fact appear to have been quite porous:

> Through the regular exchange of thousands of research reports, through the continual flow of research delegations from one company to another, and through the developing network of formal and informal communications between researchers in various divisions and laboratories of the two companies, DuPont and ICI gained an enormous amount of information that would have been impossible to obtain any other way. Researchers and research managers also learned a good deal about differing corporate research styles, differing university research traditions, and differing strategic thinking by corporate executives. (Hounshell 1996: 137)

In this regard, a deeper historical perspective would be useful in positioning Chesbrough's thesis. Moreover, Stan Metcalfe has pointed out that almost one hundred years ago, in his *Industry and Trade*, Marshall (1919: 101) talks about the 'growth of external economies associated with the wider publication of research results, providing the possibility for small(er) firms to innovate . . . if leadership is available'. He argues that external organization becomes more important as the innovation system opens up due to wider dissemination of results. Marshall links this to the division of labour and growth of knowledge, which he says does not exist in the 'ether'. Firms therefore have to organize to become embedded in the wider process of knowledge generation and knowledge capture.

Firms have often voluntarily revealed their achievements to other firms, as Allen (1983) discovered in his work on the iron production industry in England in the nineteenth century. Firms in this industry regularly shared their designs in verbal interaction as well as in published material. This resulted in collective invention where firms built upon others' work in a truly distributed and open manner. Other accounts of corporate R&D laboratories show that they are not a 'castle on the hill', but vehicles for absorbing science and methods in internal firm processes to make them better and more efficient (Freeman 1974). The SAPPHO project showed that firms are relying on many external sources of innovation (Rothwell et al. 1974), a point that is underscored in Kline and Rosenberg's (1986) paper, which highlighted complex feedback loops and interactions in innovation. Indeed, any criticism of the linear model of innovation, as practised by a wide range of innovation scholars, can be construed as an argument for open innovation. At the latest, Rothwell's fifth-generation innovation model represents a formal presentation of the idea (Rothwell 1994). As Rothwell (1994: 19) put it more than a decade ago, 'accessing external know-how has long been acknowledged as a significant factor in successful innovation'.

It has been shown that connectivity with external actors is important in order for firms to remain innovative (Freeman 1991), and in the network literature it is commonly argued that firms benefit from the social landscape in which they are embedded. Scholars in this vein have developed important findings as to how certain network structures influence firm behaviour and performance (Ahuja 2000; Baum et al. 2000; Gulati et al. 2000). Relationships with other actors helps firms to absorb technology (Ahuja 2000), improve survival rates (Baum and Oliver 1991), increase innovativeness (Baum et al. 2000; Stuart 2000), improve performance (Hagedoorn and Schakenraad 1994; Shan et al. 1994) and grow faster (Powell et al. 1996; Stuart 2000). This literature is rich with descriptions about what conditions formal and informal relationships and influences firms. Creating and sustaining ties with other actors constitutes a relational capability and creates benefits for firms that master it (Lorenzoni and Lipparini 1999). This is close to Chesbrough's idea, in which the firm's value is contingent upon its ability to create and lay claim to knowledge derived from participation in various kinds of collaborations with other actors.

Other literature takes a more micro-approach to analyse how external sources flow into the organization. This is at the heart of analysis of boundary objects and boundary spanning, where research has focused on, for example, how interactions across formal organizational boundaries facilitate the development of new innovations in high-tech industries, using individual people as the unit of analysis (Allen 1977; Tushman 1977). In this work, formal organizational boundaries are the focal point because information, resources and mindsets tend to be more alike within the same organization, or interest group. The ability to span boundaries therefore provides people with access to a broader array of ideas and opportunities. Boundary-spanning literature explores how boundary-spanning individuals facilitate the transfer to an R&D unit from the external environment (Allen and Cohen 1969). Allen and Cohen, for instance, show that central individuals in the communication patterns in R&D labs are more likely to rely on individuals outside the unit and read the literature more than other members of the lab. Within this stream of work, external communication flows are defined as information sharing between someone within the focal R&D unit and someone external to the unit. Such external relations can therefore come from the same formal organization, as well as relations with someone working for another organization (Allen 1977; Ancona and Caldwell 1992). Boundary spanners have rich contacts, both within and outside the unit. Bmundary spanners that are critical in linking the organization with the external environment are therefore considered to have gone through a multistep process, in which they have built internal and external networks in order to span the organizational boundaries (Tushman and Scanlan 1981). Through integrating and linking to external ideas, such individuals are considered to improve the R&D unit's innovative performance (Tushman and Katz 1980).

The user innovation literature also adopts a more micro-approach to innovation. Von Hippel's (1988) idea of distributed innovation is very closely related to open innovation. Von Hippel focuses on user-driven innovation, arguing that it is often users, and not manufacturers, who develop ideas for solutions to problems, sometimes customizing or modifying products and services without any involvement from the supplier. In his recent book, *Democratizing Innovation*, he suggests that distributed innovation processes are a widespread phenomenon (von Hippel 2005). He also investigates some mechanisms that firms can adopt in order to better harness distributed innovation processes. For instance, von Hippel's work on user tool-kits for innovation links into Chesbrough's work on business models (von Hippel 2001; von Hippel and Katz 2002).

DIFFERENT KINDS OF OPENNESS

Open innovation has received great attention since Chesbrough's book was published in 2003. A large number of papers are adopting this term to denote the interactive and distributed nature of innovation. A Google search for 'open innovation' indicates some 383 000 hits and a search in the title, abstract or keywords in Science Citation Index yields 27 published papers. The concept has been highly influential, among both academics and practitioners. Characterizing firms as closed or open with a binary classification, however, yields little insight into the more intricate question of how openness influences firms. We contend that the idea behind openness needs to be placed on a continuum, ranging from closed to open and covering varying degrees of openness. By adopting this reasoning, one can begin to think about the possible limitations of openness. From the review of prior literature, we know that innovation has always had *some* degree of openness, but the question is *how much*, and in *what ways*. Furthermore, open innovation is becoming an imperative for many organizations and for some, such as P&G, it is a strategic priority, and it is therefore necessary to consider under which circumstances this is a beneficial strategy.

At least three types of openness can be inferred from the literature:

1. Appropriability – different degrees of formal and informal protection
2. The number of sources of external innovation
3. The degree to which firms are relying on informal and formal relationships with other actors.

While related to some extent, these can be operationalized in fundamentally different ways, and it is not clear how or whether these types of openness are correlated.

Openness in Terms of Appropriability

Some scholars interested in open innovation have been concerned with how much appropriability is beneficial (Chesbrough 2003b; von Hippel and von Krogh 2003). In her critical examination of the book, Helfat (2006) suggests that Chesbrough does not pay sufficient attention to the potential limits of open innovation with regard to appropriability. With a more distributed and dispersed innovation process, there may be difficulties in appropriating returns. Indeed, it can be difficult for firms to detect expropriation or illegal imitation among competitors (Liebeskind 1996). In weak appropriability regimes, firms have little incentive to invest in R&D (Levin et al. 1987). Chesbrough mentions greater reliance on external actors, through licensing and collaborations and a greater decentralization of internal research and development. However, little is said about how returns are distributed among the actors involved (Gann 2005).

The appropriability regime governs an innovator's ability to capture the profits generated by an innovation (Teece 1986). Firms usually adopt both formal methods (such as patent, trademark or copyright protection) as well as informal methods (lead times, first-mover advantages, lock-ins) in their appropriability strategies. The premise is that openness caused by voluntarily or unintentionally divulging information to outsiders does not always reduce the probability of being successful (Henkel 2006; von Hippel and von Krogh 2003). Henkel (2006), for instance, suggests that firms adopt strategies to selectively reveal some of their technologies to the public in order to elicit collaboration, but without any contractual guarantees of obtaining it.

Indeed, in the absence of strong intellectual protection, there are greater chances of cumulative advancements in some cases (Levin et al. 1987). This contention is highlighted in the existence of phenomenon such as Wikipedia, free and open-source software and MySpace, where individuals collectively develop innovative solutions. This exemplifies an extreme form of organizing open innovation, where the absence of intellectual property ensures accessibility for everyone involved. In the literature on standards, it is also well known that being open and focusing less on ownership increases the opportunities to get other parties interested.

Laursen and Salter (2006), for instance, suggest that if firms place too strong an emphasis on protection of their knowledge, the result may be what they call a 'myopia of protectiveness'. Their idea is that firms may become obsessed with ownership, instead of focusing on marshalling resources and support from the external environment that is necessary to bring inventions into commercial applications and services. This suggests that firms use combinations of means of protection and thereby balance the relative inefficiency of formal protection by putting greater emphasis on alternative methods (López and Roberts 2002).

Some scholars suggest that stronger intellectual property (IP) regimes are associated with a higher reliance on external actors. Gallini (2002) predicted a relationship that with stronger IP, there would be a higher willingness to out-license. This idea is accentuated in Laursen and Salter (2006), who discovered that relying on many external sources of innovation is higher in industries with formal protection for ideas. They accentuate Chesbrough's (2003b) point that firms that have good patent protection may find it easier to engage in relationships with other organizations, as they have protection for their ideas and can respond to opportunistic behaviour among partners with normative and legal sanctions.

Openness in Terms of Relationships

When some scholars talk about openness, they refer to the vast literature on different kinds of interfirm relationships (van de Vrande et al. 2006). This is underscored in Chesbrough's (2003a) work, which portrays the trend of open innovation as progressive, where firms are becoming increasingly active in acquiring inputs from competitors, universities and other types of organizations. Chesbrough also points out that it is necessary to have some internal expertise, and that management is about leveraging internal R&D. Chesbrough et al. (2006: 290) claim that 'In the open innovation approach, firms scan the external environment prior to initiating internal R&D work. If a technology is available from outside, the firm uses it.' Following this reasoning, openness can be understood as the number and forms of external relationships with sources of technology or other actors. While acknowledging the importance of openness in terms of external sources of innovation, von Zedtwitz and Gassmann (2002) state that in order to invest in open innovation activities, firms need some degree of control over a number of the elements in their networks. Although there are many benefits of being able to buy or in-source external ideas to the organization, it requires expertise for searching and evaluating them.

Some literature underlines the need to have internal resources in order to discover developments in the external environment. Firms need to develop their absorptive capacity to assimilate developments in the external environment (Cohen and Levinthal 1990; Lane and Lubatkin 1998; Zahra and George 2002). Research has shown that firms need to have competencies in areas related to the partner's so as to assimilate external sources (Brusoni et al. 2001; Granstrand et al. 1997; Mowery et al. 1996). Conducting research is also seen as a prerequisite for being perceived as an attractive partner for others to engage in relations with (Rosenberg 1990), and in-house R&D is therefore one type of ticket to partnering with other organizations. Developing absorptive capacity through R&D to assimilate external knowledge provides an explanation for why firms invest in R&D despite the problem that findings potentially may leak to competitors. Firms therefore have a dual incentive to invest in R&D because it yields new information *and* provides the capacity to assimilate knowledge from networks. Firms vary in the extent to which they can screen, evaluate and assimilate external inputs to the innovation process. Internal capabilities and external relations are therefore to be seen not as substitutes but as complementary. The ability to absorb external inputs depends on what the firm knows. Firms presumably have to spend considerable time and resources on internal R&D, and this leads to a question of the right balance between internal and external sources of innovation. We suggest that it is the exception, rather than the rule, that firms use external R&D as a substitute for internal R&D. Indeed, there are sometimes economies of both scale and scope in R&D (Henderson and Cockburn 1996). What might be new is the balance of innovation and whether that has shifted, not least because having internal R&D capabilities can be argued to be more important when relying heavily on relationships with other actors (Helfat 2006).

Another point is related to the similarity of knowledge bases and how they facilitate the integration of ideas from distant realms (Kogut and Zander 1992), because shared language, common norms and cognitive configurations permit communication (Cohen and Levinthal 1990). In absorbing new knowledge, the firm also increases its possibility to make novel recombinations. Incorporating knowledge bases too close to what the firm already knows will hamper the positive effect of assimilating external inputs. For instance, Ahuja and Katila (2001) have suggested that knowledge relatedness between the acquiring and acquired firms is curvilinearly related to innovative performance. Too distant inputs are harder to align with existing practices, and with too similar knowledge bases it is difficult to come up with novel combinations (Sapienza et al. 2004). In other words, the effectiveness of openness is also contingent upon the resource endowments of the partnering organization.

Openness in Terms of External Sources of Innovation

Laursen and Salter (2004: 1204) define openness as 'the number of different sources of external knowledge that each firm draws upon in its innovative activities'. Their logic is that the more external sources of innovation, the more open a firm's search strategy. This is highlighted in Chesbrough's work, underscoring that innovation is often about leveraging the discoveries of others. Firms that manage to create a synergy between what the firm does and the external environment are able to benefit from the creative ideas of outsiders, while still being able to make money. Available resources become larger than a single firm can possess, allowing for innovative ways to go to the market, or creating standards in emerging markets. Such synergies can be created by relying on the external environment, by taking an active part in external developments. Rather than looking only at the benefits of openness in terms of the external sources of innovation, some scholars have begun to stress the potential limitations. Menon and Pfeffer (2003), for instance, propose that there is a managerial paradox in that firms overvalue and overuse external knowledge compared to rich internal knowledge, from which value can be captured much more easily. They argue that the preference for external knowledge is the result of managerial responses to the availability or scarcity of knowledge. Internal knowledge is more readily available and hence subject to greater scrutiny, while external knowledge is scarcer, which makes it appear more special and unique.

Simon (1947) observed that individuals can focus properly on only a few tasks at any point in time. There are cognitive limits to how much we can understand. Some organizations oversearch by spending too much time looking for external sources of innovation. Building on this reasoning, Katila and Ahuja (2002) propose that search behaviour is critical in understanding the limits and contingent effects on innovation. Based on a study of the industrial robotics industry, they suggest that some firms oversearch and that there is thus a curvilinear relationship between innovative performance and their search for new innovations. Laursen and Salter (2006) extend this reasoning by looking at external sources of innovation. Drawing on a sample of a large UK survey of 2707 manufacturing firms, they show that searching both widely and deeply for sources of innovation is curvilinearly related to innovative performance. In other words, while there may be an initial positive effect on openness, firms can oversearch or rely too heavily on external sources of innovation.

It is not obvious that all firms will rely on external partners. In fact, there are substantial variations in the degree to which firms adopt open innovation (Laursen and Salter 2004), and the degree of openness varies according to external sources of innovation as technologies mature (Christensen et al. 2005).

DISCUSSION

Implications for Theory

In spite of the upsurge of interest in the open innovation idea, there have been surprisingly few systematic reviews and critiques of it. Chesbrough et al. (2006) develop a research agenda for open innovation based on the 'paradigm shift' they argue has taken place from closed to open approaches. When reviewing earlier literature, however, it is clear that innovation has always been somewhat open.

We suggest that the open innovation literature needs to be much more precise in defining the core concepts. Chesbrough et al. (2006: 132) suggest that 'open innovation is both a set of practices for profiting from innovation, and also a cognitive model for creating, interpreting and researching those practices'. This definition is obviously quite broad and raises questions about how evidence can be gained to assess the nature of changes in innovation processes. Indeed, in this chapter we show that open innovation is not a particularly new idea and may not constitute a new paradigm in the organization of innovation. Firms have always relied on some degree of outflow and inflow of ideas, resources and individuals. A large body of literature pre-dates the recent work championed by Chesbrough and some of the closed innovators in Chesbrough's work actually shared many characteristics of open innovation (Hargadon 2003).

We suggest that making a dichotomy between closed and open approaches to innovation may be useful in explaining two 'ideal types', but this is likely to direct attention from the more relevant issues associated with the degree of openness and the extent to which firms can organize their search and development activities to benefit from multiple sources of ideas. We think that research on the degree of openness could provide richer opportunities to develop new findings about advantages and disadvantages of a variety of search strategies, collaboration arrangements and the use of new technologies to connect to ideas. In this chapter, we inferred three types of openness from the literature: (1) appropriability and the degree of formal and informal protection; (2) the number and sources of external innovation; and (3) the degree to which firms are relying on informal and formal relationships with other actors. From these it appears that researchers are discussing quite different issues under the open innovation umbrella. Although related in many ways, these three types of openness are not necessarily correlated in empirical studies. If this is not the case, openness may have different implications for firms.

In some instances, studies may be preoccupied with exploring the optimal level of openness, rather than probing how openness has changed in a qualitative sense. Openness is sometimes treated as an exogenous factor given by the environment, but through a proactive strategy firms can adopt practices to treat openness as endogenous. Indeed, perhaps openness takes different shapes today compared to the past, particularly given the availability of a new infrastructure to support innovation, what some have called 'innovation technology' (Dodgson et al. 2005). Firms may adopt new practices to cope with openness and create competitive businesses. In this regard, research could benefit from concentrating studies more explicitly on the particular nature and context of external sources of innovation (Gassmanns 2006). Implicitly, external knowledge is considered to be 'out there' to be harnessed by firms, and we have limited understanding of the process of sourcing this into corporations. This directs some of the attention away from the optimal degree of openness, to consider also new mechanisms to facilitate open innovation processes.

Implications for Practice

Although the idea of open innovation has been influential and the practices appear to be adopted by many organizations (Chesbrough et al. 2006; Gassmann and Enkel 2004), we try to highlight that the message of open innovation is far more complicated than 'the more openness, the better'. It can be costly and not always easy to have a high degree of openness (Laursen and Salter 2006). The success of open innovation can differ across technologies and industries (Christensen et al. 2005). It is therefore critical to attend to barriers and limits that can be identified, to bring credible insights to practitioners.

It may be useful to practitioners to understand the types of brokering and boundary-spanning capabilities that can support successful integration of ideas from multiple sources. There may be difficulties in evaluating external ideas compared to internal, as there is much less first-hand information available for external ideas (Menon and Pfeffer 2003). Being more involved in open innovation can therefore create tension with other practices within the organization. This begs questions about how firms might operationalize strategies for benefiting from more open approaches; what mechanisms can be implemented and how resources and capabilities are deployed to support them.

Charting Out a Research Agenda

Whilst research focuses on the motivations for openness and the implications on performance (Fey and Birkinshaw 2005; Laursen and Salter 2006), there is quite limited understanding of how external ideas travel into and out of organizations to provide beneficial outcomes. Open innovation research has recently begun to address the role of individuals, but the unit of analysis is in most instances focused on how firms compete in order to gain superior access to resources from outside firm boundaries (Chesbrough et al. 2006). Indeed, even though Chesbrough et al. discuss different units of analysis for open innovation, most of the literature is focused on how *firms* can stay ahead of competition when there is high outflow and inflow of ideas across firm boundaries. When speaking of firms as a whole, there is a risk of black-boxing interactions at the heart of innovation processes that transcend organizational boundaries. Individual networks within and across firm boundaries provide a much more fine-grained analysis of how the innovation process occurs through smaller events and social practices (Allen 1977; Tushman 1977). Rather than focusing on the structural characteristics of firms that decide to source expertise from universities (Chesbrough 2003a), one can analyse the influence of individual networks in how external ideas are brought into organizations. By doing so, it is possible to identify the adopters of external sources of innovation within the organization. The underlying issue is that individuals might not only differ with regard to their personal attributes and capabilities, but also with regard to their network positions. For instance, central individuals may have better possibilities than peripheral individuals in understanding the needs of the organization. Yet peripheral individuals may be more 'open' to counter-intuitive ideas for problem solving. In recent work on network theory, scholars have begun to explore how internal and external networks influence the probability of coming up with creative ideas (Perry-Smith 2006). While there are many descriptions of how external relations influence performance, less research focuses on the underlying decision process. This is important as firms face difficulties in maintaining a large number of relations. One could thus expect inertia in the search process, suggesting consistent patterns of collaboration over time due to socialization processes. Rather than hedging risk and avoiding redundancy through maintaining diverse relations, there may be persistent pathways in sourcing knowledge. Inertia in search processes for external expertise implies that relations that do not meet expectations might continue. This could result in over-embeddedness, with too little diversity of partners, raising another set of questions for empirical exploration.

Another theme not examined in the literature is how new technologies spur the development of open innovation processes (Gann 2005). These technologies allow for innovation to be exchanged and diffused rapidly with reduced transmission costs and a larger potential range and number of participants. Early research explored what electronic tool-kits imply for the innovation process, and later work suggests that the emergence of innovation technologies leads to an intensification of innovation and reduced uncertainty of outcomes (Dodgson et al. 2005). Acquiring technology and sharing information has spawned a range of new business opportunities over the past decade. For example, companies such as Eli Lilly have launched a knowledge brokering service, www.innocentive.com, importing and exporting intellectual capital (Lakhani et al. 2006). P&G has opened 20 web portals through which scientists work with those outside the company, including www.yourencore.com, which aims to connect to the ideas of engineers and technologists who have retired from P&G (Dodgson et al. 2006).

REFERENCES

Ahuja, G. (2000), 'Collaboration networks, structural holes and innovation: a longitudinal study', *Administrative Science Quarterly*, **45**: 425–55.

Ahuja, G. and Katila, R. (2001), 'Technological acquisitions and the innovation performance of acquiring firms: a longitudinal study', *Strategic Management Journal*, **22**(3): 197–220.

Allen, R.C. (1983), 'Collective invention', *Journal of Economic Behaviour and Organization*, **4**(1): 1–24.

Allen, T. (1977), *Managing the Flow of Technology: Technology Transfer and the Dissemination of Technological Information Within the R and D Organisation*, Cambridge, MA: The MIT Press.

Allen, T.J. and Cohen, S.I. (1969), 'Information flow in research and development laboratories', *Administrative Science Quarterly*, **14**(1): 12–19.

Ancona, D.G. and Caldwell, D.F. (1992), 'Bridging the boundary: external activity and performance in organizational teams', *Administrative Science Quarterly*, **37**: 634–65.

Arrow, K.J. (1962), 'Economic welfare and the allocation of resources for invention', in R. Nelson (ed.), *The Rate and Direction of Inventive Activities*, Princeton, NJ: Princeton University Press, pp. 609–25.

Baum, J.A.C. and Oliver, C. (1991), 'Institutional linkages and organisational mortality', *Administrative Science Quarterly*, **31**: 187–218.

Baum, J.A.C., Calabrese, T. and Silverman, B.S. (2000), 'Don't go it alone: alliance network composition and startups' performance in Canadian biotechnology', *Strategic Management Journal*, **21**: 267–94.

Brusoni, S., Prencipe, A. and Pavitt, K. (2001), 'Knowledge specialization, organizational coupling, and the boundaries of the firm: why do firms know more than they make?', *Administrative Science Quarterly*, **46**: 597–621.

Chandler, A.D. (1990), *Scale and Scope: The Dynamics of Industrial Capitalism*, Cambridge, MA: The Belknap Press of Harvard University Press.

Chesbrough, H. (2003a), *Open Innovation: The New Imperative for Creating and Profiting from Technology*, Boston, MA: Harvard Business School Press.

Chesbrough, H. (2003b), 'The logic of open innovation: managing intellectual property', *California Management Review*, **45**(3): 33.

Chesbrough, H. (2006), 'New puzzles and new findings', in H. Chesbrough, W. Vanhaverbeke and J. West (eds), *Open Innovation: Researching a New Paradigm*, Oxford: Oxford University Press, pp. 15–34.

Chesbrough, H. and Crowther, A.K. (2006), 'Beyond high tech: early adopters of open innovation in other industries', *R&D Management*, **36**(3): 229–36.

Chesbrough, H., Vanhaverbeke, W. and West, J. (eds) (2006), *Open Innovation: Researching a New Paradigm*, Oxford: Oxford University Press.

Christensen, J.F., Olesen, M.H. and Kjær, J.S. (2005), 'The industrial dynamics of open innovation: evidence from the transformation of consumer electronics', *Research Policy*, **34**(10): 1533–49.

Cohen, W.M. and Levinthal, D.A. (1990), 'Absorptive capacity: a new perspective on learning and innovation', *Administrative Science Quarterly*, **35**(1): 128–52.

Dahlander, L. and Wallin, M.W. (2006), 'A man on the inside: unlocking communities as complementary assets', *Research Policy*, **35**(8): 1243–59.

Dodgson, M., Gann, D. and Salter, A. (2005), *Think, Play, Do: Markets, Technology and Organization*, London: Oxford University Press.

Dodgson, M., Gann, D. and Salter, A. (2006), 'The role of technology in the shift towards open innovation: the case of Procter & Gamble', *R&D Management*, **36**(3): 333–46.

Fey, C. and Birkinshaw, J. (2005), 'External sources of knowledge, governance mode and R&D performance', *Journal of Management*, **31**(4): 597–621.

Fontana, R., Geuna, A. and Matt, M. (2006), 'Factors affecting university–industry R&D projects: the importance of searching, screening and signalling', *Research Policy*, **35**: 309–23.

Freeman, C. (1974), *The Economics of Industrial Innovation*, London: Pinter.

Freeman, C. (1991), 'Networks of innovators: a synthesis of research issues', *Research Policy*, **20**: 499–514.

Gallini, N.T. (2002), 'The economics of patents: lessons from recent US patent reform', *Journal of Economic Perspectives*, **16**(2): 131.

Gann, D.M. (2005), 'Review of H. Chesbrough, *Open Innovation: The New Imperative for Creating and Profiting from Technology*', *Research Policy*, **34**(1): 122–3.

Gassmann, O. (2006), 'Opening up the innovation process: towards an agenda', *R&D Management*, **36**(3): 223–8.

Gassmann, O. and Enkel, E. (2004), 'Towards a theory of open innovation: three core process archetypes', Proceedings of the R&D Management Conference (RADMA), Lisbon, Portugal, 6–9 July.

Granstrand, O., Patel, P. and Pavitt, K. (1997), 'Multi-technology corporations: why they have "distributed" rather than "distinctive core" competencies', *California Management Review*, **39**(4): 8–25.

Gulati, R., Nohria, N. and Zaheer, A. (2000), 'Strategic networks', *Strategic Management Journal*, **21**: 203–15.

Hagedoorn, J. and Schakenraad, J. (1994), 'The effect of strategic technology alliances on company performance', *Strategic Management Journal*, **15**(4): 291–311.

Hargadon, A.B. (2003), *How Breakthroughs Happen: The Surprising Truth about How Companies Innovate*, Cambridge, MA: Harvard Business School Press.

Helfat, C.E.C. (2006), 'Review of H. Chesbrough, *Open Innovation: The New Imperative for Creating and Profiting from Technology*', *Academy of Management Perspectives*, **20**(2): 86.

Henderson, R. and Cockburn, I. (1996), 'Scale, scope, and spillovers: the determinants of research productivity in drug discovery', *The RAND Journal of Economics*, **27**(1): 32–59.

Henkel, J. (2006), 'Selective revealing in open innovation processes: the case of embedded Linux', *Research Policy*, **35**(7): 953–69.

Hounshell, D.A. and Smith, J.K. (1988), *Science and Corporate Strategy: DuPont R&D, 1902–1980*, New York: Cambridge University Press.

Hounshell, D.A. (1996), 'Pondering the globalization of R&D: some new questions for business historians', *Business and Economic History*, **25**(2): 131–43.

Katila, R. and Ahuja, G. (2002), 'Something old, something new: a longitudinal study of search behaviour and new product introduction', *Academy of Management Journal*, **45**(8): 1183–94.

Katz, R. and Allen, T.J. (1982), 'Investigating the not invented here (NIH) syndrome: a look at the performance, tenure, and communication patterns of 50 R&D project groups', *R&D Management*, **12**(1): 7–19.

Kline, S.J. and Rosenberg, N. (1986), 'An overview of innovation', in R. Landau and N. Rosenberg (eds), *The Positive Sum Strategy: Harnessing Technology for Economic Growth*, Washington, DC: National Academy Press, pp. 275–305.

Kogut, B. and Zander, U. (1992), 'Knowledge of the firm, combinative capabilities, and the replication of technology', *Organization Science*, **3**(3): 383–97.

Lakhani, K.R., Jeppesen, L.B., Lohse, P.A. and Panetta, J.A. (2006), 'The value of openness in scientific problem solving', HBS Working Paper Number 07-050.

Lambert, R. (2003), *Lambert Review of Business–University Collaboration*, London: HM Treasury.

Lane, P.J. and Lubatkin, M. (1998), 'Relative absorptive capacity and interorganization learning', *Strategic Management Journal*, **19**: 461–77.

Laursen, K. and Salter, A. (2004), 'Searching high and low: what types of firms use universities as a source of innovation?', *Research Policy*, **33**(8): 1201–15.

Laursen, K. and Salter, A.J. (2006), 'Open for innovation: the role of openness in explaining innovation performance among UK manufacturing firms', *Strategic Management Journal*, **27**: 131–50.

Levin, R.C., Klevorick, A.K., Nelson, R. and Winter, S. (1987), 'Appropriating the returns from industrial research and development', *Brookings Papers on Economic Activity*, **3**: 783–831.

Liebeskind, J.P. (1996), 'Knowledge, strategy, and the theory of the firm', *Strategic Management Journal*, **17** (Winter special issue): 93–107.

López, L.E. and Roberts, E.B. (2002), 'First-mover advantages in regimes of weak appropriability: the case of financial services innovations', *Journal of Business Research*, **55**: 997–1005.

Lorenzoni, G. and Lipparini, A. (1999), 'The leveraging of interfirm relationships as a distinctive organizational capability: a longitudinal study', *Strategic Management Journal*, **20**(4): 317–38.

Marshall, A. (1919), *Industry and Trade*, London: Macmillan.

Menon, T. and Pfeffer, J. (2003), 'Valuing internal vs. external knowledge: explaining the preference for outsiders', *Management Science*, **49**(4): 497–513.

Mowery, D., Oxley, J. and Silverman, B. (1996), 'Strategic alliances and interfirm knowledge transfer', *Strategic Management Journal*, **17** (Winter special issue): 77–91.

Perry-Smith, J.E. (2006), 'Social yet creative: the role of social relationships in facilitating individual creativity', *Academy of Management Journal*, **49**(1): 85–101.

Powell, W.W. (1990), 'Neither market nor hierarchy: network forms of organization', *Research in Organizational Behavior*, **12**: 295–336.

Powell, W.W., Koput, K. and Smith-Doerr, L. (1996), 'Interorganizational collaboration and the locus of innovation: networks of learning in biotechnology', *Administrative Science Quarterly*, **41**: 116–45.

Rosenberg, N. (1990), 'Why do firms do basic research (with their own money)?', *Research Policy*, **19**: 165–74.

Rothwell, R. (1994), 'Towards the fifth-generation innovation process', *International Marketing Review*, **11**(1): 7–31.

Rothwell, R., Freeman, C., Horseley, A., Jervis, V.T.P. and Townsend, J. (1974), 'SAPPHO updated – Project Sappho Phase II', *Research Policy*, **3**: 204–25.

Sapienza, H.J., Parhankangas, A. and Autio, E. (2004), 'Knowledge relatedness and post-spin-off growth', *Journal of Business Venturing*, **19**(6): 809–29.

Shan, W., Walker, G. and Kogut, B. (1994), 'Interfirm cooperation and startup innovation in the biotechnology industry', *Strategic Management Journal*, **15**(5): 387–94.

Simon, H.A. (1947), *Administrative Behavior: A Study of Decision-Making Processes in Administrative Organizations*, Chicago, IL: Macmillan.

Stuart, T.E. (2000), 'Interorganizational alliances and the performance of firms: a study of growth and innovation rates in high-technology industry', *Strategic Management Journal*, **21**(8): 791–811.

Teece, D.J. (1986), 'Profiting from technological innovation: implications for integration, collaboration, licensing and public policy', *Research Policy*, **15**: 285–305.

Tushman, M.L. (1977), 'Special boundary roles in the innovation process', *Administrative Science Quarterly*, **22**: 587–605.

Tushman, M.L. and Katz, R. (1980), 'External communication and project performance: an investigation into the role of gatekeepers', *Management Science*, **26**: 1071–85.

Tushman, M.L. and Scanlan, T.J. (1981), 'Boundary spanning individuals: their role in information transfer and their antecedents', *Academy of Management Journal*, **24**: 289–305.

Van de Vrande, V., Lemmens, C. and Vanhaverbeke, W. (2006), 'Choosing governance modes for external technology sourcing', *R&D Management*, **36**(3): 347–63.

Von Hippel, E. (1988), *The Sources of Innovation*, New York: Oxford University Press.

Von Hippel, E. (2001), 'User toolkits for innovation', *Journal of Product Innovation Management*, **18**: 247–57.

Von Hippel, E. and Katz, R. (2002), 'Shifting innovation to users via toolkits', *Management Science*, **48**(7): 821–33.

Von Hippel, E. and von Krogh, G. (2003), 'Open source software and the "private-collective" innovation model: issues for organization science', *Organization Science*, **14**(2): 209–23.

Von Hippel, E. (2005), *Democratizing Innovation*, Cambridge, MA: The MIT Press.

Von Zedtwitz, M. and Gassmann, O. (2002), 'Market versus technology driven in R&D internationalisation: four different patterns of managing research and development', *Research Policy*, **31**(4): 569–88.

West, J. (2006), 'Does appropriability enable or retard open innovation?', in H. Chesbrough, W. Vanhaverbeke and J. West (eds), *Open Innovation: Researching a New Paradigm*, Oxford: Oxford University Press, pp. 109–33.

Zahra, S.A. and George, G. (2002), 'Absorptive capacity: a review, reconceptualisation, and extension', *Academy of Management Review*, **27**(2): 185–203.

4. Innovation policy as cargo cult: myth and reality in knowledge-led productivity growth

Alan Hughes

In the immediate post-Second World War years a series of millenarian movements known as cargo cults[1] swept through Melanesia. They emerged in the aftermath of intensive US contact in the course of the Second World War. These contacts led to a substantial increase in the material goods available to Melanesian islanders, but the end of the war meant that such material goods became less available as military withdrawal occurred. In these circumstances cargo cults emerged in which prophets would promise the return of cargoes of material goods by their ancestors (often expected to take the form of the Americans) with cargo typically shipped in the airplanes that had been such a common feature of the war experience. The means by which the return of the cargo was to be encouraged varied between different cults in different islands, but frequently involved the ritual preparation and construction of a variety of structures such as airfields, storage facilities, landing strips and associated paraphernalia. Cult members were encouraged to abandon previous cultural practices and often mimicked the behavioural characteristics of Americans (Worsley 1957; Jarvie 1964). The emergence of these cults did not lead to the return of material cargo.

There is in my view a danger today that the evolution of innovation policy structures based on copying perceived cultural characteristics and structures of the US innovation system will also fail to deliver the goods. In the case of innovation policy, the cargo is improved economic welfare through improved productivity growth based on enhanced innovation performance. The key 'ritual' structures are increased R&D expenditures; an emphasis upon the commercialization of science through university-based spin-outs and licensing routes in high-technology producing sectors; the promotion of entrepreneurship and new business entry and a supposed US entrepreneurial culture based on the subsidization of risk taking in venture capital investment and of the development of the SME sector more generally.

These perceived key elements feature centrally in policy debates. For example, in March 2000 the EU adopted the 'Lisbon' strategy to make, within the next decade, the EU the most dynamic and competitive knowledge-based economy in the world. The strategy was explicitly positioned as a response to the observed superior performance of the US economy which had in the previous decade substantially outperformed the European economies. It also explicitly accepted the view that this superior US economic performance was based on the emergence of high-technology sectors such as ICT and biotechnology as key totems of the new knowledge-based economy of the US (European Commission 2004). Despite the subsequent bursting of the dot.com bubble and an increased awareness of the emerging threat to Europe from India and China rather than the US, these key elements of the innovation and technology strategy connected with Lisbon continue to be emphasized. Thus, in 2004, it was asserted that 'There is overwhelming evidence of the vital importance of boosting R&D as a prerequisite for Europe to become more competitive. To fail to act on that evidence would be a fundamental strategic error . . . ' (European Commission 2004: 21). Similarly, it was asserted that entrepreneurship is required to take advantage of technological developments: 'Increasingly, new firms and SMEs are the major sources of growth and new jobs. Entrepreneurship is thus a vocation of fundamental importance, but Europe is not "entrepreneur-minded" enough' (ibid.: 28).

Both of these arguments were followed by calls for greater tax subsidization of high-technology investment, R&D expenditures and enhanced policies aimed at boosting entrepreneurship and new entry and reducing risk aversion and the 'stigma of failure' (ibid.).

In relation to enhancing the role of universities, the policy emphasis on spin-offs and licensing 'US style' is often noted:

> In recent years, spurred by the experience of the US in particular, policy makers, enterprises, investors and academics throughout the industrialized world have paid increasing attention to the role of universities as drivers of innovation. Many universities have established formal offices and processes for identifying promising discoveries made within their walls and turning them into revenue streams through licensing or spin-outs. (Apax 2005: 4)[2]

The belief in the centrality of university–business links to economic progress and in the commercialization of science through licensing and spin-offs is also explicit in the innovation strategies of many individual countries (OECD 2001; Yusuf and Nabeshima 2007).

In this chapter I wish to question these emphases on R&D-intensive high-technology spin-offs from the science base and entrepreneurial science. In doing so it is not my intention to argue that R&D or new entry or the growth of venture capital or university spin-offs do not matter. My contention is rather that they have been greatly exaggerated to the neglect of other key factors when one considers the innovation system as a whole. One of these factors is the importance of the diffusion and use of ICT as a general-purpose technology beyond the ICT and other R&D-intensive high-tech producing sectors. This has enabled 'unexpected' user sectors with negligible conventional R&D spend such as retailing to dominate movements in US aggregate productivity growth. A second factor is the dominant role that performance transformation in existing firms plays in driving industry-level productivity compared with the direct role of new entrants. A third is the diversified role played by universities in knowledge exchange that extends beyond a narrow focus on spin-offs and licensing to encompass the creation of human capital and a wide range of formal and informal business interactions. A further factor related to this is the pre-dominant role of customer–supplier interactions in open innovation systems (Chesbrough 2003) rather than direct university–business interactions. Finally there is the major role that public procurement policy has played in the USA in the effective provision of public rather than private sector venture capital and the high value placed by US firms on public sector sources of knowledge for innovation. The chapter attempts in the space available to provide an overview of evidence on each of these factors and to consider some broad implications for innovation policy that might be drawn on the basis of that review. In particular it concludes by arguing that the crafting of innovation policy in the context of any specific national innovation system requires a careful consideration of the structural features of that context and the particular opportunities and challenges facing policy practitioners in it. An imperfect interpretation of the experience of one country's system is unlikely to be an appropriate guide to innovation system failure or success elsewhere.

INTERPRETING US ECONOMIC PERFORMANCE

Since so much policy is linked to references to US economic performance, it is useful to begin with a brief overview of it in the recent past. Table 4.1 shows that the most dramatic feature of US performance since the Second World War is that its recent improvement is heavily concentrated at the end of the last century and at the beginning of this one, when it returned to its long-run trend performance after two decades of relatively low growth performance. The dramatic improvements in productivity growth after 1995 are now, however, due to the direct performance of R&D-intensive high-technology industries.

Table 4.1 US productivity growth 1947–2003 (real GDP per hour)

Period	Rate (%)
1947–1972	2.9
1972–1995	1.4
1995–2000	2.5
2000–2003	2.6

Sources: McKinsey Global Institute (2001); Farrell et al. (2005).

This can be seen if we decompose the aggregate performance into its components. An industry's contribution to the aggregate depends on its own change in productivity growth, and on its size, because the economy is a weighted average of the different sectors.[3]

Decomposing productivity growth in the first period from 1995 to 2000 reveals that six of 59 sectors accounted for the whole of the acceleration in productivity growth. The top three key sectors in the US economy on this basis were wholesaling, retailing, and security and commodity broking. Their joint contribution was twice as great as that of the next three, electronic and electric equipment (semiconductors), industrial machinery and equipment (computers), and telecoms (McKinsey Global Institute 2001).

None of the top three are technology-intensive sectors in any conventional sense. In the second period, the most recent years for which decomposition data are available, seven sectors accounted for 85 per cent of all the productivity growth. These were retailing, finance and insurance, computer and electronic products, wholesaling, administrative and support services, real estate, and miscellaneous professional and scientific services. None of these, with the exception of computers and electronics, are in any sense conventionally R&D intensive (Farrell et al. 2005). It's a Wal-Mart- not a Microsoft-led turnaround. The traditionally identified R&D-intensive sectors have not carried most weight.

Wal-Mart, on the back of a major IT-based business structure, has transformed – some people would argue much for the worse – a whole variety of social and economic structures in the USA and delivered enormous productivity growth in the retailing sector (McGuckin et al. 2005; Foster et al. 2002). Much of this has been linked, as in other service sectors such as transport and financial services, to the implementation of new business models based on ICT and related technologies (Hughes and Scott Morton 2005, 2006). Wal-Mart's performance is thus an example of the impact of ICT as a general-purpose technology (OECD 2003a; Helpmann 1998) in a 'user' rather than a high-tech 'producer' sector (Pilat and Lee 2001).

Microsoft, on the other hand, is a high-tech producer that contributes to the capacity for many of these changes to occur in the 'user' sectors. So in that sense Sam Walton and Bill Gates are complementary; Sam Walton and Wal-Mart are more important to productivity turnaround than Bill Gates and Microsoft, however, because of the scale of the activity that is transformed by the activities of a company such as Wal-Mart when it implements IT-linked business transformations. Differences in services productivity growth account for most of the difference in national productivity performance between the USA, the UK and Europe in the past decade, rather than differences in high-tech producing sectors (Oxford Institute of Retail Management 2004; Griffith and Harmgart 2005; Basu et al. 2003; van Ark et al. 2002).

High-technology 'producing' sectors are a small part of the economy, especially compared to the technology-using sectors and the services sector more generally. This points to the need to think extremely carefully about the mechanisms by which high-technology activity is diffused through the rest of the economy and not just the scale or productivity performances of high-technology output *per se*. A focus on high-technology production without a parallel consideration of diffusion or use throughout the innovation system, and the factors affecting that, runs a clear risk of failing to deliver the goods.

SPIN-OFFS AND NEW ENTRY

Now I want to turn to the issue of new spin-offs and their role in productivity performance; I have called this the *golden oldies* versus *the new kids on the block* debate. The new kids on the block are new high-tech spin-off firms that are often attributed such an important role in the science and innovation process. I want to present some facts about spin-offs, especially in the USA, and put them in the context of what is known about the way in which the golden oldies contribute to changes in industry structure and productivity growth.

The first thing is to get a sense of proportion. The US economy has some 500 000 firms starting up each year. That, of course, includes firms of all kinds, from small restaurants to boutique high-tech businesses, not just businesses based on the exploitation of intellectual property (IP) or new products derived from advances in scientific research. In the USA as a whole, in 2004 there were 462 IP-based start-ups where the IP was from a US university. That may be an impressive performance internationally, but its scale has to be borne in mind in interpreting claims of what might be gained in other economies from such spin-offs.

Second, although IP produced by US universities produces results in considerable patenting and licensing activity, it is insignificant numerically compared to the total amount of such research-related activity in the USA. IBM in the year 2005 alone registered 2941 patents with the US Patent office, Canon 1829, and HP 1790. The whole of the University of California (UC) state system, which is one of the most dynamic, productive and innovative university systems in the world, produced 388, MIT 136 and Stanford 90 (US Patent Office 2005). This is an impressive university performance. It is important, however, to keep it in perspective relative to corporate activity and to think of universities as a quantitatively small but qualitatively important part of a wider system.[4]

Finally, the returns from start-ups and licensing activity are enormously skewed. The following statistics illustrate just how skewed. Only 167 out of 27 322 patents held by 193 US university institutions in 2004 made over $1 million (AUTM 2005). In the case of Columbia University, Stanford and the UC system, the top five patents accounted for 65 per cent of gross licensing revenues. The chances of winning this lottery are small. That doesn't mean to say you shouldn't try to do it; you can't win unless you buy a ticket, but you have to be realistic about what the odds are. First-mover new start-ups based on radical innovations capable of transforming markets very rarely come to dominate those new markets. In the terminology of Markides and Geroski, such pioneering 'colonisers' of radical new markets rarely survive early market expansion. Fast second movers with rather different 'consolidations' skills come to scale up, dominate and capture maximum value (Markides and Geroski 2005).[5] Universities also have to be clear about the costs. The vast majority of US university technology licensing offices barely break even or don't make a profit. The gross average annual licensing revenues of the UC system in 2001–4 of $75 million cost almost $60 million per annum to maintain and manage. Thus in the period 2001–4 the net contribution of the University of California system's licensing income was $15 million annually, compared to around $235 million of commercial funding of university research (Mowery 2007).

We can now look at this in a slightly broader way. Instead of just looking at the spin-off activity by US-based universities, we can examine the impact of start-ups as a whole. A substantial amount of work has been done that attempts to decompose the change in productivity in particular industries across the OECD economies in terms of entry, exit and survivor growth (e.g. OECD 2003b; Bartelsmann et al. 2004). This work breaks down productivity growth between the gains in productivity that are made by the surviving firms that are there throughout the period studied and the transfer of activity from lower- to higher-productivity surviving firms. This is the golden oldie effect. The firms are there at the beginning and they are there at the end. Then there's the impact on productivity of firms that leave. If the worst firms drop out, there's a batting average effect and average productivity rises. Finally there is the effect of new entries, the spin-offs and new start-ups. This is the new kids on the block effect. They enter the system and either die or survive and grow over the period analysed. What is clear from this work is that the vast majority of the productivity growth that is experienced in any economy and any industry in any time period is driven by the transformation in productivity of the golden oldies; that is, it's the improvement in the performance of the firms that are there all the time. The contribution of survivors (often referred to as the 'within firms' effect) varies between 55 per cent and 95 per cent. The net effect of exits and entry accounts for 20–40 per cent, but most of this is due to the batting average effect of exits. Entry effects are small because of low entry sizes at lower average productivity than incumbents and low survival rates. Only 30–50 per cent of new entrants survive for over five years. Exit and entry rates rise and fall together across countries and over time, with high entry associated with high exit. In the case of the USA the new entry component is typically large and negative, and survival rates are low but survivors on average grow faster. Finally, it is important to note that these studies do not suggest that the USA is characterized by high net entry. Instead it appears that it is characterized by relatively rapid growth of survivors, so it is post-entry growth not entry *per se* that matters. To illustrate the effects we can look at some data from UK manufacturing for the period 1980–92 (Disney et al. 2003). The data relate to establishments that may operate a single plant and multi-plant establishments. Table 4.2 shows that net entry by

Table 4.2   Net entry, surviving firm and reallocation components of UK manufacturing establishment productivity growth 1980–92

Contributors to overall productivity growth	Singleton establishments	Group-owned establishments
Surviving establishments' productivity growth	0.6	44.6
Market reallocation between survivors with high and low productivity levels	−0.4	3.9
Market reallocation between survivors with high and low productivity growth	0.4	−2.8
Net entry productivity effect	15.9	33.2

Source: Calculated from Disney et al. (2003).

singleton establishments accounted for only 15.9 per cent of overall productivity growth, while net entry due to the closure and opening of establishments by multi-plant surviving firms accounted for over twice as much (33.2 per cent). Productivity growth within surviving establishments owned by multi-plant businesses accounted for over 44 per cent. Golden oldies, surviving firms, clearly dominate this process.[6] A policy stance that concentrates on driving innovation and productivity by looking only at new independent firms will therefore miss a very important part of the story.

There are some industries and some conditions that are relatively favourable to the success of innovative new entry (Baldwin and Gellatly 2003; Baldwin 1993; Gambardello and Malerba 1999; Audretsch 1995). The first is where the nature of the technology is constantly changing the basis on which competitiveness can be built. If there is turbulence in the technological regime and entry is relatively low cost, experimentation in new entry may be accompanied by some home runs. Also, if the incumbents – the golden oldies – in an industry are heavily committed to an existing technology, then there's a better chance of a new entity succeeding because the conservatism that goes with very heavy investment in a standard technology makes the incumbents relatively slow to react (Christensen 1997). Finally, the chances of success are higher if the resources to exploit new business ideas – complementary assets – are not owned by others. If these complementary assets, which are necessary to extract value, are owned by somebody else, it is unlikely that they can be appropriated by new independent firms going it alone (Teece 1996).

The role to be expected for new innovative entry and survival to enhance productivity performance is thus highly context specific. A blanket promotion of new start-ups in support of innovation without careful attention to industry dynamics and the ecology linking new entry and large firm success, and patterns of appropriating value, should be avoided.[7]

UNIVERSITIES AND THE INNOVATION SYSTEM

In discussing the role of universities in innovation systems I shall illustrate my argument with data from a recent survey-based comparison of the UK and US economies. The Centre for Business Research/Industrial Performance Centre (CBR/IPC) US/UK Innovation Benchmarking Survey (Cosh et al. 2006) was carried out in the period March–November 2004. The primary telephone survey covered firms of all sizes from ten employees upwards in the manufacturing and business services sectors. It achieved response rates of 18.7 per cent in the USA and 17.5 per cent in the UK. There was in addition a postal follow-up survey in both countries for firms employing more than 1000 employees. In all, the survey instrument included 200 questions, which generated over 300 variables per firm. In this chapter I shall draw only on those sections of the survey instrument that related to the interactions between universities and the firms in the survey as well as drawing on some material on the wider range of interactions that survey firms claimed were relevant to their innovation activities.

Table 4.3 shows the size distribution of the overall achieved samples in the UK and the US surveys. Approximately two-thirds of the firms in both surveys employ between ten and 99 people, around one-quarter employ between 100 and 999 people, with the remainder employing over 1000. In order to provide UK–US comparisons that are not contaminated by possible variations between countries in the distribution of responses by sector or by size of firm, I shall focus on the results that are obtained when we form a matched sample. This matched sample consists of 1149 US companies and 1149 UK companies matched by employment size and by sector, where the sectoral matching is at least at the three-digit level. Table 4.4 shows the sectoral composition of this matched sample, distinguishing between manufacturing and business services and high-tech and conventional sectors within those broad industrial groupings. The distinction between high-technology and conventional sectors is based on the R&D

Table 4.3   Size distribution of UK and US respondent firms in the CBR/IPC survey

Employment size	US	UK
10–99	62%	66%
100–999	24%	25%
1000+	14%	9%
N	1540	2129

Source: Cosh et al. (2006).

Table 4.4   The sectoral composition of a matched sample of UK and US firms (%)

Sector	High-tech	Conventional
Manufacturing	28	38
Business services	15	19

Source: As for Table 4.3.

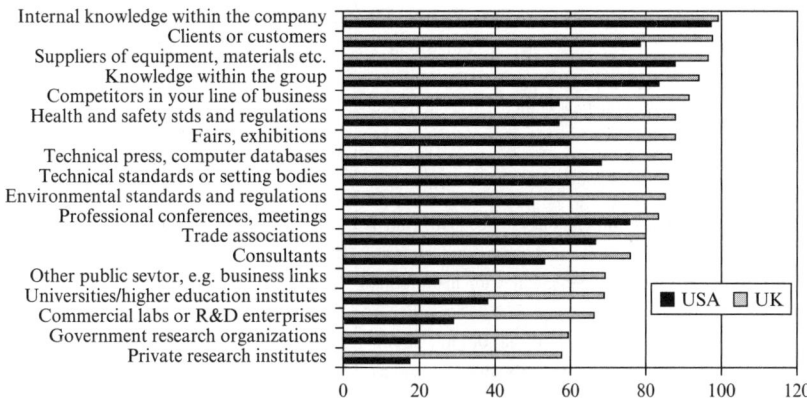

Source: Cosh et al. (2006).

Figure 4.1 Use of sources of knowledge for innovation (% companies)

intensity of their activity and the technical composition of their labour force. The survey contains a representative proportion of high-technology businesses in both countries.

One way of looking at the role of university–industry relationships is to locate universities as a source of knowledge for innovation in the wider context of the overall sources of knowledge used by innovation-active firms. The results of an analysis of this kind for firms in the UK–US matched sample are shown in Figure 4.1.[8] The picture that emerges is very clear. Customers, suppliers, competitors and the firms' own internal knowledge are the dominant knowledge sources. In both the USA and the UK, universities are relatively low in frequency of use as direct sources of knowledge for innovation. Interestingly, in terms of the proportion of firms reporting universities as a source of knowledge, the UK outstrips the USA. In both countries use is made of a very wide range of other sources. There is clearly a distributed innovation knowledge system and in terms of frequency of use universities are only a small direct part of it.[9] This does not mean that they are not important, but it does mean that their contribution has to be seen in the context of a much wider and complex system of innovation information flows. This pattern is not unique to the USA and the UK. The same is true for Australia, for instance, as is apparent from Figure 4.2. and for the EU more generally.

It is of course possible that frequency of use may not be correlated with the importance placed upon the information obtained. The survey firms were also asked to indicate the value they placed upon the sources of

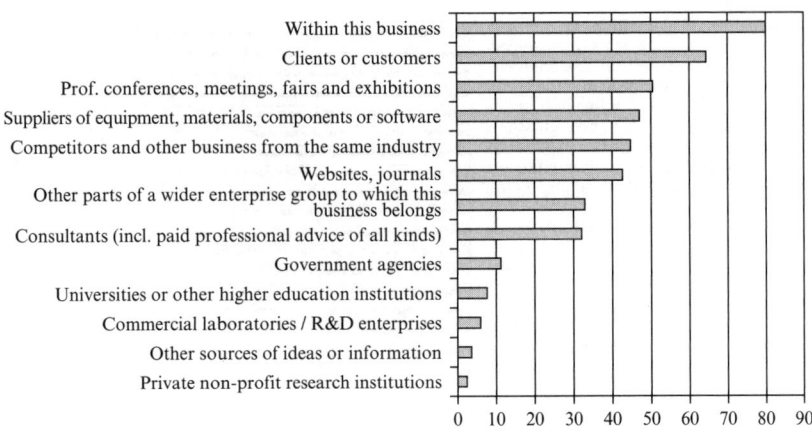

Within this business
Clients or customers
Prof. conferences, meetings, fairs and exhibitions
Suppliers of equipment, materials, components or software
Competitors and other business from the same industry
Websites, journals
Other parts of a wider enterprise group to which this business belongs
Consultants (incl. paid professional advice of all kinds)
Government agencies
Universities or other higher education institutions
Commercial laboratories / R&D enterprises
Other sources of ideas or information
Private non-profit research institutions

0 10 20 30 40 50 60 70 80 90

Source: Calculated from ABS (2006).

Figure 4.2   *Key sources of ideas or information for innovation in*
Australian innovating business 2001–3 (% companies)

knowledge as well as their use. The responses are summarized in Table 4.5, where, following Swann (2006), we group sources into three broad categories. These are the company sector, the public and private scientific knowledge base and a group of intermediating and regulatory organizations. Once again, in both countries the company sources dominate. Internal sources of knowledge plus knowledge obtained from suppliers and customers were ranked most highly as knowledge sources for innovation. In both countries they were followed by technical standards and health and safety regulations as important sources of knowledge from the intermediating and regulatory group. The need to contextualize innovation policy in the circumstances of particular countries, however, is highlighted by the fact that there are significant differences between the UK and the USA in the value placed upon knowledge from the science base, and from the intermediating organizations other than standard settings and regulators. For instance, US firms were almost twice as likely to place a high importance on knowledge gained from consultancies, government research laboratories and other public research laboratories, professional conferences and trade associations than were UK firms. Moreover, despite being more likely to cite universities as a source of knowledge, UK firms more frequently placed a lower value on it than did US firms. Another difference between the UK and the USA emerges if we probe a little more deeply into the patterns of combined use of sources of knowledge.

Table 4.5   *High importance of sources of knowledge (% of users of that source)*

	UK %	USA %	Ratio (UK/US) × 100
Company sector			
Suppliers of equipment, materials, components, or software	41.5	49.2	84.4
Internal knowledge within the company	79.9	84.5	94.6
Clients or customers	60.9	53.5	113.7
Knowledge within the group	59.4	50.7	117.1
Competitors in your line of business	27.7	20.8	132.9
Intermediating and regulatory organizations			
Consultants	12.5	26.2	47.7
Professional conferences, meetings	14.6	23.9	61.2
Trade associations	15.1	23.5	64.4
Technical/trade press, computer databases	21.5	26.5	80.8
Fairs, exhibitions	17.4	18.0	96.8
Environmental standards and regulations	31.8	46.1	69.0
Technical standards or standard-setting bodies	34.6	40.2	86.1
Health and safety standards and regulations	41.3	47.2	87.5
Other public sector, e.g. business links, government offices	10.5	38.7	27.1
Scientific knowledge base			
Government research organizations	6.6	24.7	26.6
Private research institutes	7.2	22.9	31.5
Commercial laboratories or R&D enterprises	12.2	28.4	43.0
Universities/higher education institutes	13.8	27.0	51.3

Source: As for Table 4.4.

Figure 4.3 (following Swann 2006) shows in successive quadrants the extent to which firms in the UK and the USA are specialized in their use of sources of knowledge. The first upper left quadrant simply repeats in a different form the contents of Table 4.4 with the thickness of the bands reflecting the frequency of use of each source of knowledge. The top right-hand quadrant shows the proportion of companies in each country that used at least one source from the company sector, and no other sources. This reveals immediately that although customers, suppliers and competitors and the internal knowledge base of the firm are the most frequently used (and, as we have seen, the most highly valued source), they are almost never used in isolation. When we switch to the bottom left-hand quadrant we identify

(a) Use of sources of knowledge for innovation: % companies using each source

% Firms: UK (USA)

UK	USA

Suppliers 97 (88)
Intra-Group 95 (83)
Competitors 91 (57)
H&S Reg. 88 (57)
Fairs 88 (60)
Trade Press 87 (68)
Standard setting 86 (59)
Environ. Reg. 85 (50)
Conferences 83 (76)
Trade Assocs 80 (67)
Consultants 76 (53)
Publ. Sect. Intermediary 69 (25)
Universities/HEIs 69 (38)
Commercial Labs 66 (29)
Govt Research Orgs 59 (20)
Private Research Insts 57 (18)
Customers 98 (78)

The Innovative Firm

(b) Use of at least one company source and no other source: % companies

UK	USA
0.9	2.7

Customers, Suppliers, Intra-Group, Competitors, H&S Reg., Fairs, Trade Press, Standard setting, Environ. Reg., Conferences, Trade Assocs, Consultants, Publ. Sect. Intermediary, Universities/HEIs, Commercial Labs, Govt Research Orgs, Private Research Insts

The Innovative Firm

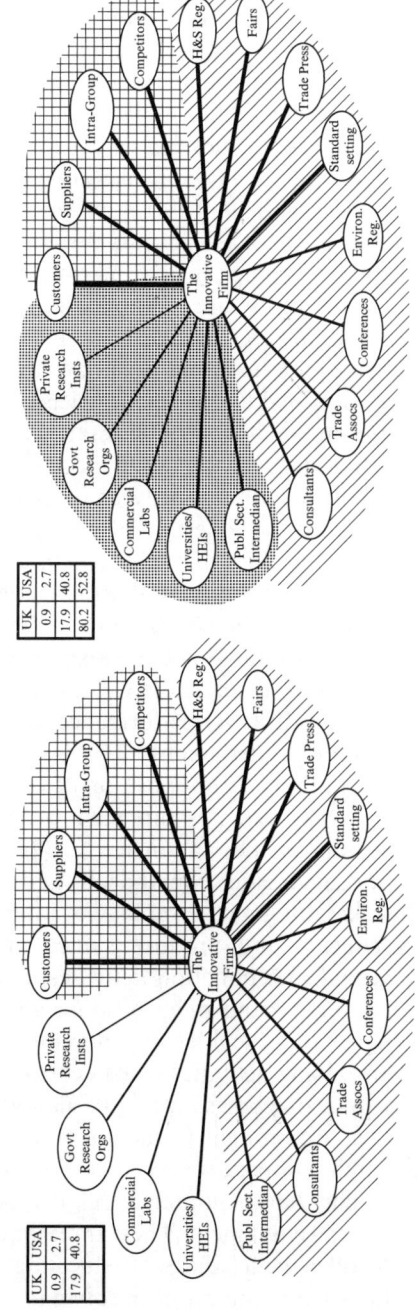

(c) Use of at least one company source and one intermediary source and no others: % companies

	UK	USA
	0.9	2.7
	17.9	40.8

(d) Use of at least one source in each group: % companies

	UK	USA
	0.9	2.7
	17.9	40.8
	80.2	52.8

Source: As for Figure 4.1.

Figure 4.3   Combined use of sources of knowledge for innovation

those firms that used at least one company source and at least one source from the intermediating and regulatory group and no others. Here a significant difference emerges between the UK and the USA. Over 40 per cent of the US firms used a company source and an intermediary source and no others, while only 17 per cent used this particular combination in the UK. When we turn finally to those companies that used at least one source in each group, we find that the UK firms are far more likely to report using a research base source in combination with the other sources of knowledge in the company and intermediating sectors. It appears therefore that US firms are much more likely to combine company and intermediating sources, while UK firms have a much more diffuse use of knowledge sources. Equally, US firms are less likely to use all three knowledge sources and have a more compact knowledge source pattern. Paradoxically, as we have already seen, when they do interact with institutions in the science base, they place a significantly higher value on the outcomes. This raises important questions about the extent to which the value placed upon the science base is enhanced by the use of intermediating institutions between the science base and companies themselves. It also raises the question of whether in the UK the use of so many sources raises difficulties of effective management and reduces their usefulness.[10] In terms of innovation policy this emphasizes the importance of paying attention to the particular structure of the innovation system in which the policy is to be introduced and an analysis of whether the particular patterns observed, for instance in the USA, are linked to a superior pattern of innovation and productivity performance. It also raises issues of depth as opposed to breadth of interactions.[11]

Once we have looked at the structural position of universities in knowledge flows in the innovation system, it is important to discuss the nature of the interactions between universities and firms. As a precursor to looking at some of the university data arising from the US–UK survey that bear on this issue, it is worthwhile setting out a typology of interactions.

First, universities educate and produce skilled graduates. Second, through their research and dissemination activities, universities increase codified knowledge. University staff publish books and scientific papers, they patent, and in engineering faculties may develop prototypes. A very wide range of problem-solving activities is also carried out – often on a regional or local basis, but sometimes on an international basis – directly addressing problems that are brought to the attention of the universities through contract research, cooperative research and faculty consulting. University laboratories may have equipment that can be used for testing various kinds of commercial equipment. These three kinds of activities are captured in Figure 4.4 under the headings of educating people, increasing the stock of codified knowledge and problem solving.

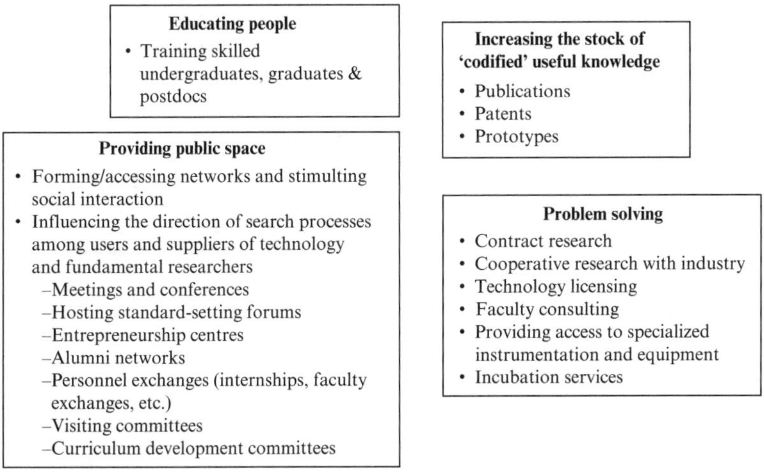

Educating people	Increasing the stock of 'codified' useful knowledge
• Training skilled undergraduates, graduates & postdocs	• Publications • Patents • Prototypes

Providing public space

• Forming/accessing networks and stimulting social interaction
• Influencing the direction of search processes among users and suppliers of technology and fundamental researchers
  –Meetings and conferences
  –Hosting standard-setting forums
  –Entrepreneurship centres
  –Alumni networks
  –Personnel exchanges (internships, faculty exchanges, etc.)
  –Visiting committees
  –Curriculum development committees

Problem solving

• Contract research
• Cooperative research with industry
• Technology licensing
• Faculty consulting
• Providing access to specialized instrumentation and equipment
• Incubation services

Source: As for Figure 4.3.

Figure 4.4   The university role is multifaceted

What tends to be less discussed is what Richard Lester and Michael Piore have called 'the public space function of universities' (Lester and Piore 2004), which is captured in the largest box in Figure 4.4. This function captures the distinctive role of universities in society and in the innovation system as public spaces in which other interested parties can 'play', if that public space is appropriately structured. This includes a range of 'soft', but nonetheless extremely important activities, to do with network forming, stimulating social interaction, influencing the direction of research processes by identifying commonly experienced problems, setting standards of a technical kind, setting up entrepreneurial centres and so on. These public space activities permit the discovery of potential complementary interests and the crafting of potential ways to develop them to mutual advantage. They also foster the role of universities as translators and providers of insights into 'new' science. For instance, in the context of the US Advanced Technology Program industrial research participants perceived that 'the university could provide research insight that is anticipatory of further research problems and that it could be an ombudsman anticipating and communicating to all parties the complex nature of the research being undertaken' (Hall et al. 2003: 491).

It is interesting to explore how these diverse public spaces and other roles are perceived by businesses, and the relative significance of licensing and spin-out formation compared to other interactions. The CBR/IPC Survey sheds some useful light here since respondent firms were asked how

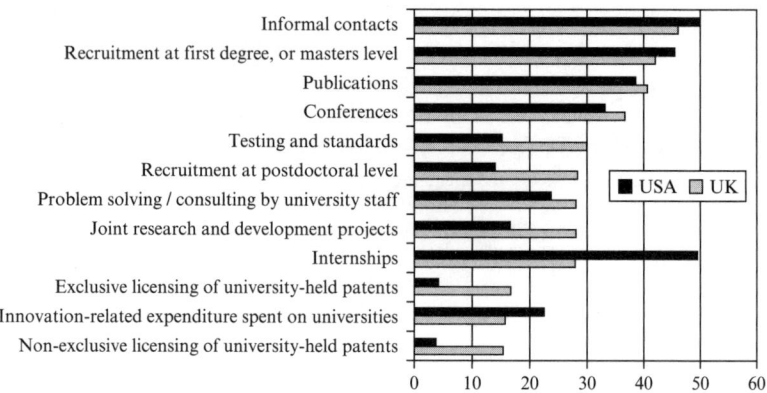

Source: As for Figure 4.3.

Figure 4.5   *Types of university–industry interaction contributing to innovation (% companies)*

they interacted with the universities in their innovation activities and what kind of emphasis they placed on different interactions. Figure 4.5 reports the results. It shows that businesses interact across the full spectrum of those elements set out in Figure 4.4. The most frequent form of interaction is via informal contacts, and it's not only the most frequent – a separate analysis (not shown here) reveals that it is also among the most highly valued (Cosh et al. 2006). All the conventional modes of university output (undergraduates and graduates, and publications and conferences) are frequently cited modes of interaction. In that sense there is no necessary conflict between how the business community says it interacts most with university activities and what academics themselves typically say they want to do.

From the point of view of differences between systems of innovation it is worth noting that US firms appear to use internships more than their UK counterparts,[12] and that they more frequently have an interaction involving innovation-related expenditure with universities. This suggests a greater depth and intensity of interaction in the USA than in the UK, even if US interaction is less frequent. US firms are, however, less, not more, likely to interact via licensing, whether exclusive or non-exclusive. However, when they interact via licensing they value it more highly (Cosh et al. 2006).

It is important to note that these are aggregate figures across manufacturing and business services. In some industries, in particular biomedical sciences, patenting and licensing are significant in terms of frequency of use and qualitative importance (Cohen et al. 2002).

From the point of view of innovation policies outside the USA, it is instructive to note that the intensification of patenting and licensing regimes in US universities has provoked a reaction. This reaction empha-sizes the threats posed to the cost and timeliness of effective knowledge exchange and exploitation. In non-biomedical sciences in particular it has been argued that the time and costs involved in negotiating IP have begun to threaten industrially funded research (Mowery 2007). Recent research suggests that major US universities are shifting knowledge exchange man-agement beyond patent and licensing to avoid possible adverse reactions on the wider range of interactions. This includes managing wider industrial liaison activities alongside patenting. It also includes negotiating royalty-free licences in some areas as part of industrial funding of research con-tracts in, for instance, electrical engineering and computer science (Mowery 2007). If the 'US model' is to guide innovation policy elsewhere, it is as well that the current evolving model rather than the 'old' one is a reference point and the full range of interactions is recognized.

PUBLIC POLICY AND VENTURE CAPITAL

I now want to turn to the issue of venture capital in the USA, and the view that what is required outside the USA is subsidization of private sector venture capital to promote a more risk-tolerant investment climate. My first point here is that in practice in the USA, one of the most powerful, proac-tive venture capital supporting activities is public R&D procurement through the Small Business Innovation Research (SBIR) programme (Connell 2006). The SBIR was established in the 1980s in the middle of the period of very low US productivity growth, when the USA experimented with a range of industrial policy mechanisms to counteract what it per-ceived, correctly, as its failure to deal with the commercial threat of Germany and Japan.

The SBIR was one of a number of initiatives taken in the course of the 1980s to address this challenge. Many were designed to encourage collabo-rative and cooperative strategies in relation to innovation policy and pro-ductivity performance (Dertouzos et al. 1989; Branscomb et al. 1999; Wessner 2003). Thus, for example, the 1984 National Cooperative Research Act relaxed anti-trust regulations to facilitate research-based joint venture collaborations. In relation to university–industry links the 1988 Omnibus Trade and Competitiveness Act established, *inter alia*, the Advanced Technology Programme to promote university–industry collaboration. In 1980 the passage of the Bayh–Dole Act was designed to enhance university patenting and licensing based on federally funded research. In the course of the 1980s several hundred university–industry research centres were also established. By 1990 such centres accounted for over $2.5 billion in academic R&D spending (see for example Branscomb et al. 1999; Mowery 2007).[13]

The SBIR as part of these policy initiatives was specifically established to support businesses with fewer than 500 employees, and provides 100 per cent funded contracts to carry out technologically intensive R&D contractual obligations for US federal agencies. The US federal agencies advertise technical or research-related problems and an open competition results in the award of a contract with potential follow-on contracts. The US government currently mandates 2.5 per cent of federal agency total R&D spend to SBIR, and that is in absolute terms a significant sum. It amounts to $2 billion annually, covering 4000 contracts (Connell 2006). The private venture capital sector in the USA, for comparison, was investing around $1 billion annually in around 200 deals per annum at the seed stage in the period 2005 to 2006 and around $4 billion annually in around 800 larger early-stage deals. This was out of a total annual amount invested in all stages of around $24 billion in those years (Money Tree 2007). The private venture capital sector in the USA is thus similar in its risk profile to private equity elsewhere, with a focus on later-stage investments and large-scale company buyouts. Only a small proportion of funding goes into seed and early-stage finance. However, the SBIR produces a situation in which much of the very risky early-stage and seed investments are supported by a public sector mandated activity. Some extremely big and successful companies have been assisted in this way. Amgen, Qualcomm and Genzyme, for instance, all have SBIR connections in their origins (Connell 2006). The SBIR effectively derisks subsequent investment by providing certification and proof of performance capacity in the earlier stages of development for small firms that win these contracts. The balance of evaluation evidence also suggests that SBIR contract winners are more likely to commercialize on the basis of their research and to grow faster than similar firms not funded through SBIR contracts (Lerner 1999; Audretsch 2002, 2003; Audretsch et al. 2002; Wessner 2001; Wallsten 2000).

OVERALL CONCLUSIONS

So, what are the overall lessons to draw from this broad overview? The first is that US productivity and growth performance is not based solely on high-tech production *per se*; it is based on the diffusion of innovations throughout the system and frequently on the transformation of what people would regard as 'low-tech' sectors (in R&D terms) by general-purpose high technologies based on ICT advances. Second, productivity gains are in general driven by firms that are in existence. Thinking about existing firms and their innovation performance is critically important in the innovation process. Innovation policy should not focus on start-ups alone. Moreover, the role that start-ups may play is conditioned by the nature of particular technological regimes and patterns of appropriability. It is better to think in terms of typologies of commercialization and knowledge exchange in which new firms' entry and independent growth is one of several potential routes. New firms and spin-offs have an important seed-bed role to play but need to be understood as part of a wider open innovation system in which the interplay between large and small firms and the transformations in large business process drive innovation and productivity. Third, public sector procurement has potentially a very powerful part to play in supporting private venture capital and bridging the highest-risk gap for early-stage development of research-intensive firms. Fourth, universities have to be seen as part of a complex system. Their direct contribution as a knowledge source is perceived in general by business as relatively small compared to other components in the innovation system. Their multifaceted role must be understood within this wider context. The mechanisms for university interaction with business are diverse and may be sector specific. Licensing and spin-offs are only one part of the story. They are significant in only some sectors and if aggressively pursued may lead to loss of other forms of research funding from business, and high rates of spin-off failure respectively. A 'one size fits all' economic development or innovation strategy for any country or any university that focuses on licensing and spin-offs alone is not appropriate. An innovation policy that promotes 'public space' interactions is likely to lead through informal and other interactions to the discovery and development of appropriate interaction modes for particular sectors and purposes.

University research is of value and interest to the business sector because it is different. Creating institutional mechanisms that promote access to the space within which this different activity is pursued in turn creates the opportunity for the translation of scientific advance, focused problem solving and the recognition and potential exploitation of commercialization opportunities. If the innovation cargo is to be delivered, the design and nature of such spaces, their adequacy and hence whether there is innovation systems failure in their provision should be high on the agenda in designing policy for 'open' innovation systems.

ACKNOWLEDGEMENTS

The author is grateful to Richard Lester, Andy Cosh and Michael Scott Morton for many stimulating discussions in this area, to Anna Bullock for help in data preparation, to the Cambridge MIT Institute for financial support for the survey research into US and UK innovation on which this chapter draws, and to the EPSRC for financial support under grant EP/EO23614/1 IKC in Advanced Manufacturing Technologies for Photonics and Electronics – Exploiting Molecular and Macromolecular Materials (which is a 'public space' experiment in fostering commercialization activities).

NOTES

1. The study of cargo cults has long engaged anthropologists and their physical manifestations are well established (Worsley 1957). There is a long and continuing controversy as to their interpretation and meaning in the cultures in which they occur (Jarvie 1964; Lindstrom 1993; Jebens 2004), and the term cargo cult is now more used outside than inside the discipline of anthropology. This is principally a result of the adoption of the term by the scientist Richard Feynman to describe as 'cargo cult science' scientific investigations that fail to deliver the scientific cargo because while apparently following all the correct forms and structures of scientific investigation they omit a key ingredient. That key ingredient is due consideration of all the evidence against as well as for a hypothesis (Feynman 1985). The argument in this chapter is in a similar spirit.

2. While noting the influence of this interpretation of the US model, the Apax report contains a good discussion of the wide range of interactions between universities and the business sector beyond licensing and spin-offs which are necessary to effect knowledge exchange. Hughes (2007) discusses these arguments in the more specific context of UK science and innovation policy.

3. More formally, the contribution of sector i to aggregate productivity growth C_i can be expressed as

$$C_i = \frac{L_0}{L_1}\left(\frac{Y_i}{Y}\dot{Y}_i - \frac{L_i}{L}\dot{L}_i\right),$$

where $\dot{Y}_i$ and $\dot{L}_i$ are sectoral output and employment growth rates over the period 0 to 1, Y_i/Y, and L_i/L are the sectors' shares in output and employment in period 0 and L_0 and L_1 are levels of national employment in time periods 0 and 1 (McKinsey Global Institute 2001).

4. Patent statistics are subject to a number of problems in assessing performance. Companies may patent for strategic reasons, and this strategic significance varies across sectors (see for example Hall 2004). The broad university–industry picture is, however, clear enough. It is less clear whether the quality of university patents has risen or fallen as their numbers have risen (Sampat et al. 2003; Henderson et al. 1998).

5. From an innovation system point of view this points to the importance of understanding the interactions between types of firms and the complementarity between spin-offs as a seed-bed of new ideas and the role of subsequent acquisition or replacement by fast second movers.

6. It should be noted that the interpretation in the text is rather different from that drawn by the authors who emphasize new entry effects. They choose to regard as 'new entry' new plants introduced by existing multi-establishment businesses. This is clearly not new entry in the sense of new independent firms. Most new plants which open and survive are built by surviving multi-plant firms (the golden oldies).

7. It has been argued that focus on independent growth by new start-ups rather than their acquisition and integration by established firms is also questionable, given the relative

strengths of large firms in exploiting or scaling up radical innovations pioneered by new firms (Markides and Gersoki 2005).

8. The 18 sources identified are consistent with a number of previous innovation surveys including the European Community Harmonised Innovation Survey and the periodic survey of the Small Business Sector in the UK carried out by the CBR since 1991.

9. These results are similar to those obtained for the USA in the well-known 1994 Carnegie Mellon survey (see for example Cohen et al. 2002).

10. It is interesting to note that an analysis of European Community Harmonised Innovation data shows an inverted U-shaped relationship between innovation performances and the number of knowledge sources used (Laursen and Salter 2006).

11. The CBR/IPC Survey also reveals that US firms support these university interactions with a greater commitment of resources than is the case in the UK (Cosh et al. 2006).

12. They also value them more highly (Cosh et al. 2006).

13. In addition to specific policy initiatives there is also abundant evidence that points to the important role played by federal expenditures, foreign policy related military expenditures generally and the (Defense) Advanced Research Project Agency (DARPA) in particular. This includes, for example, their role in emergence of the Internet, computing and IT as a general-purpose technology (Flamm 1987; Segaller 1998; Mowery and Rosenberg 1998); the development of Silicon Valley (Lécuyer 2006) and the impact of defence expenditure more generally on the structure and funding of basic applied science (Stokes 1997).

REFERENCES

ABS (2006), *Innovation in Australian Business 2003 (Reissue)*, Canberra: Australian Bureau of Statistics.

Apax (2005), *Understanding Technology Transfer*, London: Apax Partners Ltd.

Audretsch, D.B. (1995), *Innovation and Industry Evolution*, Boston, MA: The MIT Press.

Audretsch, D.B. (2002), 'Public/private technology partnerships: evaluating SBIR-supported research', *Research Policy*, **31**(1): 145–58.

Audretsch, D.B. (2003), 'Standing on the shoulders of midgets: the U.S. Small Business Innovation Research Program (SBIR)', *Small Business Economics*, **20**(20): 129–35.

Audretsch, David B., Link, Albert N. and Scott, John T. (2002), 'Public/private technology partnerships: evaluating SBIR-supported research', *Research Policy*, **31**(1): 145–58.

AUTM (2005), *US Licensing Survey FY 2004*, Northbrook, IL: AUTM.

Baldwin, J.R. (1993), *The Dynamics of Industrial Competition: A North American Perspective*, Cambridge: Cambridge University Press.

Baldwin, J.R. and Gellatly, G. (2003), *Innovation Strategies and Performance in Small Firms*, Cheltenham, UK and Northampton, MA, USA: Edward Elgar.

Bartelsman, E., Haltiwanger, J. and Scarpetta, S. (2004), 'Microeconomic evidence of creative destruction in industrial and developing countries', Policy Research Working Paper Series 3464, The World Bank.

Basu, S., Fernald, J.G., Oulton, N. and Srinivasan, S. (2003), 'The case of missing productivity growth: or does information technology explain why productivity accelerated in the United States but not in the United Kingdom?', Federal Reserve Bank of Chicago WP8, June.

Branscomb, L.M., Kodama, F. and Florida, R. (eds) (1999), *Industrializing*

Knowledge: University Industry Linkage in Japan and the United States, Boston, MA: MIT Press.

Chesborough, H. (2003), *Open Innovation: The New Imperative for Creating and Profiting from Technology*, Boston, MA: Harvard Business School Press.

Christensen, C.M. (1997), *The Innovator's Dilemma. When New Technologies Cause Great Firms to Fail*, Boston, MA: Harvard Business School Press.

Cohen, W.M., Nelson, R.R. and Walsh, J.P. (2002), 'Links and impacts: the impact of public research on R&D', *Management Science*, **48**(1): 1–23.

Connell, D. (2006), *Secrets of the World's Largest Seed Capital Fund*, Cambridge, UK: Centre for Business Research, University of Cambridge.

Cosh, A.D., Hughes, A. and Lester, R. (2006), *UK PLC: Just How Innovative Are We?*, Cambridge: Cambridge MIT Institute, University of Cambridge, http://www.cbr.cam.ac.uk/news/160206_Report_only.htm.

Dertouzos, M.L., Lester, R.K. and Solow, R.M. (1989), *Made in America: Regaining the Productive Edge*, Boston, MA: The MIT Press.

Disney, R., Haskel, J. and Heden, Y. (2003), 'Restructuring and productivity growth in UK manufacturing', *The Economic Journal*, **113**: 666–94.

European Commission (2004), *Facing the Challenge: The Lisbon Strategy for Growth and Employment. Report from the High Level Group Chaired by Wim Kok*, Luxembourg: Office of Official Publications of the European Communities, November.

Farrell, D., Baily, M.N. and Remes, J. (2005), 'US Productivity after the Dot Com Bust', McKinsey and Company.

Feynman, R.P. (1985), *Surely you're joking, Mr Feynman! Adventures of a Curious Character*, New York: W.W. Norton.

Flamm, K.S. (1987), *Targeting the Computer: Government Support and International Competition*, Washington, DC: The Brookings Institution.

Foster, L., Haltiwanger, J. and Krizan, C.J. (2002), 'The link between aggregate and microproductivity growth: evidence from the retail trade', National Bureau of Economic Research NBER Working Paper 9120, August.

Gambardello, A. and Malerba, F. (eds) (1999), *The Organization of Economic Innovation in Europe*, Cambridge: Cambridge University Press.

Griffith, R. and Harmgart, H. (2005), 'Retail productivity', The Institute for Fiscal Studies Working Paper WP05/07, London: IFS, December.

Hall, B.H. (2004), 'Exploring the patent explosion', CBR Working Paper, WP 291, Cambridge: Centre for Business Research, University of Cambridge, September.

Hall, B.H., Link, A.N. and Scott, J.T. (2003), 'Universities as research partners', *The Review of Economics and Statistics*, **85**(2): 485–91.

Helpmann, E. (ed.) (1998), *General Purpose Technologies and Economic Growth*, Cambridge, MA: The MIT Press.

Henderson, R., Jaffe, A.B. and Trajtenberg, M. (1998), 'Universities as a source of commercial technology: a detailed analysis of university patenting 1965–1988', *Review of Economics and Statistics*, **80**(10): 119–27.

Hughes, A. (2007), 'University industry links and UK science and innovation policy', in S. Yusuf and K. Nabeshima (eds), *How Universities Promote Economic Growth*, Washington, DC: World Bank, pp. 71–90.

Hughes, A. and Scott Morton, M.S. (2005), 'ICT and productivity growth – the paradox resolved', CBR Working Paper, WP 316, Cambridge: Centre for Business Research, University of Cambridge, December.

Hughes, A. and Scott Morton, M.S. (2006), 'The transforming power of complementary assets', *MIT Sloan Management Review*, **47**(4): 50–58.
Jarvie, I.C. (1964), *The Revolution in Anthropology*, London: Routledge and Kegan Paul.
Jebens, H. (ed.) (2004), *Cargo, Cult and Culture Critique*, Honolulu, HI: University of Hawaii Press.
Laursen, K. and Salter, A. (2006), 'Open for innovation: the role of openness in explaining innovations performance among UK manufacturing firms', *Strategic Management Journal*, **27**(2): 131–50.
Lécuyer, C. (2006), *Making Silicon Valley: Innovation and the Growth of High Tech, 1930–70*, Cambridge, MA: The MIT Press.
Lerner, J. (1999), 'The government as venture capitalist: the long-run impact of the SBIR program', *Journal of Business*, **72**(3): 285–318.
Lester, R.K. and Piore, M.J. (2004), *Innovation: The Missing Dimension*, Cambridge, MA: Harvard University Press.
Lindstrom, L. (1993), *Cargo Cult: Strange Stories of Desire from Melanesia and Beyond*, Honolulu, HI: University of Hawaii Press.
Markides, C.C. and Geroski, P.A. (2005), *Fast Second: How Smart Companies Bypass Radical Innovation to Enter or Dominate New Markets*, San Francisco, CA: Jossey Bass/Wiley.
McGuckin, R.H., Spiegelman, M. and van Ark, B. (2005), 'The US advantage in retail and wholesale trade performance: how can Europe catch up?', The Conference Board Working Paper 1358, New York, March.
McKinsey Global Institute in association with Solow, R.M., Bosworth, B., Hall, T. and Triplett, J. (2001), *US Productivity Growth 1995–2000: Understanding the Contribution of Information Technology Relative to Other Factors*, McKinsey Global Institute.
Money Tree (2007), 'Money tree report', PricewaterhouseCoopers, www.pwcmoneytree.com/moneytree/.
Mowery, D. (2007), 'University–industry research collaboration and technology transfer in the United States since 1980', in S. Yusuf and K. Nabeshima (eds), *How Universities Promote Economic Growth*, Washington, DC: The World Bank, pp. 164–81.
Mowery, D. and Rosenberg, N. (1998), *Paths of Innovation: Technological Change in 20th-Century America*, Cambridge: Cambridge University Press.
OECD (2001), 'Fostering hi-tech spin offs: a public strategy for innovation', *OECD Science Technology Industry Review*, Special Issue, No. 26, Paris.
OECD (2003a), *ICT and Economic Growth: Evidence from OECD Countries, Industries and Firms*, Paris: OECD.
OECD (2003b), *The Sources of Economic Growth in OECD Countries*, Paris: OECD.
Oxford Institute of Retail Management (2004), *Assessing the Productivity of the UK Retail Sector*, Templeton College, Oxford, April.
Pilat, D. and Lee, F.C. (2001), 'Productivity growth in ICT and ICT using industries: a course of growth differentials in the OECD?', STI Working Papers 2001/4, OECD, June.
Sampat, B.N., Mowery, D.C. and Ziedonis, A.A. (2003), 'Changes in university patent quality after the Bayh–Dole Act: a re-examination', *International Journal of Industrial Organization*, **21**(9): 1371–90.
Segaller, S. (1998), *Nerds 2.0.1: A Brief History of the Internet*, New York: TV Books.

Stokes, D.E. (1997), *Pasteur's Quadrant: Basic Science and Technological Innovation*, Washington, DC: The Brookings Institution.

Swann, G.M.P. (2006), 'Innovators and the research base: an exploration using CIS4', in *Report for the Department of Trade and Industry/Office for Science and Innovation*, London.

Teece, D.J. (1996), 'Competition, cooperation and innovation: organizational arrangements for regimes of rapid technological progress', *Journal of Economic Behaviour and Organization*, **8**: 1–26.

US Patent Office (2005), 'Patenting by organisations', Washington, DC: US Patent Office.

van Ark, B., Inklaar, R. and McGuckin, R.H. (2002), 'Changing gear: productivity, ICT and service industries: Europe and the United States', Research Memorandum GD-60 Gröningen Growth and Development Centre, University of Gröningen.

Wallsten, S.J. (2000), 'The effects of government–industry R&D programs on private R&D: the case of the Small Business Innovation Research program', *RAND Journal of Economics*, **31**(1): 82–100.

Wessner, C.W. (ed.) (2001), *The Small Business Innovation Programme SBIR: Challenges and Opportunities*, Washington, DC: National Research Council and National Academy Press.

Wessner, C.W. (2003), *Government–Industry Partnerships for the Development of New Technologies: Summary Report*, Washington, DC: National Research Council and National Academy Press.

Worsley, P. (1957), *The Trumpet Shall Sound: A Study of 'Cargo' Cults in Melanesia*, London: MacGibbon and Kee.

Yusuf, S. and Nabeshima, K. (eds) (2007), *How Universities Promote Economic Growth*, Washington, DC: World Bank.

5. New innovation models and Australia's old economy

Mark Dodgson and John Steen

INTRODUCTION

This chapter addresses some issues of open innovation in Australia, a relatively small and remote economy with significant resources industries. To date there has been little examination of what open innovation means for resource-based industries like mining and agriculture, nor indeed for 'non-high-tech' sectors in general. By providing a very different research setting to previous studies, this chapter helps to contribute to the process of testing and developing the concept of open innovation.

In the open innovation model the old internalized and linearly driven innovation process in firms and national innovation systems is replaced by a new innovation paradigm characterized by the presence of non-linear, iterative interactions within open systems of firms and other actors (Rothwell 1992; Chesbrough 2003; Gassmann 2006; Chesbrough 2006). In this model, several points of departure have been noted from previous thinking on innovation, including an equalization in the importance of internal and external knowledge, purposive outbound streams of knowledge and technology, and the rise of 'knowledge brokers' – or intermediaries – as important creators of connections and facilitators of knowledge flows. The contribution of knowledge or innovation brokers provides the focus of this chapter.

Managing the flows of knowledge within economic systems is especially important in a small economy on the periphery of world technological developments. We start from the position that the best policy for a country like Australia is to maximize its openness to innovation from multiple sources and to build on its core strengths in industries where comparative advantage already exists. Not only would those industries such as agriculture and mining become more competitive but they would create knowledge platforms for new industries. This form of related diversification can be seen in the evolution of Pharmacia from Swedish forestry and agricultural science (Waluszewski 2004) and the embryonic stages of a new geothermal power technology in Australia that has evolved from geophysics developed through oil and gas exploration (Geodynamics 2006).

Study of knowledge brokers provides a means of capturing the evolution of the institutions supporting knowledge flow within and between sectors. The way that these institutions develop – old institutions evolve and new ones are created – is a good indicator of how responsive innovation systems are to openness. This chapter looks at two case studies – AMIRA International as an evolving institution in the minerals industry and Innovation Xchange (IXC) as a new one working across sectors – that indicate how institutions designed to manage flows through an innovation system can work effectively to act as brokers and connectors. It is of course unwise to generalize on the basis of two case studies, and we see the need for an extensive research agenda on 'openness' in Australia's national innovation system.

Following from the discussion of these cases in the Australian context, we conclude with more general theoretical concerns about the open innovation model. There is a risk that without locating open innovation within existing theory it could become a transitory fad, like knowledge management and total quality management, believed to have broader and greater implications than it deserves. If open innovation holds real prospects for rethinking innovation at both the government policy and firm strategy level, then it will need to stand up to analysis from the existing literature on national innovation systems, corporate strategy and the theory of the firm.

INNOVATION IN AUSTRALIA

Innovation is critical to the success of all elements of the Australian economy, including those that are significantly resource-based. As Australia's foremost economic historian, Geoffrey Blainey, has observed, Australia's powerful mining industries today would have been insignificant without a long series of innovations, especially in mining and metallurgical methods, transport and marketing, and the wool industry, which for more than a century produced about half of Australia's export revenue, would have been of little importance without a long series of major and minor innovations.[1]

The historical importance of innovation in the resources sector in Australia is also captured by the Australian chapter in Nelson's 1993 collection on national innovation systems, which concentrates almost exclusively on resource industries (Gregory 1993).

Innovation in Australia is not, of course, limited to the resources sector. There are, for example, a large number of government and business reports on innovation covering all sectors (Commonwealth of Australia 2003). There is significant innovative activity in manufacturing (Dodgson and Innes 2006) and R&D spending in the service industries increased from £200 million in 1985 to just below £1 billion in 2003.

R&D spending in Australia was £5.5 billion in the year 2002–3, with the government share 44.7 per cent and business spending at 46.7 per cent.[2] Businesses performed the largest amount of R&D in Australia in 2002–3, worth £2.9 billion (53 per cent), followed by universities at £1.4 billion (25.5 per cent), government research institutes (18.5 per cent) and private non-profit intermediaries (2.6 per cent). Business R&D expenditure on R&D increased fourfold between 1985 and 2003, but at 0.89 per cent of GDP, remains low by OECD standards, positioning Australia at fifteenth on the OECD rankings in 2004 (Australian Bureau of Statistics 2004).

Universities receive half of the Australian government's £2.5 billion funding for research from the Australian government. Just over 30 per cent goes to other government research institutes, such as the Commonwealth Scientific and Industrial Research Organization (CSIRO), and 13 per cent of the R&D spend is allocated to business. The bulk of university funding (88 per cent) is provided by the government, with only 0.05 per cent of university research funded by business.

The challenges confronting innovation in Australia, a relatively small country, are immense. They include the sheer scale of investment and complexity of new innovation capabilities in the USA, Europe and Asia. Spending, as it does, less than 2 per cent of the OECD's total R&D expenditure, Australia will always remain at the periphery of global developments in competition based on innovation. This provides policy makers and managers with substantial challenges when considering innovation policy and practice, and places particular importance on the need to respond quickly and well to changing models of innovation that might encourage better use of limited resources.

The challenges for Australia are argued by some knowledgeable observers to be exacerbated by a lack of clarity and tension between education, science and industry policy; unclear roles of state governments in a federal system; and structural impediments to innovation, such as a predominance of small firms in the industrial structure and a high reliance on overseas multinational companies in high-tech sectors (which explain the low levels of business expenditure on R&D) (Cutler 2006). These challenges remain for those who retain a more complacent view of Australia's innovation performance (Barlow 2006). Despite the improvement in Australia's terms of trade since 2003 due to rising commodity prices, the rate of productivity growth has fallen compared to the 1990s, and the economy is now exposed to the vagaries of international commodity markets and assumptions about the unproblematic growth of China and India as consumers of these commodities (Eslake 2007).

Efforts have been made to achieve a more interactive model of innovation in various government initiatives to improve innovation performance. The significance of networking between firms and the universities; the importance of private intermediary knowledge brokers in the commercialization of university research; maximizing the utilization of infrastructure support for innovation; and the greater role of funding bodies in reforming the dissemination of information on innovation have been appreciated by government.

The Australian Research Council's (ARC) 'Linkage' schemes and the government's cooperative research centres (CRC) programme are examples of public–private partnerships designed to enhance the flow of knowledge between the public and private sectors. A total of 615 Australian and 166 foreign companies and industry bodies were partner organizations with universities in proposals submitted under ARC's 'Linkage Projects' in 2005. The ARC funded a total of A$261 million to 'Linkage' schemes in 2005–6, with cash and in-kind contributions from participating partner organizations more than doubling this amount (ARC 2006). Although this scheme appears attractive to businesses, in practice difficulties are often experienced in their management, especially concerning research priorities and ownership of IP rights.

The CRC programme was conceived in 1989 with the first centres starting in 1991. The Minister for Science and Technology claimed at their launch that 'The cooperative research centres will help Australia to achieve closer linkages between science and the market'. The programme has continued with successive governments and had nine selection rounds by 2006. Centres have a seven-year funding cycle. There are currently 57 CRCs operating. In total, stakeholders have committed £4.6 billion in cash and in-kind contributions: £1.1 billion from the CRCs, £1.2 billion from universities and £0.9 billion from industry.

Scientific outputs over the first ten years of the CRC programme have been impressive, including: 1400 PhDs; 15 839 papers accepted for publication and 549 overseas patents. Some CRCs have become well known – e.g. Photonics, Cochlear and Beef, but there is not much evidence about widespread adoption of research outcomes. Only £13 million has been produced in technology licences and £20 million raised in contract research over a ten-year period, and only £2.7 million in revenues have been created from spin-out companies back to CRCs between 2000–2001 and 2001–2 (Howard Partners 2005). While some have been funded for three rounds (21 years), only one or two CRCs have continued without government support. There is industry concern about their lack of commercial impact, and while CRCs have generated clear economic benefits (Insight Economics 2006), these are geared towards larger-scale, longer-term arrangements that are suited to big research users, requiring extensive management. They have also been used inappropriately, for example in an abortive attempt to recreate the Australian space industry (Moody and Dodgson 2006).

As the CRC example shows, creating the institutions for interaction does not guarantee the formation of effective links between elements of the innovation system. There remain significant problems with knowledge flow due to weak or non-existent linkages between research organizations and firms in Australia. This poses challenges for government, universities, business and intermediaries.

A major challenge for government is the need to complement existing cooperative research arrangements with other smaller, flexible and shorter collaborative research arrangements between groups of firms, either independently or in conjunction with universities and public sector research agencies. There is also a concern to strike a better balance between policies for knowledge production and technology invention, and policies for knowledge flow and technology diffusion and adoption.[3]

There are major challenges for universities. Only large universities are equipped with the necessary scale and expertise to develop effective commercialization units. Intellectual property management in research organizations and universities is inconsistent (Johnston et al. 1999), and involve high transaction costs for businesses that deal with them, resulting in less incentive for business to collaborate (as seen in the tiny proportion of university research funding received from business). Businesses also lack commercialization capabilities, especially smaller firms that have little experience of collaboration (Institution of Engineers Australia 2002).

These issues facing government, universities and business pose challenges and raise opportunities for intermediaries. As mentioned above, the R&D contribution by private sector intermediaries is very low at £150 million (2.6 per cent of total R&D expenditure) and this suggests that these private and other research intermediaries need to have a more proactive role. There need to be more flexible arrangements, including the use of private sector intermediaries that can allow universities to draw on their commercial expertise in a cost-effective way.

Open Innovation

The open innovation model is described elsewhere in this volume and will not be revisited here. As traditionally inward-looking firms refocus their activities to interact with knowledge created outside the firm, this leads to new opportunities and renders innovation processes used in the past sub-optimal. The simple implication of this is that it becomes necessary to create new capabilities (Cohen and Levinthal 1990) and innovation brokerage roles (Dodgson and Bessant 1996; Hargadon 2003) to access and utilize external knowledge.

In practice some organizations have gained competitive advantage by employing this approach to innovation for many years. Organizations such as Porsche have outmanoeuvred their much bigger competitors, such as BMW and Mercedes Benz, to bring innovations such as the revolutionary ceramic brake system to the market (Harryson and Lorang 2005). Porsche has done this by collaborating with universities whenever it lacks the expertise to develop a specific innovation. Various other organizations such as Intel and Xerox have also used similar open innovations systems to ensure that they stay ahead of competitors (Allio 2005). Chesbrough quotes the examples of Intel, Microsoft, Sun, Oracle, Cisco and Amgen, and explains how these organizations have achieved success with the discoveries of others (Chesbrough 2006). This approach means that complementarities exist between the creation and use of knowledge, leading to a tighter link between creators and users of knowledge. It suggests that innovative firms have changed the way they search for new ideas, adopting open search strategies that involve the use of a wide range of external actors and sources to achieve and sustain innovation (Laursen and Salter 2006). Far removed from the early Schumpeterian model of the lone entrepreneur bringing innovations to the market, these newer models highlight the interactive nature of innovation, stressing the importance of networks and the reliance of firms on external suppliers of knowledge.

This approach, typified by Procter & Gamble's 'Connect and Develop' (Dodgson et al. 2006), represents a new business model for integrated innovation processes. It also suggests a new style of relationship with suppliers, research providers and customers assisted by new kinds of technology.

Despite its large-firm and high-tech focus, the open innovation model potentially has important implications for the Australian economy and Australian firms. The obvious step here is to point to the high-tech sector of the Australian economy and recommend initiatives to help these firms become more open in their innovation practices. However, we wish to avoid this direction for a number of reasons. First, we know from several studies that it is often not the high-tech sector *per se* that drives mature economies. A study by Hughes and colleagues of US growth in the 1990s showed that performance and wealth generation were dominated by retailing and financial services (Cosh et al. 2006). This is not to say that technology does not matter, but instead it shows that the way backbone industries innovate and use technology is of vital importance. A similar conclusion was reached by a European Commission study of growth per-formance and technology use in EU firms (Hirsch-Kreinsen et al. 2003). The fastest-growing industries in Europe were not biotech or ICT; instead the data showed how frequently dismissed and 'boring' industries such as food packaging and heavy transport were in fact leaders in wealth creation and innovation. Marx was off target when he talked about the success of new industries being a life-or-death matter for the prosperity of nations. Perhaps what is more important is the way that old industries use innova-tion to reinvent themselves in a competitive international economy. If open innovation models *do* indeed hold prospects for the management of innovation in firms and industries, then this has implications for estab-lished Australian sectors such as mining, agriculture, fisheries, financial services and manufacturing.

To help assess the extent to which 'openness' is an existing or potentially important feature of the knowledge flow within Australia's innovation system, we turn to the evolution of two intermediary, or broker, institu-tions: AMIRA International and the InnovationXchange (IXC). These cases were chosen as exemplar existing and emerging organizations. Knowledge of AMIRA International was derived from interviews with its CEO, published material and information in Vandermark (2003). Research into IXC included interviews and discussions with its CEO and another board member and published material.

AMIRA INTERNATIONAL

AMIRA International, a not-for-profit private sector company, was estab-lished in 1959 as 'an independent association of mineral companies to facil-itate the technical advancement of its members in the mineral, coal, petroleum and associated industries'. Based in Melbourne, it has offices in Perth, Cape Town, Johannesburg, Denver and Santiago, and affiliations with similar bodies in North America and Europe. It was created to 'develop, broker and facilitate collaborative research projects, with a focus on assisting with the uptake of leading-edge science and technology by members to strengthen their business development'.

AMIRA quickly established itself as a broker and manager of multi-sponsor collaborative research projects, acting as the agent for industry sponsors for research in universities, CSIRO or other institutions. In 1998, AMIRA opened its membership to foreign companies, and offices were incorporated overseas. It reconstituted itself as AMIRA International in 2000. It continues to attract new members from North America, Latin America and South Africa. Many of its projects have had multiple extensions. P9 is a mineral processing modelling project that has been subject to ever-advancing research for 45 years.

AMIRA operates by developing and managing jointly funded research projects on a fee-for-service basis. Project sponsors are required to be AMIRA members, whose primary benefit is the output from the research. AMIRA estimates companies sponsoring research projects enjoy financial leverage of 10 to 20 times the funds contributed, simply by sharing the costs and benefits of the research. AMIRA draws on a wide range of internationally recognized research providers. Currently there are 81 minerals research organizations participating in active or developing projects in Australia, New Zealand, Canada, the USA, Brazil, Chile, the UK, Germany, the Netherlands, Russia and South Africa. In the year 2005–6, 14 new projects were begun and 11 were completed, bringing total projects under management to 47, valued at £20 million of member funds. These are leveraged by supporting grants from government and institutional sources, bringing the total research funding to over £33 million.

Australia's world-class minerals research is reflected in the fact that while 47 per cent of AMIRA's funding in the year ending June 2002 was sourced from companies outside Australia, over 95 per cent of the research was still conducted in Australian institutions.

Spillover effects extend into other sectors of the economy. 'R&D in the minerals industry does not stop at mineral processing, it spills over into all sorts of unrelated industries, supporting start up ventures, new products and services' (Dick Davies, CEO, AMIRA International, 2006). An example is project P266D, 'Improving Thickener Technology', a continuing collaborative research project for companies involved in extractive metallurgy. As part of the project, CSIRO fluid dynamics specialists applied their knowledge to creating a better design for split-feed inlets to thickener feedwells. By manipulating this design, they improved the distribution of liquid coming into the thickener and kept it at the sheer rate needed to improve the coagulation/flocculation process. The design changes were taken up by equipment manufacturers and are being used in mineral processing plants for significant productivity improvements. CSIRO's expertise has contributed to other resource sector projects, such as improving mixing and agitation in alumina and nickel production sectors, and improving electrostatic precipitator performance. The underlying skills needed for this project have also been utilized in life-saving surgical techniques, for example in determining the restraining forces required to keep in place a graft within the aorta of the heart.

AMIRA's members range from the world's biggest resource companies to small producers, suppliers and consultants. Members are structured into three groups:

- 'Group' members are large multinational resource companies with many wholly owned subsidiaries. By paying for group membership, intellectual property (IP) can be used in operations throughout the company.
- 'Mid-cap' group members are firms with a market capitalization of under £170 million, with a reduced group membership fee reflecting the fact that they tend to have fewer subsidiaries through which they can exploit any IP they gain.
- 'Member' companies pay a smaller fee but can use any IP accessed through AMIRA only in the operations of the member company.

AMIRA claims to offer four key advantages to its members.

1. Leverage on research funds combined with reduced risk by helping a number of like-minded companies share the cost of a research project and access all IP benefits.
2. Access to the world's best researchers, expertise and relationships, with leading mineral research institutions and individuals accumulated over half a century.
3. Providing proven operating protocols and service, developed to ensure research delivery.
4. Networking, that is, the advantage of being part of a network, access to contacts with the global research community, and opportunities to stay abreast of technology developments.

AMIRA claims it allows companies to concentrate research resources on their core projects, critical to their competitive success.

Since its formation there have been a number of setbacks in AMIRA International's evolution. Challenges to its model of collaborative research come from several quarters, including changes to investments in innovation in the minerals industry, shifts in Australian innovation policy, operational difficulties and rising expense of individual projects.

The research environment has recently been affected by two major developments in the minerals industry: a period of price and demand strength, and continuing industry consolidation through acquisition and mergers. While firms have reported stronger results, there are competing demands for these funds as in a period of high demand there is pressure to divert available funds to immediate capacity increases. Industry consolidation has meant fewer organizations exist to share research costs. Minerals companies have been restructuring, cutting back internal research groups and devolving responsibility for research to company business units. As a result, AMIRA, which previously worked through 10 or 20 head offices, now has to deal with mine sites spread across the globe with inherently unstable commitments to R&D. Industry changes have led to a high degree of mobility of technicians and an overall decline in on-site expertise, making it hard to achieve the goal of successful exploitation of research. In this setting it is becoming increasingly difficult to target research towards long-term step-generation advances in the industry rather than to short-term commercialization of identified technologies.

Shifts in Australian innovation policy in the 1990s and other changes in government policy, a restraint on university funding and initiatives aimed at improving relations between research providers and industry needs all meant that AMIRA faced increasing new challenges. Furthermore, some new players – such as a number of cooperative research centres, 22 of which have been in the minerals and energy sector – received matching government funding for industry sponsorship. In recent times AMIRA has faced a turbulent operating environment. There was decline in research funds in three consecutive years from 2002 until 2004. New research contract value for 2005/6 was £6 million, a 7.5 per cent decrease compared to 2004/5. Income from operations at £1.8 million was 2 per cent lower than the previous year, while the operating expenses at £1.8 million were 15.7 per cent higher. AMIRA has to deal with the basic problem of generating the quantum of support for projects and project generation.

In the wake of many setbacks, AMIRA's strategies and outlook have responded to these developments in the industry. It states that its emphasis in the near- to medium-term strategy is to consolidate and further penetrate existing markets, to expand into new areas with the existing customer base and to continuously engage opportunities as members expand their operations globally. There have been initiatives to create new services, support infrastructure development that promotes opportunities for members, and to pursue additional sources of funding. AMIRA's long-term strategy is to provide a 'genuine global network strategy for its members and research providers'. A range of important steps has been taken towards implementing this strategy. They include producing technology road maps – such as for alumina and copper – that build on the vision for the future agreed by an industry sector, and chart broad priorities for subsequent research programmes and projects over the next 10–20 years. AMIRA reorganized its business units in 2002, in response to member's suggestions, and a new 'Sustainable Development Business Unit' was established with a principal focus on eco-efficiency, energy efficiency and other environmental issues.

Innovation and breakthrough technologies depend upon a solid and growing science base. To help ensure industry researchers have access to such a resource, AMIRA has set up the Australian Mineral Science

Research Institute and is responsible for coordinating its industry support and monitoring its effective applications. Generation of increased business in new regions is a major strategy, resulting in offices being opened in Denver, USA and Santiago, Chile. The geographical spread of its staff has resulted in limited face-to-face knowledge sharing of markets and activities. To help overcome this disadvantage AMIRA has established a collaborative web service for sharing information. It is further being expanded by linking it to AMIRA's core network file service in Melbourne, enabling greater information access.

AMIRA has transformed itself from an Australian-based association with fierce independence and little government interaction and sponsorship. Many of its recent projects have attracted leveraged funding from government schemes in Australia and overseas, including THRIP in South Africa; NSERC in Canada and ARC and CRC programmes in Australia. The growing level of international membership and their active partnership in the make-up of its Council, and receipt of overseas government funds, is evidence of AMIRA's success in internationalization, without threatening or reducing investments in the Australian R&D community.

INNOVATIONXCHANGE

IXC Australia was initially established in 2003 as the 'Australian Industry InnovationXchange Network', to provide an online 'open network' to improve communication across the boundaries of industry, government and academia. It was created with the support of the Australian Industry Group (an industry association), the Australian government and three Australian state governments. The federal government's Department of Industry, Tourism and Resources (DITR) provided £0.5 million from its Innovation Access Program to help establish the network.

In May 2004, IXC began to trial the use of its employees as 'trusted intermediaries' to facilitate the secure and managed exchange of sensitive knowledge between its members. This intermediary service pilot scheme, which initially focused on life sciences and IT, was successfully reviewed and the 'IXC Intermediary Service' became IXC's core business activity. After an initial period of intense engagement when IXC intermediaries attempt to absorb client needs, capabilities and strategic intentions, they maintain regular online contact supported by quarterly site visits.

Sensitive client information is shared among IXC intermediaries using a secure database called 'the Vault'. Client needs and capabilities are continuously updated and connections are sought by matching these logged intentions with those of others. Confidential information is not seen by other client–members, but when an opportunity is found, IXC intermediaries help the members engage directly through a step-by-step disclosure process.

IXC Australia Limited was officially formed in July 2006, headquartered in Melbourne as a fully independent, not-for-profit company limited by guarantee. Its objective is to help potential business partners such as growing companies, universities and research institutes look for opportunities to collaborate. IXC's model aims to provide value to various stakeholders in different ways. SMEs with limited internal resources gain access to external technology, research and complementary resources. Large companies are directed to pockets of international research excellence. Research institutes are assisted with commercialization.

In its initial year, IXC focused on technology-related industries. It has assisted the establishment of a life sciences cluster with members including IBM Healthcare, Johnson & Johnson Research, Walter & Eliza Hall Medical Research Institute, the University of New South Wales and seven Melbourne-based listed companies. Within months of its establishment, IXC claims the cluster is already generating the following outcomes:

- the creation of new valuable inventions for a Melbourne-based company,
- the creation of seven significant, but confidential, opportunities for direct collaboration amongst members,
- 28 other opportunities that are presently under investigation,
- the aim to double the size of the consortium with the support of ResMed and IBM and connect it to consortia of firms with converging interests in the medical devices, smart building and smart manufacturing sectors.

IXC's attention has subsequently been directed towards Australia's strength in mining, agriculture and other resource sectors. The Victorian government has provided up to £0.4 million (over three years) to IXC Australia, to supply intermediary services to strategic Victorian industry clusters including food and agriculture and smart manufacturing sectors. By September 2006, IXC intermediaries had completed 155 investigations, identified 115 opportunities and secured 20 engagements for collaboration (DITR 2006). The collaborative opportunities and commercial engagements created by IXC are expected by the organization to be worth many tens of millions of dollars within the next three to five years.

IXC has attracted early international interest. In May 2006, the UK government, through its Higher Education Innovation Fund, committed £3.6 million to the Birmingham Business School to partner with the Birmingham Chamber of Commerce and IXC. IXC UK has been operational since the end of 2006. IXC aims to further expand its cross-country international linkages. It recognized early on that intermediary services are a unique vehicle at a very practical level to promote international collaboration. According to IXC board member Terry Cutler, 'in a small country like Australia, where both industry and research are looking in and looking out, it is vital to be international'. IXC has helped facilitate the process of local research tapping into international networks. An example of such cross-country networking is seen in the two-way transfer of railway technology between Australia and the UK. It is also an example of IXC's intermediary function, where underlying technology transfer opportunities were identified.

IXC has the goal of establishing a consortium of regional hubs around the world. In a bid to establish a presence in South America, IXC created a regional hub in Chile at the end of 2006 and will be working towards implementing it in 2007. It is in active discussion with the Malaysian government to establish a South East Asian hub. The deployment of IXC intermediaries internationally will be through licensing its model to local organizations established along similar lines, i.e., independent, not-for-profit. Under the IXC banner, these nationally based IXCs are supported by IXC Australia, with its associated ethics, methodologies and standards. Linked in this way to IXCs internationally, clients could potentially benefit from the shared knowledge of an expanding network of IXC intermediaries. The IXC now has 25 member organizations, including a number of multinational pharmaceutical companies, with IXC intermediaries located in the USA, New Zealand and Europe. There are plans to establish IXC International in Melbourne by 2008. This body will guide the development of global networks of IXC operations and will be governed by the representatives of the participating countries.

IXC's major challenges include recruiting more staff – it currently has fewer than ten employees – and creating market awareness to develop increasing participation in the IXC networks. IXC stresses the significant role of the government in achieving this. In particular, the Commonwealth government has facilitated SME participation through various subsidies and programmes.

IXC's financial model uses a flexible fee system to increase membership. It operates:

- a subscription model – where IXC Intermediary Services are delivered on a fee-for-service basis by annual contract. At the same time shorter, project-based contracts are also available. The IXC Access Service is provided at a lower cost and geared towards very small organizations. It can vary the intensity of its engagement as

necessary – from two full days a week of the IXC Intermediary
Service, to simply maintaining a 'watching brief';
*   a research contract model – where IXC attempts to help solve a par-
    ticular research problem, i.e. on a project-by-project basis.

As IXC is structured as a not-for-profit organization, and deals with
commercially sensitive and valuable information, it has to ensure that its
dealings with various partners operate with integrity. Standard terms of
engagement and confidentiality agreements with no unauthorized disclosure are strictly maintained.

In its latest initiative, IXC will collaborate with the Australian Institute
for Commercialization (AIC) to cross-refer candidate technologies and
companies with its flagship innovation programme – the AIC TechFast
Program. The programmes are geared to work in complementary ways,
but at different points in the stages in the development of an innovation or
business.

CONCLUSION AND A FUTURE RESEARCH AGENDA

Numerous obstacles can be identified in Australia that constrain an
efficient innovation system. These include an industrial structure that comprises small firms; a preponderance of overseas multinational companies in
high-tech sectors and large resources industries – features that limit the
demand for innovation from business; the structure, multiplicity and
administration of public support programmes; a lack of personnel in business and universities skilled at facilitating knowledge flows; deficiencies in
levels of funding for research and innovation; and a need to better balance
support for basic and applied R&D and technology creation and diffusion
policies.

Australia also possesses many features of a successful innovation
system, including a strong research base and policies, such as CRCs,
directed towards encouraging the stream of knowledge from it into business. The two case studies analysed above reveal that Australia also possesses intermediary institutions that play an important role in facilitating
and managing the connections and flow of knowledge. The first case,
AMIRA International, has a long history of operating as a research
manager and broker for the minerals industry, working with firms of all
sizes. AMIRA has been very successful from an innovation systems perspective, in enabling and strengthening links between public sector minerals research in the knowledge base and industry needs. It has produced

spillovers into other industries. It has faced numerous difficulties due to changes in the industry and in government policy, yet it has continued to change and evolve to deal with the problems it faces. AMIRA is using new tools, structures and strategies to extend its effectiveness as a broker. It has become a much more open institution, engaging with many more players on an international level. It has become much more engaged with government than in the past, helping build better connections in the system. The second case, IXC, is a new organization that has grown quickly and immediately linked into international networks. It has received initial backing from many parts of the Australian innovation system and operates as a trusted broker, connecting its various elements and assisting even the smallest organizations. Both institutions are using the new innovation-supporting technological infrastructure to support and develop their knowledge brokering roles.

More research is needed into the complexity of the contributions and interrelationships of the major stakeholders that facilitate such knowledge brokering in Australia. Investigation of the intricacies of the linkages between these stakeholders, including research institutes, small and large firms and governments will improve our understanding of the link between knowledge creation and its commercialization. In a world where major firms aspire to acquire more than half of their innovations from outside the firm (Dodgson et al. 2005), such an understanding may be pivotal to ensuring the successful application of new knowledge. From the perspective of the research sector and government, the quick and broad dissemination of research results increases the efficiency by which they are accessed by users, thereby increasing total returns on research investment. The concept of open or distributed innovation provides a focal analytical lens for such future study.

The idea that such interaction should exist between government, public R&D and business is not new, and policies like the CRCs have been formulated to stimulate it. However, open innovation takes this model of interaction another step further with its focus on structure, connections and evolution of the innovation system that is nested within other international systems of innovation. The strategic aspects of open innovation represent a challenge for managers and academics alike. Innovation creates new capabilities, as knowledge and other resources inside and beyond the organization are recombined, but what are the antecedents of these new capabilities? Do particular forms of organization develop capabilities for using open innovation better than others? Also, having created these new capabilities in an open system, how does the organization capture the value created by them for competitive advantage? The continuing conflict between openness and the barriers to competition that are sources of competitive advantage warrants significantly more research (Helfat 2006).

A core proposition of a future research agenda is that firms are more likely to innovate openly when the capabilities that are being developed have natural barriers to competition associated with them. Intellectual property strategies involving patents and trade secrets have restricted application in open innovation because they can erect barriers preventing collaboration and the flow of information (Laursen and Salter 2006). Capabilities that are embedded within the organization and supported by clusters of intra-organizational weak ties in a systemic fashion, however, can remain appropriated by firms exposed to partners (Black and Boal 1994). Australia has deep, historical competitive and comparative advantages in the resources sectors. These are based on the availability of fabulously rich raw materials, an excellent research base, and strong internationally leading firms. The industry has also traditionally been marked by a significant element of inter-organizational trust in the industry (Quinn 1992), helping deal with the appropriability problem in an open innovation model.

The specificity and the idiosyncracy of the Australian industrial context – seen particularly clearly in the resources sectors – shows how the efficacy of the open innovation model will depend upon its tailoring to specific circumstances. Australian resources firms can well afford to be open with their intellectual property because their primary sources of competitive advantage are operating mines, tenements and logistics. In economic terms, these are inimitable factors of production as they tend to be geographically specific to a firm's operations. High levels of cooperation and trust may exist in the industry because there is little chance of effective opportunism from competitors.

This brings us to our final remarks on the emerging open innovation approach. There is a tendency to focus on the capabilities-building dimension of openness as firms use markets and alliances to develop knowledge and find new uses for their existing knowledge base. An agenda that examines open innovation purely from a capabilities perspective, however, will result in incomplete theoretical understandings. Open innovation represents an organizational form and therefore must be considered in the light of established literature on the theory of the firm that endeavours to explain firm boundaries (e.g. Coase 1937; Cyert and March 1963; Chandler 1992; Conner 1991; Madhok 1996; Kay 1997). To use the AMIRA example, perhaps open innovation in the form of alliances and markets in the mining industry is possible due to lowered transaction costs created by reduced levels of opportunism and asset specificity (Williamson 1999). Transaction costs matter and without considering them we may be reduced to overly socialized explanations of governance in systems of open innovation (Foss 1996, 2003). Chesbrough's (2006) account of the IXC case, and the role that 'beer, bonding and being there' plays in increasing the levels of trust within the network, signals a warning to researchers to take governance and opportunism seriously in the open innovation research model. As Foss (1996) has remarked, social cohesion and moral communities are insufficient to explain organizational forms.

Perhaps one of the most fruitful directions for open innovation research lies in attempts to reconcile capabilities and transaction costs explanations of organizational forms (Madhok 1996; Combs and Ketchen 1999; Williamson 1999; Colombo 2003; Steen et al. 2008). Firms may be compelled to use open innovation rather than the old internalized models of the innovation process because they can create more value, more rapidly. However, they do this in the face of increased transaction costs incurred through using an open organizational form, especially when operating internationally. The strategic choice of whether to be open or not is therefore a dynamic reconciliation of these opposing considerations. If we are to believe Langlois's (2003) hypothesis of the vanishing hand of management in the late twentieth century, where general lowering of transactions costs triggers the disintegration of large corporations, then perhaps we can see the shift towards open innovation as part of this broader process. More buyers and sellers of innovation serve to increase the efficiency of the marketplace and innovation technologies may further improve these market mechanisms (Chesbrough 2006). Open innovation may be a new and emerging literature, but it needs to review its significance for 'old' economies and should also be reconciled with 'old' theories of corporate strategy and organizational forms and firm boundaries.

NOTES

1. Speech at the National Innovation Summit, Melbourne, 6 December 2006.
2. We have, of course, to be cautious about the use of old innovation performance measures such as R&D expenditure and patenting. It simply does not tell us anything about how well a firm, or an industry, is working as an open system of knowledge flows, especially one like Australia's which is predominantly 'low-tech' and resource-based. Always a proxy and piecemeal indicator, it is useful only as a broad indicator of scale of investment in one, and occasionally subsidiary, element of innovation investments.
3. Manifesto for the Promotion of an Innovative Australia (www.mindsharing.com.au).

REFERENCES

Allio, R. (2005), 'Interview with Henry Chesbrough: innovating innovation', *Strategy and Leadership*, **33**(1): 19–24.

AMIRA (1997), Review of Business Programs, 'Industry leads: Government Follows', March.

AMIRA International (2002), National Research Priorities – Submission 105, June.

ARC (2006), 'Response to Productivity', Commission Draft Report.

Australian Bureau of Statistics (2004), *Research and Experimental Development, 2003–2004*.

Barlow, T. (2006), *The Australian Miracle: An Innovative Nation Revisited*, Sydney: Picador.

Black, J.A. and Boal, K.B. (1994), 'Strategic resources: traits, configurations and paths to sustainable competitive advantage', *Strategic Management Journal*, **15**: 131–48.

Chandler, A.D. (1992), 'What is a firm? A historical perspective', *European Economic Review*, **36**(2–3): 483–92.

Chesbrough, H. (2003), *Open Innovation: The New Imperative for Creating and Profiting from Technology*, Boston, MA: Harvard Business School Press.

Chesbrough, H.W. (2006), *Open Business Models: How to Thrive in the New Business Landscape*, Boston, MA: Harvard Business School Press.

Coase, R.H. (1937), 'The nature of the firm', *Economica*, **4**: 386–405.

Cohen, W. and Levinthal, D. (1990), 'Absorptive capacity: a new perspective on learning and innovation', *Administrative Science Quarterly*, **35**(2): 128–52.

Colombo, M.G. (2003), 'Alliance form: a test of the contractual and competence perpsectives', *Strategic Management Journal*, **24**: 1209–29.

Combs, J.G. and Ketchen, D.J. (1999), 'Explaining interfirm cooperation and performance: toward a reconciliation of predictions from the resource-based view and organizational economics', *Strategic Management Journal*, **20**: 867–88.

Commonwealth of Australia (2003), *Backing Australia's Ability 2*, Canberra: Australian Government, December.

Conner, K.R. (1991), 'A historical comparison of resource-based theory and five schools of thought within industrial organization economics: do we have a new theory of the firm?', *Journal of Management*, **17**(1): 121–54.

Cosh, A., Hughes, A. and Lester, R. (2006), *UK PLC: Just How Innovative Are We?* Cambridge, MA: Cambridge/MIT Institute, pp. 1–24.

Cutler, T. (2006), 'Innovation – where do we go from here?', *FastThinking Magazine*.

Cyert, R.M. and March, J.G. (1963), *A Behavioral Theory of the Firm*, Englewood Cliffs, NJ: Prentice Hall.

DITR (2006), *Submission from the Department of Industry, Tourism and Resources to the Productivity Commission Study into Science and Innovation*, September.

Dodgson, M. and Bessant, J. (1996), *Effective Innovation Policy*, London. Routledge.

Dodgson, M. and Innes, P. (2006), *Australian Innovation in Manufacturing: Results from an International Survey*, Report to the Australian Business Foundation.

Dodgson, M., Gann, D. and Salter, A. (2005), *Think, Play, Do: Technology, Innovation, and Organization*, Oxford: Oxford University Press.

Dodgson, M., Gann, D. and Salter, A. (2006), 'The role of technology in the shift towards open innovation: the case of Procter & Gamble', *R&D Management*, **36**(6): 329–42.

Eslake, S. (2007), *The Australian Economy: Annual Presentation to Students from the University of Delaware*, ANZ Bank, January.

Foss, N.J. (1996), 'Knowledge-based approaches to the theory of the firm: some critical comments', *Organization Science*, **7**(7): 470–76.

Foss, N.J. (2003), 'The strategic management and transaction cost nexus: past debates, central questions, and future research possibilities', *Strategic Organization*, **1**(12): 139–69.

Gassmann, O. (2006), 'Opening up the innovation process: towards an agenda', *R&D Management*, **36**(3): 223–6.

Geodynamics Limited (2006), *Annual Report 2006*.

Gregory, R.G. (1993), 'The Australian innovation system', in R.R. Nelson (ed.), *National Innovation Systems: A Comparative Analysis*, New York: Oxford University Press, pp. 324–52.

Hargadon, A. (2003), *How Breakthroughs Happen. The Surprising Truth about How Companies Innovate*, Cambridge, MA: Harvard Business School Press.

Harryson, S. and Lorange, P. (2005), 'Bringing the college inside', *Harvard Business Review*, **83**(12): 30–32.

Helfat, C.E. (2006). 'Open innovation: the new imperative for creating and profiting from technology', *Academy of Management Perspectives*, **29**(2): 86–8.

Hirsch-Kreinsen, H., Jacobsom, David, Laesadius, Staffan and Smith, Keith (2003), 'Low-tech industries and the knowledge economy: state of the art and research challenges. Pilot: Policy and Innovation in Low-Tech', Working Paper.

Howard Partners (2005), *Knowledge Exchange Networks in Australia's Innovation System: Overview and Strategic Analysis*, Report of a Study Commissioned by the Department of Education, Science, and Training, June.

Insight Economics (2006), *Economic Impact Study of the CRC Program*, prepared for the Australian Government Department of Education, Science and Training, October.

Institution of Engineers Australia (2002), *Business R&D in Australia*, Submission to the House of Representatives Standing Committee on Science and Innovation Inquiry into Business Commitment to Research and Development.

Johnston, R., Matthews, M. and Dodgson, M. (1999), *Enabling the Virtuous Cycle. Evaluations and Investigations Programme*, 00/14 Higher Education Division, Canberra.

Kay, N. (1997), *Pattern in Corporate Evolution*, Oxford: Oxford University Press.

Langlois, R.N. (2003), 'The vanishing hand: the changing dynamics of industrial capitalism', *Industrial and Corporate Change*, **12**(2): 351–85.

Laursen, K. and Salter, A. (2006), 'Open for innovation: the role of openness in explaining innovation performance among UK manufacturing firms', *Strategic Management Journal*, **27**(2): 131–50.

Madhok, A. (1996), 'The organization of economic activity: transaction costs, firm capabilities and the nature of governance', *Organization Science*, **7**(5): 577–90.

Moody, J. and Dodgson, M. (2006), 'Managing complexity in complex product systems: lessons from the development of a new satellite', *Journal of Technology Transfer*, **31**(5): 567–88.

Quinn, B.J. (1992), *Intelligent Enterprise: A Knowledge and Service Based Paradigm for Industry*, New York: The Free Press.

Rothwell, R. (1992), 'Successful industrial innovation: critical factors for the 1990s', *R&D Management*, **22**(3): 221–39.

Steen, J., Liesch, P., Matthews, M. and Thorburn, L. (2008), 'Balancing flexibility and control: international technology and market development in Australian biotechnology firms', *International Journal of Technology Management*, **41**(1/2): 203–22.

Vandermark, S.E. (2003), 'An exploration of innovation in Australian minerals industry: an innovation systems approach', PhD thesis, Australian National University.

Williamson, O.E. (1999), 'Strategy research: governance and competence perspectives', *Strategic Management Journal*, **20**: 1087–108.

Waluszewski, A. (2004), 'A competing or co-operating cluster or seven decades of combinatory resources? What's behind a prospering biotech valley?', *Scandinavian Journal of Management*, **20**(2): 125–50.

6.   Evolution of UK government support for innovation

Tim Minshall

INTRODUCTION

The aim of this chapter is to summarize the main stages in the evolution of the UK government's support for innovation in the period 1945 to mid-2007.

The UK government's view of the role of innovation within the UK's economic performance has been based on the perceived impact of innovation upon productivity. For HM Treasury, 'productivity growth relies on a continual stream of inventions and innovations of both new technologies and improved working practices' (HM Treasury 2000: 12). For the DTI: 'In the past, many UK-based businesses have prospered even when selling in low value markets, but today British industry faces a new challenge: how to raise its rate of innovation?' (DTI 2003: 8).

The perception of the role of innovation in relation to the performance of the UK economy (and the consequent attitudes of successive governments) has passed through a series of clear phases since the 1950s (see Figure 6.1). A description of each of these phases and examples of government policies to support the different needs is given in the following sections.

1945–1960S: REBUILDING THE ECONOMY

In the immediate post-Second World War years, the UK economy faced huge challenges in recovering from the extreme short-term demands that had been placed upon industry, together with a lack of investment. This was coupled with the UK facing the 'end of Empire' and needing to reposition itself as an economy operating in a world very different from that which existed before the Second World War. Examples of schemes that were initiated by the government around this time to help manage the transition to peace and to support the rebuilding of the economy included the following:

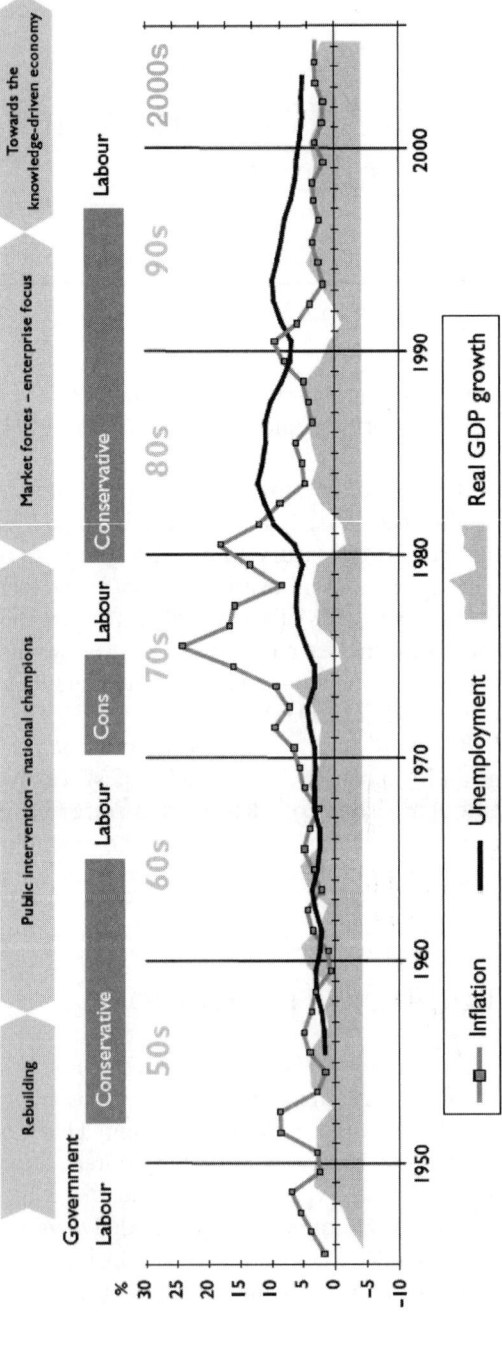

Source: Gill et al. (2007).

Figure 6.1 Economic performance and government policy themes 1945–2006

- Council of Industrial Design (established 1944): precursor to the Design Council that aimed to 'promote by all practicable means the improvement of design in the products of British industry' (Design Council 2007).
- Industrial and Commercial Finance Corporation (ICFC) (established 1945): ICFC was formed by the UK clearing banks, the Bank of England, the Board of Trade and the Treasury to meet the needs of smaller companies and address the shortage of long-term capital available to them for development. The idea for this corporation stemmed back to the findings of the Macmillan Committee in the 1930s (Coopey 1994). The ICFC, and its subsidiary function named Technology Development Capital (TDC), eventually became part of the UK venture capital firm Investors in Industry (3i).
- Finance Corporation for Industry (FCI) (established 1945): FCI was established by the same group that established ICFC but aimed to provide finance for large-scale, long-term investments needed for industrial rationalization (Coopey 1994). The FCI also eventually became part of the basis for 3i.
- National Research Development Corporation (NRDC) (established 1948): set up by the government to commercialize British publicly funded research. NRDC was involved in the exploitation of technologies including interferon, the hovercraft and the continuously variable transmission (CVT).

These organizations provided a foundation upon which the government could target its resources for rebuilding the economy, but also to support innovation under the coordination of, predominately, the Board of Trade (the predecessor of the Department of Trade and Industry) and the Department of Scientific and Industrial Research.

1960S AND 1970S: PUBLIC INTERVENTION – NATIONAL CHAMPIONS

Among the many strands of the government's industrial policy in these two decades, three themes can be identified clearly.

1. There was a strong move to support (or even create) larger firms through actively encouraging mergers between UK companies to form more substantial opponents for overseas (mostly Japanese and American) rivals (Owen 2004).

2. There was an articulation of the view that innovation which exploited scientific excellence was to be a critical element of the UK's future success. The Labour government of the mid- to late-1960s saw science and technology as a key foundation of a 'new' Britain (as summed up in Harold Wilson's speech on the 'scientific and technological revolution'[1]).
3. There was an emerging recognition that small firms were not only an important driver of change and innovation, but that this sector had been chronically undersupported since 1945.

Examples of public bodies created during this period to support the achievement of these three objectives include:

●   Ministry of Technology (established 1964): the Ministry of Technology was formed partly from the Department of Scientific and Industrial Research and partly from the Board of Trade.
●   Industrial Reorganisation Corporation (IRC) (established 1966): the IRC was set up to promote mergers in fragmented industries, working in partnerships with the newly formed Ministry of Technology.
●   National Enterprise Board (NEB) (established 1975): the remit of the NEB was to help regenerate British industry. It did this by (a) actively supporting the formation of new firms in 'strategic sectors' and (b) providing support for smaller firms with the aim of accelerating their growth.

The activities of the IRC in merging smaller UK firms to be stronger national champions resulted in the formation of companies such as ICL (Owen 2004). ICL was the result of the bringing together of a number of smaller UK computer companies to form one organization that it was hoped would have the potential to rival IBM in the USA. ICL, though achieving some success, particularly in the design and implementation of IT systems for the public sector, never became a serious rival for IBM and was in 1990 bought by Japan's Fujitsu.

It has been observed that this policy of providing public support and encouragement for the development of large, vertically integrated companies at this time was 'out of tune with the new economic realities' (Owen 2005). The support for the creation of new firms in selected industry sectors by the NEB helped the development of biotechnology and semiconductor activities in the UK. The NEB's role in the helping smaller, high-potential firms, especially in computing, had some successes (such as Sinclair Research).

Another example output of this approach of attempting to pick national innovation champions and supporting them through direct public funding is that of the Anglo-French Concorde project. Though widely acknowledged to be an extraordinary technological achievement, the project proved to be a long, drawn-out commercial failure (Spufford 2003).

At the end of the 1970s, with the UK economy struggling with high unemployment and high inflation, it became clear that the aims of this strong interventionist approach to supporting innovation had not been achieved.

1980S AND 1990S: MARKET FORCES – ENTERPRISE FOCUS

With the election of the Conservatives to power in 1979, the emphasis of government policy was firmly on the free market. Under the Thatcher government, the agencies (such as NEB) who had been tasked with 'picking winners' were closed down; numerous state-owned agencies were privatized; and the emphasis of government support shifted strongly to creating the climate for enterprise to flourish.

Examples of public bodies created during this period to support the free-market-focused economy included:

- British Technology Group (BTG) (established 1981): BTG was formed by the merger of the National Enterprise Board and the National Research Development Corporation. BTG went on to be privatized in 1992 and listed on the London Stock Exchange in 1995.
- Investors in Industry (3i) (established 1983): 3i was formed from the Industrial and Commercial Finance Corporation. Its activities are now focused around buy-outs, development capital and venture capital.

Although the Conservative government was taking a largely non-interventionist stance to support for innovation, it launched a number of schemes to provide targeted funding at stimulating activity. These schemes included:

- Small Firms Merit Award for Research and Technology (SMART): SMART drew upon US experience to provide small firms with relatively modest amounts of funding to test the commercial and technical feasibility of an idea, and to help these firms attract further funding.
- Small Firm Loan Guarantee (SFLG): provided government guarantees to allow small firms to borrow from commercial banks.
- Business Start-up Scheme/Business Expansion Scheme/Enterprise Investment Scheme: these schemes allowed investors to claim tax relief on income from investments in unquoted companies and, in particular, start-ups.

- Alternative Investment Market (AIM): although this was not a government scheme *per se*, the government strongly encouraged the London Stock Exchange to establish a separate market for smaller and younger firms.

In 1993, a government document was published that led to the reorganization of public support for science and innovation. *Realising Our Potential: A Strategy for Science, Engineering and Technology* communicated a significant shift in government thinking. Science policy was conceived within the broader framework of innovation policy, and this led to a complete overhaul of the organization of support for science and technology in the UK, including the move of Office of Science and Technology (OST) to the DTI.

LATE 1990S AND 2000S: TOWARDS THE KNOWLEDGE-DRIVEN ECONOMY

In response to the changing pattern of international competition, New Labour under Blair saw the future of the UK as based around the development of a strong 'knowledge-driven' economy; and innovation and entrepreneurship were key strands in the achievement of that objective. The evolution of the current government's approach to developing the mechanisms to use innovation to drive the UK towards a knowledge economy can be tracked in the key documents as detailed in Table 6.1.

The policy documents outlined in Table 6.1 have underpinned the implementation of a range of initiatives to support innovation, examples of which are shown in Table 6.2.

There have also been a series of initiatives targeted at increasing the supply of investment available to young firms seeking to exploit new technologies. These include:

- Regional Venture Capital Funds (RVCFs): since 2000, over £120 million has been committed by the Small Business Service and the European Investment Fund in creating a network of nine regional venture capital funds (each with, typically, £20 million to £50 million under management).
- Enterprise Capital Funds (ECFs): an adaptation of the equity investment activities of the US Small Business Innovation Company (SBIC) for the UK environment.
- Venture Capital Trusts (VCTs): the Enterprise Investment Scheme (EIS) was a means by which individual 'informed investors' could

Table 6.1   Selected key innovation policy documents[2]

Policy document	Summary
Our competitive future: Building the knowledge-driven economy, 1998	Set a 'co-ordinated and coherent programme of action' to close performance gap with competitors. Actions focused around capabilities, collaborations and competition
Excellence and opportunity: A science and innovation policy for the 21st Century, 2000	Set framework for government's role as key investor in science base; facilitator for collaboration between HEIs and business; and the regulator for innovation
Opportunity for all in a world of change: A white paper on enterprise, skills and innovation, 2001	Emphasized importance of innovation to regional and national growth, with policy objectives for: skills; building strong regions; investment in innovation; fostering enterprise and growth; and strengthening international links
Science and Innovation Strategy, 2001	Outlined DTI's aims, objectives and science and innovation priorities
Innovation Report, 2003	Outlined direct measures to be taken in seven key areas to ensure that the UK would be a 'key knowledge hub in the global economy'. Recommended the establishment of a Technology Strategy Board
Lambert Review, 2003	Analysed the specific role of university–industry collaborations in supporting innovation
Science and innovation investment framework 2004–2014	Set qualitative attributes of a successful system to support improvements in UK innovative performance over medium to long term
Business Support Solutions: A new approach to business support, 2004	Defined the new approach to business support in the light of the review of DTI activities
DTI: Five Year Programme: Creating Wealth from Knowledge, 2004	Outlined the key challenges facing the UK economy and the role that the 'new' DTI would play in addressing these challenges
Technology Strategy Board – Annual Report 2005	Summarized the activities of the board since inception in 2004 and outlined next stages of activities 'to deliver a technology strategy for wealth creation and to position the UK as a global leader in innovation'

Table 6.2   Examples of direct and indirect public support for innovation

Main title	Managed by	Broad purpose	Target companies	Amount per company	Programme budget
Grants for investigating an innovative idea	DTI	Reimbursed consultancy to help businesses get advice on the steps needed to implement their ideas. The grant will cover 75 per cent of the costs of the mentor and expert consultant, up to a ceiling of £12 000	Businesses with fewer than 250 employees wishing to exploit an innovative idea	<£12k	£1.2 m–£1.5m for pilot year (currently on hold)
Grants for R&D (research)	DTI/RDAs	Aim to investigate the technical and commercial feasibility of innovative technology. Up to £75k	Businesses with fewer than 50 employees	<£75k	£100 m for 2004–7
Grants for R&D (exceptional)	DTI/RDAs	Projects that involve a significant technological advance and are strategically important for a particular technology or industry sector. Grants of up to £500k	Any business	<£500k	
Grants for R&D (development)	DTI/RDAs	Aim to develop a pre-production prototype of new product or process that involves a significant technological advance. Up to £200k	Businesses with fewer than 200 employees	<£200k	
Grants for R&D (micro)	DTI/RDAs	Simple, low-cost development projects lasting no longer than 12 months. Grant of up to £20k	Businesses with fewer than 10 employees	<£20k	

Knowledge transfer partnerships	DTI	A grant to cover part of the cost of using a person to transfer and embed knowledge into a business from the UK knowledge base via a strategic project	All businesses needing expert help to innovate
Knowledge transfer networks	DTI	A grant to an intermediary to set up a network in a priority technology area, bringing together businesses, universities and others with an interest in technology applications	All businesses wanting to grow by exploiting technology
Collaborative R&D	DTI	Funding for collaborative research and development projects between businesses, universities and other potential collaborators. The level of grant support will vary from between 25 per cent and 75 per cent of R&D costs	All UK-based businesses wishing to exploit technology

Not specified

£370 m for 2005–8

Source: Livesey et al. (2006).

133

make investments directly into new, privately held companies and be exempt from certain capital gains tax payments. The VCTs widen the opportunities by allowing individuals to invest in a fund which then itself invests in new, privately held companies.

The government was also keen to support universities in taking on a greater role at all stages of the innovation process, and not just to be focused on the traditional university activities of teaching and research. To that end, the government has implemented a number of university-focused programmes (see Table 6.3).

The New Labour government has been able to build upon a number of platform activities developed over the last 20 years to focus new initiatives upon increasing innovation to support higher national productivity. The direct effectiveness of these activities on improving the UK's productivity is clearly hard to measure, although data from Livesey et al. (2006) provide some insight into this linkage.

The evolution of the consultations and analysis listed in Table 6.1 has led the government to view innovation as one of the five main drivers of improved UK productivity; the others are skills, investment, competition and enterprise. The DTI developed a strategy and selected five areas to focus efforts where they believed they have the greatest impact on improving the UK's productivity:

- Transferring knowledge
- Maximizing potential in the workplace
- Extending competitive markets
- Strengthening regional economies
- Fostering stronger partnerships.

The approach taken specifically to support innovation encompassed a series of direct (e.g. awards paid directly to companies to 'do' innovation, such as grants for R&D) and indirect support measures (e.g. fiscal measures to give incentives to companies to undertake various innovation-related activities, such as R&D tax credits).

HOW HAS GOVERNMENT STRUCTURED ITS SUPPORT?

The number of schemes initiated as a result of the various policy documents outlined earlier, or which were initiated through other activities but which are aimed at improving the productivity of UK industry

Table 6.3   *Example UK government funding streams to support 'third stream' activities*

Start year	Initiative	Purpose	Details
1998	Higher Education Reach Out to Business and the Community (HEROBaC)	Funding to support activities to improve linkages between universities and their communities	£20 m per year allocated to provide funding for the establishment of activities such as corporate liaison offices
1999	University Challenge Fund (UCF)	Seed Investments to help commercialization of university intellectual property rights	£45 m allocated in the first round of the competition in 1999 (with 15 seed funds being set up) and £15 m in October 2001. 57 HEIs how have access to this funding
1999	Science Enterprise Challenge (SEC)	Teaching of entrepreneurship to support the commercialization of science and technology	SEC provided £44.5 m through two rounds of funding. There are now over 60 HEIs participating in SEC-funded activites
2000	Higher Education Innovation Fund (HEIF)	Single, long-term commitment to a stream of funding to 'support universities' potential to act as drivers of growth in the knowledge economy'	HEIF was launched in 2000 to bring together a number of previously independently administered third stream funding sources. This was then extended with HEIF2 in 2004 and HEIF3 funding in 2006
2000	Cambridge–MIT Institute	A range of research projects and education activities to drive improvements in the UK's competitiveness, productivity and entrepreneurship	£65 m for a five-year programme of activities

under the broad headings of 'business support', had reached over 3000 by 2005. A review of DTI business support activities in 2004 (DTI 2004a, 2004b) highlighted concerns at lack of coherence and clear communication of business support offerings. The DTI's response to this was to:

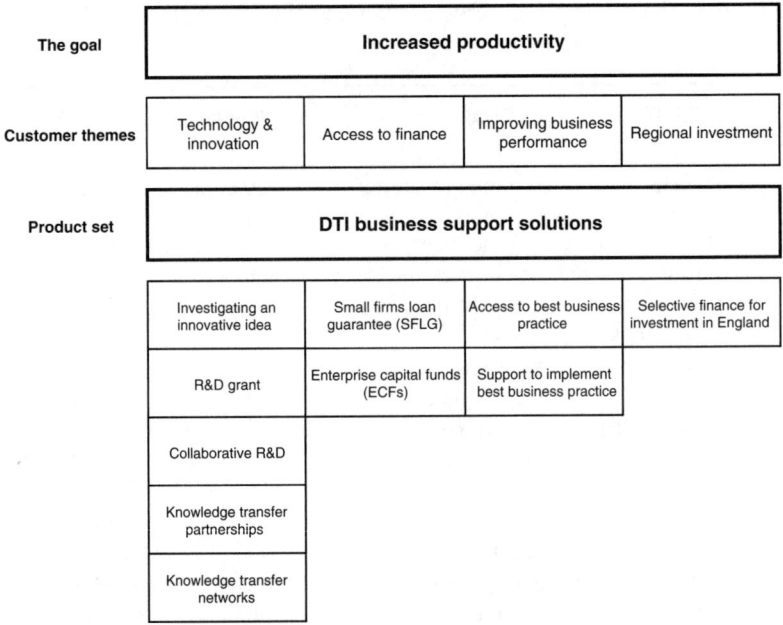

Figure 6.2   DTI's structuring of innovation support activities as of 2005

- make strategic investments in business support to drive up productivity;
- simplify through focus on customers and their relationships with
  DTI; and
- increase accessibility through improving the delivery of business
  support.

One output of the review and refocusing of activities was the winding
down of numerous business support schemes and the structuring of activities around customer themes. On the specific goal of improving the UK's
ability to innovate, five core products were identified following the 2004
review: knowledge transfer networks; collaborative R&D; grants for investigating an innovative idea (since discontinued); grants for R&D; and
knowledge transfer partnerships.

Information for firms on what practical support is available to help them
innovate is now channelled through a number of sometimes overlapping
online sources provided by the DTI, Business Link, HM Treasury and HM
Customs and Excise. This reflects the fact that while the core offerings have
been rationalized and simplified, the message to companies seeking to

know more about how to innovate was still somewhat complex. It was also clear that while there were many routes to finding out about the core offerings to directly support innovation, information on the indirect sources of support will come from a wide range of sources. For example there are also additional sources of funding to support innovation available via organisations such as NESTA[3] and the Carbon Trust.

To simplify the way in which companies access the help they need, there has been a clear desire by the government to encourage firms to access support via the Business Link network, delivered at a regional level. A potentially powerful resource to link individual company need is that of the Business Link Grants and Support Directory.[4] This provides information on national and region-specific assistance available to help firms access funding and direct business support, including innovation-related activities. However, using such a system will raise many options but there needs to be intelligent interpretation of what the diverse funding schemes identified as a result of this actually mean. Users are then encouraged to contact their regional Business Link to receive advice and support for accessing the funds they need.

While efforts have continued to clarify the routes for firms to access public support for innovation, there has also been a recent significant internal restructuring at the national level. The DTI's Innovation Group (IG) was merged with the Office of Science and Technology (OST) to form the Office of Science and Innovation (OSI).

EMERGING ISSUES

Changing Nature of the Process of Innovation

The policies described in the first part of this chapter show the clear recognition by various governments of the importance of innovation for the economic growth of the UK. However, the nature of innovation has changed and continues to change. Shortening product life cycles, intensification of competition, the speed at which discontinuous innovations may disrupt markets, and increased product complexity are all factors driving the need for companies in many industries to change the innovation process – to move beyond relying upon the exploitation of ideas generated through internal R&D to a more open model of innovation.[5]

Government support for innovation is reflecting this change to some degree. For example, the government can play a useful role in supporting collaborations between firms, particularly when one firm is a resource-limited start-up and the other is a large, established firm. The smaller firm is typically in (or perceives itself to be in) a weak position when seeking to negotiate with a larger firm. Programmes such as collaborative R&D or knowledge transfer networks can place smaller firms on a more equal footing with large partners. Larger firms get access to a new source of ideas, and the small firm gets access to a wide range of complementary assets needed to bring its invention to market.

Increased Importance of Different Types of Innovation

Government support for innovation in the UK has been, to a certain extent, blurred with that of support for invention and scientific discovery. Focusing on the latter has led to a strong emphasis on increasing the percentage of GDP that is spent on R&D, in both the public and private sectors. R&D spend as a proportion of GDP has become the main yardstick of 'innovation' performance for Europe in its efforts to deal with the continued innovation dominance of the USA and Japan, and emerging innovation challenge of countries such as India and China. One instance is the EU-wide Lisbon target of 3 per cent of GDP invested in R&D by 2010 (EC 2007). The UK has itself set a lower (and probably more realistic) target of 2.5 per cent by 2014.

Innovation needs to be considered in all its manifestations – only a few of which may be directly linked to the level of national R&D – if its real potential to support improvements in productivity is to be recognized. Innovations not just in terms of bringing new technologies to market, but also in business models, organizational structures, production processes, business processes, service and support, etc., all have the potential to increase the productivity of UK firms. Government support can be targeted at activities to stimulate innovation in all these forms.

The Emerging Role of the Regions

The increasing role of the regions as the focus of public support for innovation presents both opportunities and challenges. On the positive side, innovation support can be tailored to particular regional needs. On the downside, it may lead to rationing of support to regions with higher levels of innovative activity.

One regional innovation support instrument that is proving popular is that of the 'enterprise hub'. Though many different definitions exist for these, most have the common characteristic of being a regionally funded collaboration around innovation and enterprise. They typically bring together a range of independent support activities around a particular theme, be it type of firm (e.g. high growth potential) or industry sector (e.g. biotechnology) or both, to provide an 'ecosystem' of support for innovation.

The Role of Universities

As shown in Table 6.3, the New Labour government has seen universities as playing a number of important roles in supporting innovation. First, they provide a significant proportion of the world-class research that feeds IP into the innovation system. Second, they provide people with the skills to bring ideas to market. Third, they have been increasingly expected not just to be passive suppliers of IP, but also to be active in the generation of commercial value from their IP. Encouraging universities to become directly involved in innovation activities helps them to bring ideas out of the labs and into market application. However, a blurring of the boundaries between research and innovation may present challenges in the future, as exemplified by the current debate in the USA around royalty-free use of patented IP by universities.

Responsiveness versus Consistency

With the nature of innovation changing, the ability of government to be responsive and develop programmes appropriate for emerging needs is clearly critical. However, a lack of consistency in programmes can lead to confusion and poor uptake among the very organizations the government is seeking to support. The earlier sections of this chapter reflect the almost constant state of flux within which many innovation support programmes have been developed, implemented, adapted and terminated.

Spreading the Ability to Innovate More Widely

Capturing and creating value from leading-edge research outputs is one clear thrust of programmes to support innovation. However, the government also recognizes the needs of the large group of 'non-innovating' firms, and it is these firms that need support not only to recognize the importance of innovation, but also direct assistance in implementing appropriate tools and techniques to support innovation within their organizations. For these firms, leading-edge research outputs may not be what is required; it may be more mature, standardized approaches that just need to be effectively implemented that will allow them to create more value.

Integration of Activities and Flows of Knowledge

With innovation identified as a key driver of productivity growth, support for innovation is unsurprisingly a theme in the strategies of a wide range of public agencies. Although having the innovation agenda spread widely is positive, it also presents significant challenges in integrating – or at least linking – support for all the different types and dimensions of innovation. Although innovation itself relies upon multiple flows of knowledge from around the world connecting and fusing to form new outcomes, individual programmes to support innovation may be forced to focus on discrete, measurable, local outcomes. A result of this can be the focusing of programmes whose outcomes can be measured but whose real effectiveness at supporting innovation more broadly may be missed.

The development of the knowledge infrastructure to underpin increased innovation also presents policy challenges. A knowledge infrastructure (such as the development of technology intelligence activities) typically requires sustained support over a significant period without there being directly attributable near-term improvements in innovation. The DTI's Global Watch programme is one example of such an infrastructure-building activity which has recently closed.

Appropriate Lessons from Overseas

There are always lessons to be learned from the ways in which other nations support innovation. However, successful programmes may be highly context-specific. The UK has often looked to the USA for models of ways in which it could improve its ability to innovate. While there are successful programmes from the USA that may be equally effective in the UK (such as the US SBIR programme), factors related to the scale of the US economy when compared to the UK may affect the transferability of US programmes. Observation of activities within smaller economies such as Israel and Singapore may provide alternative valuable insight into practical options for the UK.

CONCLUSIONS

UK government support for innovation has passed through a series of phases over the past 50 years, and these phases have also reflected, to different degrees of success, the changing nature of the innovation process, and the widening of types of innovation able to drive productivity improvements. Drawing upon the themes presented in this chapter, the following conclusions can be drawn:

- Public support for innovation needs to be responsive to the changing, and increasingly open and international, nature of innovation in all its forms. Although there is clearly recognition of the open, international and diverse nature of innovation, placing primary emphasis on increasing R&D expenditure may not be the most effective route to improving UK productivity in the face of global competitors.

- Recognition of the importance of knowledge flows to the innovation process is required to ensure that an appropriate infrastructure for innovation is developed for the UK in the twenty-first century. Such knowledge flows may be between firms (be they start-ups, SMEs or multinationals), universities, the public sector, regions and nations.

- There are lessons in the public support of innovation that can be drawn from the experience of other nations, but these lessons should be drawn from many sources and not just the USA. Not only do smaller nations such as Singapore and Israel present interesting lessons, but there are novel approaches to innovation and its support that can also be observed, for example, in the rapidly developing BRIC economies (Brazil, Russia, India and China) as they adapt their roles within the knowledge economy.

NOTES

1. 'The Britain that will be forged in the white heat of this [scientific and technological] revolution will have no place for restrictive practices and outdated measures on either side of industry.' Harold Wilson, Speech to Labour Party conference, 1 October 1963.
2. These reports can be downloaded via http://www.ifm.eng.cam.ac.uk/cep/govpolicy.html.
3. National Endowment for Science, Technology and the Arts – www.nesta.org.uk.
4. http://www.businesslink.gov.uk/bdotg/action/gsdt.
5. Open innovation has been described as 'the use of purposive inflows and outflows of knowledge to accelerate internal innovation, and expand the markets for external use of innovation, respectively' (Chesbrough et al. 2006). This is contrasted with the 'closed' model of innovation where firms typically generate their own ideas which they then develop, produce, market, distribute and support.

REFERENCES

Chesbrough, H., Vanhaverbeke, W. and West, J. (eds) (2006), *Open Innovation: Researching a New Paradigm*, Oxford: Oxford University Press.
Coopey, R. (1994), 'The first venture capitalist: financing development in Britain after 1945: the case of ICFC/3i', *Business and Economic History* **23**(1): 262–71.
Design Council (2007), website: http://design-council.org.uk/en/Design-Council/1/Our-history/.
DTI (2003), *Innovation Report: Competing in the Global Economy: The Innovation Challenge*, London: Department of Trade and Industry, http://www.berr.gov.uk/publications.

DTI (2004a), *Business Support Solutions: A New Approach to Business Support*, London: Department of Trade and Industry, http://www.berr.gov.uk/publications.

DTI (2004b), *Five Year Programme: Creating Wealth from Knowledge*, London: Department of Trade and Industry, http://www.berr.gov.uk/publications.

EC (2007), 'Europa: Growth and Jobs', European Commission website: http://europa.eu.int/growthandjobs/index_en.htm.

Gill, D., Minshall, T.H.W., Pickering, C. and Rigby M. (2007), *Funding Technology: Britain Forty Years On*, Cambridge: St John's Innovation Centre and University of Cambridge Institute for Manufacturing, ISBN 1-902546-50-4, http://www.fundingtechnology.org.

HM Treasury (2000), *Productivity in the UK: The Evidence and the Government's Approach*, London: HMSO.

Livesey, F., Minshall, T.H.W. and Moultrie, J. (2006), 'Investigating the technology-based innovation gap for the United Kingdom: A report for the UK Design Council', Cambridge: University of Cambridge Institute for Manufacturing, ISBN 1-902546-49-0.

Owen, G. (2004), 'Where are the big gorillas? High technology entrepreneurship in the UK and the role of public policy', *The Diebold Project UK*, The Diebold Institute, www.dieboldinstitute.org.

Spufford, F. (2003), *Backroom Boys: The Secret Return of the British Boffin*, London: Faber and Faber.

PART II

Firm Development through Knowledge

7. Entrepreneurship in the knowledge economy

Erik Stam and Elizabeth Garnsey

INTRODUCTION

If the industrial economy ran on coal and iron ore, the fuel of today's economy is knowledge.[1] Technologies have always been underpinned by knowledge, but an economy run on knowledge is characterized by a critical role for information and communication technology (ICT), a high proportion of knowledge-intensive activity and intangible capital that amounts to more than tangible capital in the economy's capital stock (Atkinson and Court 1998; Foray 2004).[2] The emergence of the knowledge economy is not confined to high-technology and ICT services; it has spread across all sectors of market economies since the 1970s.[3] Wealth creation increasingly depends on the generation and exploitation of knowledge involving not only science and technology but also knowledge of practice required to create economic value (Gibbons et al. 1994). That knowledge plays an important role in the economy is not a new idea or finding. Every economy is based on knowledge of farming, mining and construction (Mokyr 2002). Knowledge, embodied in people and technology, has always been central to economic development. But only over recent decades has its importance received so much emphasis (Harris 2001). The OECD economies are more than ever dependent on the production, distribution and use of knowledge (OECD 1996).

While scholars in the 1950s and 1960s pointed to the economic importance of large firms (Galbraith 1956; Servan-Schreiber 1968), more recently a shift from the managed economy to the entrepreneurial economy in OECD countries has been identified (Audretsch and Thurik 2000, 2001). Radical changes in ICT and biotechnology have created market opportunities that are more effectively developed by new firms than by established companies. The shift to knowledge-based economic activity is said to be the driving force underlying the emergence of the entrepreneurial economy (Audretsch and Thurik 2001).[4] There is an emphasis on individual motivation, new ideas and risk taking, which render small and new flexible firms critical to economic success.[5] In the entrepreneurial economy, flexibility and innovation are more important than stability and control. Policy makers are counting on entrepreneurial initiative to address contemporary economic problems associated with structural change, including unemployment and industrial stagnation. Several studies have found that (especially ambitious) new firms have a positive effect on economic growth in advanced capitalist economies and to a marked extent in transition economies (Van Stel et al. 2005; Wong et al. 2005; Acs and Mueller 2006; Stam et al. 2007; Bosma et al. 2006). This is not the case in developing countries, where other mechanisms are currently more important for economic growth (cf. Bwalya 2006).

Recently comparative data have been amassed on entrepreneurship and young firm growth in different countries. Figure 7.1 shows indicators of entrepreneurship and young firm growth in several knowledge-based economies. The indicators of ambitious entrepreneurship and total entrepreneurial activity vary to a large degree (with Japan having values that are about five times smaller than those of the USA). Within the UK, about 5 per cent of the adult population is actively preparing a start-up or owns a recently started (<42 months) business. Slightly more than 1 per cent of the adult UK population has the ambition to start a business that will expand beyond 20 employees. The indicator of young firm growth shows extensive variation: about 9 per cent of the US young firms has grown by more than 60 per cent while this is only the case for about 1 per cent of the German firms. Japan has the lowest values on most indicators, and the USA has the highest values. The UK and Italy also have high shares of high-growth young firms, while the Netherlands, Denmark, Norway and Germany reveal relatively low rates of realized growth. Ambitious entrepreneurs in the latter four countries face either well-paid competing opportunities or severe constraints that frustrate the realization of their growth ambitions.

In the international comparisons, the USA stands out on all indicators of entrepreneurship and firm growth. This is usually attributed to cultural factors, which are doubtless present. Historically, innovations have been pioneered in the USA since the nineteenth century. But the USA benefits from three further advantages: a large and competitive domestic market; a highly developed financial system; and a high level of long-term government support for basic science (Owen 2004). The disadvantage of small home markets has been only partially offset by the removal of trade barriers within the EU, given strong differences in language and national practice. The scale of US government support for scientific research, defence-related and health-related research remains above that in the EU as a whole (Shahid and Kaora 2007). Moreover, government support to small business has been very extensive in the USA (Connell 2006). Policy

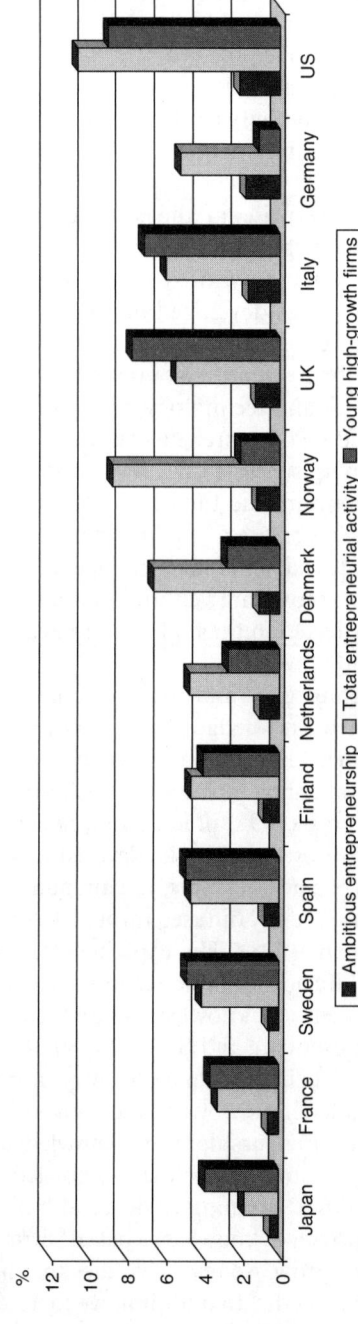

Sources: Ambitious entrepreneurship (20+ employees) and total entrepreneurial activity (nascent entrepreneurs + owners of young (<42 months) businesses) in 2001 (Stam et al. 2007, based on GEM data); high-growth firms (at least 60 per cent employment growth over three years of young (<5 years) firms with 15–200 employees, on average in 1999–2001) (Hoffmann and Junge 2006, based on Bureau van Dijk data).

Figure 7.1   Entrepreneurship, growth ambitions and high-growth firms: an international comparison

in the UK and other European countries has been influenced by 'level playing field' considerations which would view US levels of support to companies by governments in member states as anti-competitive practices (cf. Dosi et al. 2006). Another remarkable fact is that countries that score high on R&D indicators and productivity levels, such as Japan and Sweden, seemingly realize this without the 'intervention' of high levels of entrepreneurship.

Attention to the role of enterprise in the knowledge economy has resulted in policy makers around the globe counting on entrepreneurship to provide the engine of economic growth. A major source of entrepreneurial innovation is the knowledge developed in scientific organizations and private research labs. Entrepreneurs build on significant discoveries and emergent knowledge in their scientific communities (Merton 1993). Extensive investments in science and technology have given rise to opportunities for innovation pursued by entrepreneurs (Baumol 2002). Whether or not opportunities are taken up successfully, however, depends on entrepreneurial behaviour in an economy and the way in which new businesses are managed. New firms have to create a resource base in order to commercialize knowledge on a scale that can make an impact. New firms excel at detecting opportunities and resourcing ventures, but matching these up is a delicate process. This is reflected in the highly non-linear growth paths of new firms and high exit rates.

As knowledge-based firms enter, grow and exit the economy, they demonstrate the economic value of new knowledge. An economy with a high proportion of knowledge-based firms is building the knowledge and expertise required for the future when emerging technologies will diffuse into other parts of the economy. As Penrose (1995) pointed out, new firms are often the lead innovators in new fast-growth industries devoted to the production of new goods and services (e.g. since her time, in communications). Their emerging technologies are subsequently diffused through intersectoral flows (see e.g. Mowery and Rosenberg 1998). The experimental nature of new firms and their role in the diffusion of new technologies are important reasons why the encouragement of knowledge-based entrepreneurship should be a policy objective (Rosenberg and Birdzell 1986; Rosenberg 1992; Eliasson 1998). Such firms may benefit from early entry into new markets and establish technological leadership (Mowery and Nelson 1999).

This chapter deals with entrepreneurship in the knowledge economy. It reviews the literature on scientific and technological knowledge as a source of entrepreneurial opportunities, examining evidence at both the regional and organizational levels. Because knowledge-based firms have more impact on the economy if they grow, we review studies that have analysed the determinants of new firm growth.[6] In addition, we focus on the causal mechanisms of new firm growth, discussing longitudinal case study research on problem solving and competence creation in new growing firms. We end with a summary and a discussion on policy issues related to entrepreneurship in the knowledge economy.

KNOWLEDGE AS A SOURCE OF ENTREPRENEURIAL OPPORTUNITIES

Entrepreneurship necessarily involves individuals and their response to economic opportunities (Shane and Eckhardt 2003). Not only is the source of opportunities important, but so is the nature of the individual recognizing and commercializing these opportunities. Studies have shown that entrepreneurial opportunities are not exogenously given but rather endogenously and systematically created under certain conditions. They are the outcome of investments in new knowledge and ideas (Schumpeter 1942) on the one hand, and the accumulation of knowledge in individuals (Shane 2000) and firms (Cohen and Levinthal 1989, 1990) on the other hand. Prior knowledge enables certain entrepreneurs to be alert to new opportunities (Shane 2000; Kirzner 1973).

Prevailing theories of entrepreneurship have revolved around the ability of individuals to recognize opportunities and act on them by starting a new firm. This has generated a literature asking why entrepreneurial behaviour varies across individuals with different characteristics, while implicitly holding constant the external context in which individuals find themselves. Here the source of opportunities is implicitly taken as given. However, it is unlikely that individual personality is the key factor in view of evidence that the entrepreneurial role is often set aside when individuals risk losing what they have gained from being entrepreneurial. Entrepreneurial start-ups often become conservative small businesses.

There has been much empirical research showing that firms located near knowledge sources introduce innovations at a faster rate than rival firms located elsewhere. These studies frequently invoke the existence of localized knowledge spillovers as an explanation for this correlation (see Breschi and Lissoni 2001 for a critical review). Agents investing in research or technology development often end up facilitating other agents' innovation efforts, either unintentionally, as when inventions can be imitated, or intentionally, as when scientists report on their research. Economists have termed this non-rival characteristic of knowledge 'knowledge spillovers' (Arrow 1962; Nelson 1959). Knowledge spillovers have been defined as 'any original, valuable knowledge generated somewhere that becomes accessible to external agents, whether it be knowledge fully characterizing an innovation or knowledge of a more intermediate sort. This knowledge is absorbed by an individual or group other than the originator' (Foray 2004: 91).

The generation of knowledge in centres of research does not automatically lead to new economic value. New ideas and knowledge embodied in goods and services need to reach markets and meet demand. Routes to market can be established by corporate marketing and, less readily, by technology transfer units of public organizations. The knowledge spillover theory of entrepreneurship (KSTE) suggests that entrepreneurship provides a crucial mechanism in translating knowledge into new value, and ultimately economic growth (Acs et al. 2005; Audretsch and Lehmann 2005b; Audretsch et al. 2006). In contrast with entrepreneurial traits theories,[7] knowledge spillover theory takes individual characteristics as given and examines variation in context. One justification is that the same individuals move in and out of entrepreneurial roles over time. Moreover, the knowledge context is held to influence cognitive processes. Endogenous entrepreneurship is said to occur when knowledge workers respond to opportunities by starting a new firm. In this view entrepreneurship is a rational choice made by economic agents who seek to appropriate the value they attribute to knowledge endowments, whether their own or their employers'. This theory does not claim that entrepreneurship is the only mechanism for turning formal knowledge into economically valuable knowledge, but attempts to throw light on this particular mechanism. (See Chapter 12 in this volume by Perkmann and Walsh for other theories of knowledge-based wealth generation.)

In principle, established companies are better placed to exploit opportunities as they have more resources to deploy than new ones. But as Penrose pointed out in 1959, the established company faces constraints in perceiving and responding to new opportunities. Its managers 'will be guided in its expansion programmes as much by the nature of its own resources as by market demand, for every firm is . . . a more or less specialised collection of resources and cannot move with equal ease in every direction' (Penrose 1995: 224). Penrose wrote that there are in consequence opportunities for small firms to arise in the interstices neglected by large companies. Potential entrepreneurs may recognize opportunities in new knowledge and ideas that are not recognized as valuable by the originating organization. Famous examples of companies developing resources which they did not exploit are Bell and Xerox, private companies that incubated emerging technologies. During the emergence of the semiconductor industry, the growth of knowledge developed at the Bell Labs and the Bell System provided more opportunities for new semiconductor firms than the Bells could exploit (Holbrook et al. 2000: 1037; cf. Moore and Davis 2004 for a similar situation at Fairchild Semiconductors). The diversity of start-ups based on newly developed knowledge of semiconductor electronics ensured that much of the opportunity space presented by the transistor's invention was explored and exploited. It has been claimed that roughly half the population of Silicon Valley semiconductor manufacturers can be traced back to the Bell Labs (Rogers and Larsen 1984: 43–5). Another well-known source of entrepreneurial opportunities was Xerox Corporation. In the 1960s and 1970s managers at Xerox who understood the potential of digital electronics and computing set up Xerox PARC near Stanford University. PARC (its employees aided by Pentagon funding) created many of the key technologies of the PC industry, but failed to take advantage of their opportunities (Smith and Alexander 1988). Xerox's innovations in computing were largely underexploited because its business model was based on developing copier systems in house with proprietary standards. PARC employees were alert to business opportunities neglected by Xerox and chose to leave to found new companies based on novel business models (Chesbrough and Rosenbloom 2002). Large firms in new technologies are often repositories of unused ideas: big firms have natural diseconomies of scope that a cluster of start-ups does not have (Moore and Davis 2004; cf. Nooteboom 2000).

As regards university-based spin-offs, the incidence of this activity has increased considerably in the last decades, not only in the USA but also elsewhere (Shahid and Kaora 2007). These companies explore applications of knowledge beyond the academic remit, which established firms find commercially uncertain or which conflict with their current activities (Pavitt 2001). The pioneer in Europe among centres of high-tech activity was the University of Cambridge. The first spin-out company from the university was the Cambridge Scientific Instrument Company, founded in 1881 by Horace Darwin, the youngest son of Charles Darwin. The cluster of high-tech activities resulted from multigenerational spin-out from the university (Garnsey and Heffernan 2005; Library House 2006).[8]

Regions without larger firms at the technological frontier, or sizeable research organizations, will probably have fewer spin-off firms, because of both a lack of technically trained people and a shortage of ideas (Moore and Davis 2004). A mix of large and small knowledge-based organizations is thus a better starting point for the exploration and exploitation of new ideas than a concentration of small entrepreneurial firms alone (Moore and Davis 2004; Rothwell and Dodgson 1994; Nooteboom 1994).

EMPIRICAL STUDIES ON KNOWLEDGE AS A SOURCE OF ENTREPRENEURIAL OPPORTUNITIES

How does the creation of new knowledge stimulate high-tech enterprise? Two different mechanisms are found to be relevant for high-growth technology-based start-ups: research and human capital (Audretsch et al. 2005; Audretsch and Lehmann 2005b). The latter mechanism involves embodied knowledge flows via highly educated entrepreneurs (Colombo and Delmastro 2002) and the recruitment of students (Mian 1996). Research excellence is a critical factor for high-growth technology-based firms. Technical universities are not necessarily more successful in facilitating the spillover and commercialization of knowledge (Audretsch and Lehmann 2005a).[9]

Continued access and absorption of external (scientific) knowledge can also be achieved via the attraction of managers and directors with an academic background. Audretsch and Lehmann (2006) showed – based on board composition of 295 high-technology firms – that there is a strong link between both geographic proximity to research-intensive universities and board composition, and firm performance. Scientists who act as board members facilitate the access to and absorption of firm-external knowledge (Audretsch and Stephan 1996).

Knowledge-rich regions are found to generate more entrepreneurial opportunities than knowledge-poor regions (Audretsch et al. 2005; Link and Scott 2005). Even though advances in information technology have increased the access to information, geography still matters. Because knowledge is more easily shared through close interaction in local networks, proximity and locality feature prominently in a knowledge economy, with knowledge-rich regions operating as centres of new activity. Geographic proximity between nascent entrepreneurs and knowledge sources is very important for the emergence of new firms. For example, Zucker et al. (1994) show that in biotechnology, an industry based almost exclusively on new formal knowledge, the firms tend to cluster in just a handful of locations. This finding is supported by Audretsch and Stephan (1996), who examine the geographic relationships of scientists working with biotechnology firms. The importance of geographic proximity is clearly shaped by the role played by the scientist, who is more likely to be located in the same region as the firm when the relationship involves the transfer of new economic knowledge (cf. Egeln et al. 2004). Knowledge spillovers are localized and tend to decay rapidly with transmission across geographic space (Audretsch and Feldman 1996; Jaffe et al. 1993). Regions integrate the various agents – individuals, networks, firms and other organizations – involved in the innovation process in a regional innovation system (Garnsey 1998a; Cooke 2001; Breschi and Lissoni 2001; Asheim and Gertler 2005).

The empirical evidence suggests that entrepreneurs' perceptions of opportunity are affected by their location in a knowledge-rich region. This is consistent with the knowledge spillover theory of entrepreneurship, as explained above. Those contexts with greater investment in knowledge experience a higher incidence of knowledge-based entrepreneurship, *ceteris paribus*. However, institutions, capital markets and other factors affect the level of entrepreneurship in knowledge-rich regions. Thus it is only recently that knowledge-rich regions such as Cambridgeshire and the Washington DC area have become centres of enterprise and growth (Garnsey and Heffernan 2005; Feldman and Francis 2003).

The growth of new firms in the knowledge economy has received extensive attention in the literature to which we now turn. The experience of start-up is very different from that of growth; that few start-ups reach a substantial size continues to be of interest. Fast-growth companies that become major players in their sector are the main job providers among any cohort of new firms (Storey 1997; Davidsson 2005).

GROWTH OF NEW KNOWLEDGE-BASED FIRMS

The emerging structure of opportunities in the economy is among the key factors explaining the emergence and growth of firms. However, entrepreneurs and firms differ in their ability to absorb and act on perceived opportunities. Beyond knowledge as the source of entrepreneurial opportunities, a matching of opportunities and resources to create value through new activity must take place (Garnsey 1998b; Hugo and Garnsey 2005; Stam and Garnsey 2006).

Ownership of or a licence to IP originating in a university can endow the start-up with a unique resource. It has been argued that valuable, rare, inimitable and non-substitutable resources may endow a firm with a competitive advantage that translates to superior performance (Barney 1991). This does not automatically lead to a competitive advantage, just as knowledge spillovers cannot be absorbed by all firms. A key element is absorptive capacity: a firm's ability to recognize, value and assimilate new external information (Cohen and Levinthal 1989, 1990). The increased absorptive capacity of new firms interacting with academic institutions may provide advantages for developing new products and making alliances with other firms, and ultimately improve firm performance. Empirical studies have shown a lack of (direct) positive effects of these university–industry flows on the post-entry performance of knowledge-based firms; an indirect effect via increased absorptive capacity may be more important (Rothaermel and Thursby 2005; Roper et al. 2006). Cockburn and Henderson (1998) demonstrate that firms must exhibit substantial absorptive capacity to capture and appropriate rents from publicly available knowledge.

In order to commercialize technical intellectual property, the new firm needs organizational knowledge. This latter type of knowledge is the fundamental source of competitive advantage of firms (Grant 1996). Opportunities must be identified by entrepreneurs and resources must be accessed, secured and mobilized in order to generate returns. Key problems facing the start-up venture must be solved by developing a repertoire of problem-solving skills or competence. As learning is built up to overcome these problems, competences and dynamic capabilities are developed (Hugo and Garnsey 2005). Competences can be viewed as individual and team-based knowledge and skills which yield economic benefit. By accessing, developing and integrating new and existing knowledge, firms will be able to reconfigure the nature of their resource base, which is necessary to achieve sustainable competitive advantage in a technologically dynamic environment (Teece et al. 1997; Eisenhardt and Martin 2000). The way firm growth is managed affects whether internal resources are developed and successfully matched to opportunities (Penrose 1995; Garnsey 1998b; Kogut and Zander 1992). In the case of the young knowledge-based firms, the key dynamic capability is their ability to detect opportunities for their new technologies and to use their competence to sustain innovation.

In the next section we present an overview of empirical studies on the determinants of new firm growth, which gives insight into the favourable start-up conditions for the growth of new firms.

EMPIRICAL STUDIES ON EMPLOYMENT GROWTH IN NEW FIRMS

There is continuing interest in identifying the key factors shaping new firm growth, but the answers are elusive. Several empirical studies have sought to examine the determinants of employment growth in new firms using correlation analysis. These studies are summarized in Table 7.1 [10] This table does not give an exhaustive overview of all independent variables ('determinants') analysed in these studies but of those featuring in at least two studies. We have categorized the determinants of the growth in employee numbers in new firms into three sets of factors. Personal-level determinants include human capital, social capital and ambitions of the entrepreneur; firm-level determinants include organizational capital and financial capital; variables related to the business environment of the firm are industry or geographical location. Table 7.1 shows that the outcomes of these studies are scattered: hardly any study takes a similar set of determinants into account, and when the same determinants are taken into account, contrasting outcomes may result.

Consensus is clearest for personal-level determinants in Table 7.1, where effects are described in terms of statistical associations. The human capital

Table 7.1 Empirical studies on employment growth of new firms*

	Cooper et al. (1994)	Vivarelli and Audretsch (1998)	Brüderl and Preisendörfer (1998)	Almus and Nerlinger (1999)	Dahlqvist et al. (2000)	Schutjens and Wever (2000)	Bosma et al. (2004)	Colombo and Grilli (2005)	Stam et al. (2006)
Personal									
Human capital									
Education level	+		0			0	0	+	0
Immigrant	−	0	0		−				0
Self-employed parents	0	+	0					0	0
Management experience	0	+	0			0		0	
Unemployment		0			0				
Self-employment/start-up experience			0		+		0	+	0
(Long) work experience			−			−			
Industry experience			0			0	+	+	0
Technical experience				+		0	+	+	0
Male founder	+		+		+		+		+
Age of entrepreneur						0	0	+	−
Social capital									
Entrepreneurial networks			0				+		+
Emotional support from spouse			0				0		
Business partners	+		0	0		+		+	0
Ambitions									
Start-up motivation: market need/niche		0			0				
Start-up motivation: realize innovation		+			0			+	

Table 7.1 (continued)

	Cooper et al. (1994)	Vivarelli and Audretsch (1998)	Brüderl and Preisendörfer (1998)	Almus and Nerlinger (1999)	Dahlqvist et al. (2000)	Schutjens and Wever (2000)	Bosma et al. (2004)	Colombo and Grilli (2005)	Stam et al. (2006)
Goal: sales growth									
Goal: employment growth							+		+
Start-up motive: higher income		+					0		
Firm									
Fin. capital — Start-up capital	+		+			0			
Organizational capital — Incorporation			+	+	+	0		+	0
Start-up size: sales			0			+			+
Start-up size: employees			−	−		+			
Start-up of take-over	−		0						
Industry: retail or personal services					−	0			0
Environment									
Industry: manufacturing/construction		+	0			+			
Industry: business services		0				+	0		
Metropolitan/urban location				0	0/+	0			−

Note: * See the appendix for an overview of the characteristics of the samples on which these studies are based; '0' = no relation; '+' = positive relation; '−' = negative relation.

variables of educational level, start-up experience, industry experience and technical experience have generally been found to be positively associated with firm growth. Positive effects of the entrepreneurial, managerial and technical skills of the entrepreneur on sales growth in new firms were also found by Chandler and Jansen (1992), as was industry experience on sales growth by Siegel et al. (1993). In contrast, Stuart and Abetti (1990) found that only entrepreneurial experience (previous new venture involvements) and not managerial and technical experience were important determinants for a composite indicator of new firm growth (based on sales, employment, profits and productivity growth). The transfer of market experience from the parent had a positive effect on the growth of corporate spin-offs according to Tübke and Empson (2002), while the transfer of technical experience was revealed to have a negative effect. Demographic characteristics of founders including gender (female) and belonging to an immigrant group are negatively correlated with firm growth. Social capital measured by such factors as starting a firm with business partner(s) has a consistent positive effect on measures of subsequent firm growth. The motivation at start-up to realize an idea or launch an innovation is also positively associated with firm growth.

Among firm-level determinants, two factors have a consistent positive correlation: the level of start-up capital and the firm's incorporation. High levels of start-up capital provide the means to invest in resources that enable growth in the longer term, and act as a buffer against external shocks. Being incorporated provides legal status that reduces risks and supports potentially high gains. Research has shown that firms under limited liability are more likely to become insolvent, but also more likely to exhibit high growth (Harhoff et al. 1998). Among business environment determinants, starting in retail/personal services has a negative association with firm growth, while starting in manufacturing or construction seems to have a positive association with growth. This may reflect the minimum efficient size required by different industries or the business cycle in construction.

There is controversy on the effect of work experience and of the initial (employment) size of the firm on growth prospects. On the one hand work experience can provide on-the-job learning, leading to valuable knowledge for managing a growing business. However, this depends on type of activity and type of organization in which experience has been gained. Gompers et al. (2005) show that young venture-capital-backed firms are a fertile breeding ground for new venture-capital-backed firms. In these types of organizations, employees learn from their co-workers about what it takes to start a successful new firm and are exposed to a network of suppliers and customers who are used to dealing with start-up companies. Entrepreneurs with lengthy work experience are likely to become more cautious and conservative than entrepreneurs with shorter work experience.

Inconsistent evidence has been found on the effect of the initial employment size on subsequent firm growth. In the industrial economic literature it is a stylized fact that young and small firms grow relatively fast, because they have to achieve the minimum efficient size (MES) in their industry (Mansfield 1962; Audretsch et al. 2004). Initial size has been found to have a negative effect on firm growth in these studies (Audretsch et al. 1999; Lotti et al. 2001). Smaller start-ups thus have a higher need to grow (Davidsson 1991). On the other hand, relatively large start-ups have more human resources at hand to realize growth and are more likely to attract financial capital and human resources, which enables them to grow more rapidly than small start-ups (cf. Westhead and Cowling 1995). These large start-ups may also be more ambitious regarding future growth. There is not much evidence on a discriminating effect of a metropolitan or rural location on new firm growth. There are large international differences between entrepreneurship and firm growth, ambitions as well as realizations, but as yet there is no empirical international comparative research on the determinants of employment growth in (new) firms.[11]

In brief, analysis of factors associated with firm growth are dominated by the analysis of variance using cross-sectional measures to compare the attributes or conditions of new firms in samples. Attributes of firms with a successful growth record provide a guide to desirable attributes of new firms. However, inferences from the founding attributes to firm growth are not robust, and are plagued by methodological and conceptual weaknesses. They are unable to trace the feedback effects between dynamic business conditions and the responses of entrepreneurs, employees and managers that underlie firm growth (Garnsey et al. 2006).

ENTREPRENEURIAL MANAGEMENT

The identification of opportunities is critical for the emergence of new businesses. However, businesses do not grow unless opportunities are realized. This requires firm building, i.e. the creation of a multi-person organization with a distinctive organizational capability. Long-term growth is determined to a large extent by the interplay of opportunity perception and the ability to realize opportunities (Penrose 1995; Ghoshal et al. 1999).

One of the key assets of new and small firms is their flexibility (Piore and Sabel 1984; Yu 2001). Key elements of the flexibility of firms in general are so-called dynamic capabilities, i.e. 'the firm's ability to integrate, build and reconfigure internal and external competences to address rapidly changing environments' (Teece et al. 1997: 516). Dynamic capabilities are the organizational and strategic procedures (referred to as routines) by which firms achieve new resource combinations (Eisenhardt and Martin 2000: 1107). They include the capacity to undertake specific and identifiable processes such as R&D, interfirm alliances, new product development, and exporting. With knowledge creation routines (R&D) new knowledge is built within the firm that is of particular strategic relevance in high-tech industries. Alliancing routines bring new resources into the firm from external sources, also often essential in high-tech industries (Powell et al. 1996; Baum et al. 2000; Tapon et al. 2001). With new product development routines the varied skills and backgrounds of firm members are combined to create revenue-producing goods and services. Strategic decision making, for example regarding the entrance into new (international) markets, is a dynamic capability in which firm members pool their various business, functional and personal expertise to make the choices that shape the major strategic moves of the firm.

Quantitative studies such as those cited above have limitations. In particular, they lack a conceptually grounded explanatory model in terms of which to make sense of findings. The authors have proposed one such model, based on Penrosian theory, that explains some of the puzzles associated with stage-based models of firm development (Garnsey et al. 2006).

This model differs conceptually from stage models of firm growth. Stages are states or phases that characterize a system (here a firm) at a point in time. In contrast, processes are a related set of events, actions and outcomes that induce change in a system. There may be several change processes under way at a time and interacting with each other, but system states or phases are consecutive. The developmental processes found among new firms as they mobilize and build capabilities to generate market returns can only be fully illuminated by micro-data from case histories. Companies that go through similar developmental processes because they build similar capabilities in a common sequence may exhibit some measurable evidence of similar phases (e.g. of opportunity search, fund raising, recruitment etc.). This may give the appearance of regular junctures at which growth problems can be said to have been overcome (Vohora et al. 2004). But phases are not universal since new firms differ in the resource endowments they inherit and need different resource bases for different types of output. There is no critical juncture at which sustained growth is assured since young firms that have achieved a period of sustained growth often hit setbacks that, at best, require them to repeat earlier developments. Different types of resource base are built and used in different ways, depending on the activity and business model of the firm (Chesbrough and Rosenbloom 2002). Developmental processes may occur in parallel when firms build capabilities for one product or service while being at an early stage in developing other planned offerings, as in the 'soft to hard' strategy of product development funded on early service provision (Bullock 1983).

In brief, we find that while each firm is unique, there are common processes that bring about development and common problems that have to be resolved. Common processes include opportunity recognition and resource matching, resource mobilization, resource generation and resource accumulation. These make possible the development of competences and capital in a base made up of productive, commercial and financial resources. Problems originating within or outside the firm may deplete this resource base, leading to a turning point in the life course of these firms. These have negative consequences when problems are not solved, but positive consequences when they lead to new solutions and the development of new competences that extend the firm's resource base.

Evidence supporting this model is to be found not only from case evidence but also when we examine the unfolding processes through which firm growth takes place. This throws more light on the constraints and success attributes cited in the quantitative studies on new firm growth. Continuous linear growth is the exception rather than the rule. A study by Garnsey et al. (2006) found that only 6 per cent of the new high-tech firms in Cambridgeshire continuously grew over their early life course, while 37 per cent faced severe setbacks (see Figure 7.2).[12]

Hugo and Garnsey (2005) showed that the difficulties faced by new firms provide a stimulus to creating technological competence and marketing capability that propel their growth. Initial disadvantages are addressed by mobilizing resources in new ways, by resource economy ('asset parsimony': Hambrick and Macmillan 1984), resource leverage ('bootstrapping': Bhide 2000) and by creating new resources (e.g. 'bricolage': Baker and Nelson 2005). These efforts are linked in a dynamic process of problem solving that requires strategic relations with others. Resource economy is achieved internally by rearranging the firm's activities and resources in order to produce more with less. New growing firms use their initial resources to gain further leverage. When faced with a resource deficit that cannot be remedied externally, the firms set out to build their own proprietary resources. Cooperative interactions with other parties, including funders, regulators and suppliers, are used to mobilize resources and open further opportunities. When market solutions prove unavailable, this barrier to the pursuit of the original business idea may be an opportunity to develop a new business idea. A key feature of entrepreneurial responses to adversity is cognitive. Entrepreneurs view the situation they face as a soluble problem which they can address proactively and on which they can have some impact. They reconsider their situation and find ways to turn obstacles to their advantage by rerouting the firm. Recurrent problem solving of this kind enables these new firms to build capability on a cumulative basis. As Penrose (1995) anticipated, to succeed they have to match their resources

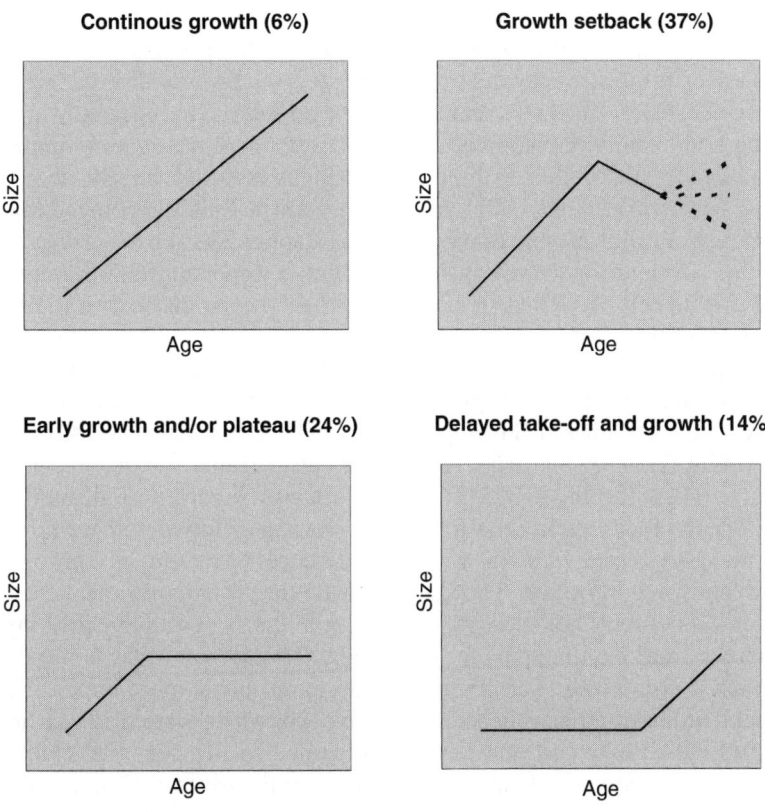

Source: Garnsey et al. (2006).[13]

Figure 7.2 Turning points among Cambridgeshire high-tech firms founded in 1990, surviving ten years

(in particular the competence they had developed) to shifting opportunities. Information asymmetries, technologies advancing ahead of market provision and government regulation are examples of sources of opportunity. Entrepreneurial opportunities often emerge when leads and lags in market needs and provision create asynchronies between supply and demand and stimulate innovative responses to 'market failure' (cf. Metcalfe 2004) that drive entrepreneurial activity.

Hugo and Garnsey (2005) do not suggest that any and every deficiency can be transformed by entrepreneurial problem solving into an asset. The cliché that every problem is an opportunity does not recognize that problems can combine in such a way as to close off opportunities and crush motivation. Undoubtedly early endowments (financial and human capital) are facilitating and attract other favourable attributes in a self-reinforcing process. Timing also plays a large part in securing favourable outcomes for these new firms. But more than good luck is involved in repeatedly identifying and exploiting resources and timely opportunities so as to improve productivity and build capability. Alliancing is essential, but the new firm must have something to offer partners in return. Building competence in response to problems makes it possible to establish useful partnerships that further increase the firms' capability. Not only opportunities but impending threats can be turned to advantage when they spur creative thinking about objectives and new strategic moves. Although most new firms are held back by the continual difficulties besetting growth, those that find their way around these problems grow to be major players in their industry.

A study by Stam and Garnsey (2006) revealed that even in an elite sample of young fast-growing firms, most firms face turning points in their life course. These turning points often constrain growth for a period, and force the firm to focus again after a resource shortage. However, these turning points also enable growth when competence is developed through a problem-solving process. The study showed that there are endemic asynchronies between constituents of the new firm's resource base, input resources and requirements for expansion. This explains why continuous growth is so unlikely. Certain growth mechanisms are more important in certain industries than others. For example, knowledge-based service firms require close ties to customers, while for biomedical firms growth is initially realized by acquiring financial resources from investors (Pisano 2006). However, in the long run the biomedical ventures also have to generate adequate resources from a product market to avoid being taken over or abandoned by their investors.

Entrepreneurial founders do not necessarily have the problem-solving skills required by good entrepreneurial managers. People with the right combination of skills and experience are scarce and the assimilation and motivation of staff can create serious difficulties (Witt 1998, 2000). As the firm grows, management information becomes increasingly complex (Greiner 1972). The difficulty for decision makers in assimilating and making considered judgements increases under conditions of rapid growth. Where reserves have been run down, delays and ill-judged decisions can bring growth to a halt. As new firms grow, they face increasing organizational complexity; according to some authors this will require periodic restructuring (Greiner 1972; Romanelli and Tushman 1994; Vohora et al. 2004). Competence-based scholars have pointed to benefits of paced growth (Penrose 1995; Hugo and Garnsey 2005; Teece et al. 1997), while organizational ecologists have undertaken studies that show why radical organizational changes impair growth prospects and even survival in young technology-based firms (Baron and Hannan 2002; Hannan and Freeman 1984).

How the founders of new technology-based firms approach organizational and HR challenges in the early days of building their firms may have enduring effects on the firms (Baron and Hannan 2002: 8–9). This is the issue addressed in several papers based on the Stanford Project on Emerging Companies (SPEC) (see for example Baron et al. 1999, 2001; Burton 2001; Baron and Hannan 2002). This study found an important determinant of growth of technology-based firms to be organizational models or blueprints that entrepreneurs use in launching their new ventures. These blueprints guide entrepreneurs' thinking about how to organize employment and manage personnel. If the origin of the firm is formative for its subsequent development (Hannan and Freeman 1977, 1984), blueprints are likely to be enduring in the life course of new fast-growing firms. Baron and Hannan (2002) showed that changes in organizational blueprints are in general very destabilizing to young technology firms, adversely affecting employee turnover, financial performance and even survival. These findings suggest that disruption may be considerable when investors replace technical founders who have had a formative role in the company. Selecting people who fitted into the organization and coordination via peer control and/or culture was more commonly found among firms that achieved an IPO (Baron and Hannan 2002). Selection based on exceptional talent, intrinsic work attachment, and professional standards of coordination, most often found in biotech firms, was common in firms that fared best in the post-IPO phase (Baron and Hannan 2002). These findings demonstrate the importance of dedicated people and a sense of community for the longer-term success of the firm.

CONCLUSIONS

This chapter has provided an overview of studies on scientific and technological knowledge as a source of business opportunities, and on the emergence and growth of new firms in the knowledge economy. We have shown that new knowledge in science and technology is an important and localized source of entrepreneurial opportunities. Public and corporate sector players do not necessarily commercialize this knowledge because they lack the vision or incentives. New firms arise that seek to do so, but the recognition of emerging opportunities and the mobilization of the necessary resources in order to create new economic value is achieved by only a few. These few high-growth start-ups are of greater importance for economic growth than new firms in general. Corporate spin-offs are more likely to turn into these high-growth firms than university spin-offs. The international variation in realized firm growth is far greater than the variation in ambitious entrepreneurship, suggesting that entrepreneurs in certain countries face severe constraints preventing their firms from realizing intended growth. The empirical studies on new firm growth showed that high levels of human, social and financial capital are enabling endowments, facilitating the growth of new business. Despite or because of the many problems facing new firms, among the successful innovations that are achieved by dedicated entrepreneurial teams, there are some that have a very major impact on their firm and industry. The fact that these firms more often originate from the USA than from any other country can partly be traced back to the huge (direct and indirect) government support to new technology-based firms in the USA.

ACKNOWLEDGEMENTS

We are grateful for helpful comments on an earlier draft of this chapter from Pablo D'Este and Geoffrey Owen. This research was carried out in the Innovation and Productivity Grand Challenge with financial support from the Engineering and Physical Sciences Research Council and the Economic and Social Research Council through the AIM initiative.

NOTES

1. In policy terms this is reflected in respectively the European Coal and Steel Community (the forerunner of the current European Union) and the current Lisbon Strategy of the EU.
2. The OECD (2005) uses the 'knowledge based economy' to describe trends in advanced economies towards greater dependence on knowledge, information and high skill levels, and the increasing need for ready access to all of these by the business and public sectors.
3. Ironically, this has gone hand in hand with a labour productivity growth slowdown in Europe ever since, and initially also in the USA (Van Ark 2000).
4. A reverse causality has also been suggested: entrepreneurship driving the transition towards a knowledge-based economy (DTI 1998).
5. In the knowledge economy (innovative) new firms might be more important than small firms in general (cf. Parker 2001).
6. It has been said that the UK has been 'producing' too many university spin-offs, but not enough growing spin-offs, which are needed to commercialize new knowledge on a sufficiently large scale (Lambert 2003). In a similar vein, Owen (2004) showed that in comparison with the USA the UK does not lag behind in the number of high-tech start-ups, but that the UK has a lack of technology-based firms which grow very fast from start-up into major international corporations (see also Bartelsman et al. 2005 for a lack of post-entry growth of successful entrants in European countries in general).

7. The psychological traits associated with entrepreneurial behaviour – independence, achievement needs, tolerance of ambiguity, persistence etc. – help people to survive and get on everywhere, as shown by the success of micro-credit schemes in developing countries (Prahalad 2004).
8. A similar but larger-scale process took place at other high-tech clusters like Boston's Route 128 and California's Silicon Valley (Saxenian 1994). A study by BankBoston (1997) revealed that MIT graduates have founded 4000 currently active companies. Another study suggests that nearly 2000 of the current San Francisco Bay Area's high-tech firms (like HP, Varian Associates, Cisco Systems, Silicon Graphics, Sun Microsystems, Google and Yahoo) were founded by Stanford alumni or faculty (Byers et al. 2000).
9. In a study in the 50 largest metropolitan areas in the USA, Rosenbloom (2007) found a positive effect of the number of science and engineering doctorates on SBIR/STTR grant funds, venture capital investments in a community, and the number of IPOs. Armington and Acs (2002) found that firms are more likely to form in US labour market areas (LMAs) that have a high percentage of college graduates than in those LMAs with high concentrations of less skilled workers.
10. All these studies address moderate growth in new firms, which creates at most tens of jobs. This is the most common type of firm growth. Those creating hundreds of jobs, such as firms in the INC.500 (see Bhide 2000; Markman and Gartner 2002) or Europe's 500 lists (see BCG 2002) are the exceptions that alter industries (see Owen 2004).
11. See Autio (2005) for one of the few international comparative studies on high-growth entrepreneurship.
12. In a sample of Netherlands start-ups covering all industries and size classes only 0.3 per cent of the firms reveal a continuous growth path, while 68.6 per cent had a plateau growth path (Stam et al. 2006).
13. This is derived from bi-annual employment data at the firm level, with a threshold of at least 5 per cent change.

REFERENCES

Acs, Z. and Mueller, P. (2006), 'Employment effects of business dynamics: mice, gazelles and elephants', Discussion Papers on Entrepreneurship, Growth and Public Policy No. 2306, Jena: Max Planck Institute of Economics.

Acs, Z., Audretsch, D., Braunerhjelm, P. and Carlsson, B. (2005), 'The Knowledge Spillover Theory of Entrepreneurship', CEPR Discussion Paper No. 5326, London: CEPR.

Almus, M. and Nerlinger, E.A. (1999), 'Growth of new technology-based firms: which factors matter?', *Small Business Economics*, **13**: 141–54.

Armington, C. and Acs, Z.J. (2002), 'The determinants of regional variation in new firm formation', *Regional Studies*, **36**: 33–45.

Arrow, K.J. (1962), 'Economic welfare and the allocation of resources for invention', in *The Rate and Direction of Inventive Activity*, Princeton, NJ: Princeton University Press, pp. 609–26.

Asheim, B.T. and Gertler, M.S. (2005), 'Regional innovation systems and the geographical foundations of innovation', in J. Fagerberg, D. Mowery and R. Nelson (eds), *The Oxford Handbook of Innovation*, Oxford: Oxford University Press, pp. 291–317.

Atkinson, R.D and Court, R.H. (1998), *The New Economy Index: Understanding America's Economic Transformation*, Washington, DC: Progressive Policy Institute Technology, Innovation, and New Economy Project.

Audretsch, D. and Feldman, M. (1996), 'R&D spillovers and the geography of innovation and production', *American Economic Review*, **86**(4): 253–73.

Audretsch, D.B. and Lehmann, E. (2005a), 'Do university policies make a difference?', *Research Policy*, **34**: 343–7.

Audretsch, D.B. and Lehmann, E. (2005b), 'Does the knowledge spillover theory of entrepreneurship hold for regions?', *Research Policy*, **34**: 1191–202.

Audretsch, D.B. and Lehmann, E. (2006), 'Entrepreneurial access and absorption of knowledge spillovers: strategic board and managerial composition for competitive advantage', *Journal of Small Business Management*, **44**(2): 155–66.

Audretsch, D.B. and Stephan, P.E. (1996), 'Company–scientist locational links: the case of biotechnology', *American Economic Review*, **86**(3): 641–52.

Audretsch, D.B. and Thurik, A.R. (2000), 'Capitalism and democracy in the 21st century: from the managed to the entrepreneurial economy', *Journal of Evolutionary Economics*, **10**(1–2): 17–34.

Audretsch, D.B. and Thurik, A.R. (2001), 'What's new about the new economy? Sources of growth in the managed and entrepreneurial economies', *Industrial and Corporate Change*, **10**(1): 267–315.

Audretsch, D.B., Keilbach, M.C. and Lehmann, E.E. (2006), *Entrepreneurship and Economic Growth*, Oxford: Oxford University Press.

Audretsch, D.B., Lehmann, E. and Warning, S. (2005), 'University spillovers and new firm location', *Research Policy*, **34**: 1113–22.

Audretsch, D.B., Santarelli, E. and Vivarelli, M. (1999), 'Start-up size and industrial dynamics: some evidence from Italian manufacturing', *International Journal of Industrial Organization*, **17**: 965–99.

Audretsch, D.B., Klomp, L., Santarelli, E. and Thurik, A.R. (2004), 'Gibrat's Law: are the services different?', *Review of Industrial Organization*, **24**(3): 301–24.

Autio, E. (2005), *Global Entrepreneurship Monitor 2005: Report on High-Expectation Entrepreneurship*, London: GEM.

Baker, T. and Nelson, R.E. (2005), 'Creating something from nothing: resource construction through entrepreneurial bricolage', *Administrative Science Quarterly*, **50**(3): 329–66.

BankBoston (1997), *MIT: The Impact of Innovation*, Boston, MA: BankBoston Economics Department.

Barney, J.B. (1991), 'Firm resources and sustained competitive advantage', *Journal of Management*, **17**(1): 99–120.

Baron, J.N. and Hannan, M.T. (2002), 'Organizational blueprints for success in high-tech start-ups: lessons from the Stanford Project on Emerging Companies', *California Management Review*, **44**: 8–36.

Baron, J.N., Burton, M.D. and Hannan, M.T. (1999), 'Engineering bureaucracy: the genesis of formal policies, positions, and structures in high-technology firms', *Journal of Law, Economics, and Organization*, **15**(1): 1–41.

Baron, J.N., Hannan, M.T. and Burton, M.D. (2001), 'Labor pains: organizational change and employee turnover in young, high-tech firms', *American Journal of Sociology*, **106**(4): 960–1012.

Bartelsman, E., Scarpetta, S. and Schivardi, F. (2005), 'Comparative analysis of firm demographics and survival: evidence from micro-level sources in OECD countries', *Industrial and Corporate Change*, **14**(3): 365–91.

Baum, J.A.C., Calabrese, T. and Silverman, B.S. (2000), 'Don't go it alone: alliance network composition and startups' performance in Canadian biotechnology', *Strategic Management Journal*, **21**: 267–94.

Baumol, W.J. (2002), *The Free-Market Innovation Machine: Analyzing the Growth Miracle of Capitalism*, Princeton, NJ and Oxford: Princeton University Press.

BCG (2002), *Setting the Phoenix Free*, Munich: Boston Consulting Group GMBH.

Bhide, A. (2000), *The Origin and Evolution of New Businesses*, New York: Oxford University Press.

Bosma, N.S., Praag, C.M. van, Thurik, A.R. and Wit, G. de (2004), 'The value of human and social capital investments for the business performance of startups', *Small Business Economics*, **23**(3): 227–36.

Bosma, N., Stam, E. and Schutjens V. (2006), 'Creative destruction and regional competitiveness', SCALES-paper H200624, Zoetermeer: EIM Business and Policy Research.

Breschi, S. and Lissoni, F. (2001), 'Knowledge spillovers and local innovation systems: a critical survey', *Industrial and Corporate Change*, **10**(4): 975–1005.

Brüderl, J. and Preisendörfer, P. (1998), 'Network support and the success of newly founded businesses', *Small Business Economics*, **10**: 213–25.

Brüderl, J. and Preisendörfer, P. (2000), 'Fast growing businesses: empirical evidence from a German study', *International Journal of Sociology*, **30**: 45–70.

Bullock, M. (1983), *Academic Enterprise, Industrial Innovation and the Development of High-Technology Financing in the United States*, London: Longman.

Burton, D.M. (2001), 'The company they keep: founders' models for organizing new firms', in C.B. Schoonhoven and E. Romanelli (eds), *The Entrepreneurship Dynamic: Origins of Entrepreneurship and the Evolution of Industries*, Stanford, CA: Stanford University Press, pp. 13–39.

Bwalya, S.M. (2006), 'Foreign direct investment and technology spillovers: evidence from panel data analysis of manufacturing firms in Zambia', *Journal of Development Economics*, **81**: 514–26.

Byers, T., Keeley, R., Leone, A. and Parker, G. (2000), 'The impact of a research university in Silicon Valley: entrepreneurship of alumni and faculty', *Journal of Private Equity*, **4**(1): 7–15.

Chandler, G.N. and Jansen, E. (1992), 'The founder's self-assessed competence and venture performance', *Journal of Business Venturing*, **7**: 223–36.

Chesbrough, H. and Rosenbloom, R.S. (2002), 'The role of the business model in capturing value from innovation: evidence from Xerox Corporation's technology', *Industrial and Corporate Change*, **11**(3): 529–55.

Cockburn, I. and Henderson, R. (1998), 'Absorptive capacity, coauthoring behavior, and the organization of research in drug discovery', *Journal of Industrial Economics*, **46**(2): 157–82.

Cohen, W.M. and Levinthal, D.A. (1989), 'Innovation and learning: the two faces of R&D', *The Economic Journal*, **99**: 569–96.

Cohen, W.M. and Levinthal, D.A. (1990), 'Absorptive capacity: a new perspective on learning and innovation', *Administrative Science Quarterly*, **35**: 128–52.

Colombo, M. and Delmastro, M. (2002), 'How effective are technology incubators? Evidence from Italy', *Research Policy*, **31**(7): 1103–22.

Colombo, M.G. and Grilli, L. (2005), 'Founders' human capital and the growth of new technology-based firms: a competence-based view', *Research Policy*, **34**(6): 795–816.

Connell, D. (2006), *'Secrets' of the World's Largest Seed Capital Fund: How the United States Government Uses its Small Business Innovation Research (SBIR) Programme and Procurement Budgets to Support Small Technology Firms*, Cambridge, UK: Centre for Business Research, University of Cambridge.

Cooke, P. (2001), 'Regional innovation systems, clusters, and the knowledge economy', *Industrial and Corporate Change*, **10**(4): 945–74.

Cooper, A.C., Gimeno-Gascon, F.J. and Woo, C.Y. (1994), 'Initial human and financial capital as predictors of new venture performance', *Journal of Business Venturing*, **9**(5): 371–95.

Dahlqvist, J., Davidsson, P. and Wiklund, J. (2000), 'Initial conditions as predictors of new venture performance: a replication and extension of the Cooper et al. study', *Enterprise & Innovation Management Studies*, **1**(1): 1–17.

Davidsson, P. (1991), 'Continued entrepreneurship: ability, need and opportunity as determinants for small firm growth', *Journal of Business Venturing*, **6**: 405–29.

Davidsson, P. (2005), *Researching Entrepreneurship*, New York: Springer.

Dosi, G., Llerena, P. and Labini, M.S. (2006), 'The relationships between science, technologies and their industrial exploitation: an illustration through the myths and realities of the so-called "European Paradox"', *Research* Policy, **35**: 1450–64.

DTI (Department of Trade and Industry) (1998), *Our Competitive Future: Building the Knowledge-Driven Economy*, London: HMSO.

Egeln, J., Gottschalk, S. and Rammer, C. (2004), 'Location decisions of spin-offs from public research institutions', *Industry and Innovation*, **11**(3): 207–23.

Eisenhardt, K.M. and Martin, J.A. (2000), 'Dynamic capabilities: what are they?', *Strategic Management Journal*, **21**: 1105–21.

Eliasson, G. (1998), 'The nature of economic change and management in the knowledge-based information economy', DRUID working paper 1998-05.

Feldman, M. and Francis, J. (2003), 'Fortune favours the prepared region: the case of entrepreneurship and the capitol region biotechnology cluster', *European Planning Studies*, **11**(7): 765–88.

Foray, D. (2004), *The Economics of Knowledge*, Cambridge, MA: MIT Press.

Galbraith, J.K. (1956), *American Capitalism*, Boston, MA: Houghton Mifflin.

Garnsey, E. (1998a), 'The genesis of the high technology milieu: a study in complexity', *International Journal of Urban and Regional Research*, **22**(3): 361–77.

Garnsey, E. (1998b), 'A theory of the early growth of the firm', *Industrial and Corporate Change*, **7**: 523–56.

Garnsey, E. and Heffernan, P. (2005), 'High tech clustering through spin out and attraction: the Cambridge case', *Regional Studies*, **39**(8): 1127–44.

Garnsey, E., Stam, E. and Hefferman, P. (2006), 'New firm growth: exploring processes and paths', *Industry and Innovation*, **13**(1): 1–24.

Ghoshal, S., Hahn, M. and Moran, P. (1999), 'Management competence, firm growth and economic progress', *Contributions to Political Economy*, **18**: 121–50.

Gibbons, M., Limoges, C., Nowotny, H., Schwartzman, S., Scott, P. and Trow, M. (1994), *The New Production of Knowledge: The Dynamics of Science and Research in Contemporary Societies*, London: Sage.

Gompers, P., Lerner, J. and Scharfstein, D. (2005), 'Entrepreneurial spawning: public corporations and the genesis of new ventures, 1986 to 1999', *Journal of Finance*, **60**(2): 577–614.

Grant, R.M. (1996), 'Toward a knowledge-based theory of the firm', *Strategic Management Journal*, **17**: 109–22.

Greiner, L.E. (1972), 'Evolution and revolution as organizations grow', *Harvard Business Review*, July–August: 37–46.

Hambrick, D.C. and Macmillan, I.C. (1984), 'Asset parsimony – managing assets to manage profits', *Sloan Management Review*, **25**: 67.

Hannan, M.T. and Freeman, J. (1977), 'The population ecology of organizations', *American Journal of Sociology*, **82**: 929–64.

Hannan, M.T. and Freeman, J. (1984), 'Structural inertia and organizational change', *American Sociological Review*, **49**: 149–64.

Harhoff, D., Stahl, K. and Woywode, M. (1998), 'Legal form, growth and exit of West-German firms – empirical results for manufacturing, construction, trade and service industries', *Journal of Industrial Economics*, **46**(4): 453–88.

Harris, R. (2001), 'The new economy: intellectual origins and theoretical perspectives', *International Journal of Management Reviews*, **3**(1): 21–40.

Hoffmann, A.N. and Junge, M. (2006), 'Documenting data on high-growth firms and entrepreneurs across 17 countries', Mimeo, Fora Copenhagen.

Holbrook, D., Cohen, W.M., Hounshell, D.A. and Klepper, S. (2000), 'The nature, sources, and consequences of firm differences in the early history of the semiconductor industry', *Strategic Management Journal*, **21**: 1017–41.

Hugo, O. and Garnsey, E. (2005), 'Problem-solving and competence creation in the early development of new firms', *Managerial and Decision Economics*, **26**(2): 139–48.

Jaffe, A., Trajtenberg, M. and Henderson, R. (1993), 'Geographic localization of knowledge spillovers as evidenced by patent citations', *Quarterly Journal of Economics*, **63**: 577–98.

Kirzner, I.M. (1973), *Competition and Entrepreneurship*, Chicago, IL: University of Chicago Press.

Kogut, B. and Zander, U. (1992), 'Knowledge of the firm, combinative capabilities, and the replication of technology', *Organization Science*, **3**(3): 383–97.

Lambert, R. (2003), *Lambert Review of Business–University Collaboration*, London: HM Treasury.

Library House (2006), *The Cambridge University Economic Impact Study*, Cambridge: Library House Ltd.

Link, A.L. and Scott, J.T. (2005), 'Opening the ivory tower's door: an analysis of the determinants of the formation of U.S. university spin-off companies', *Research Policy*, **34**(7): 1106–12.

Lotti, F., Santarelli, E. and Vivarelli, M. (2001), 'The relationship between size and growth: the case of Italian newborn firms', *Applied Economics Letters*, **8**: 451–4.

Mansfield, E. (1962), 'Entry, Gibrat's Law, innovation, and the growth of firms', *American Economic Review*, **52**: 1023–51.

Markman, G.D. and Gartner, W.B. (2002), 'Is extraordinary growth profitable? A study of *Inc. 500* high-growth companies', *Entrepreneurship Theory and Practice*, **27**(1): 65–75.

Merton, R.K. (1993), *On the Shoulders of Giants*, Chicago, IL: University of Chicago Press.

Metcalfe, J.S. (2004), 'The entrepreneur and the style of modern economics', *Journal of Evolutionary Economics*, **14**: 157–75.

Mian, S. (1996), 'Assessing value-added contributions of university technology business incubators to tenant firms', *Research Policy*, **25**: 325–35.

Mokyr, J. (2002), *The Gifts of Athena: Historical Origins of the Knowledge Economy*, Princeton, NJ: Princeton University Press.

Moore, G. and Davis, K. (2004), 'Learning the Silicon Valley way', in T. Bresnahan and A. Gambardella (eds), *Building High-tech Clusters: Silicon Valley and beyond*, Cambridge: Cambridge University Press, pp. 7–39.

170   *Firm development through knowledge*

Mowery, D.C. and Nelson, R.R. (1999), *Sources of Industrial Leadership: Studies of Seven Industries*, New York: Cambridge University Press.
Mowery, D.C. and Rosenberg, N. (1998), *Paths of Innovation. Technological Change in 20th-Century America*, Cambridge: Cambridge University Press.
Nelson, R.R. (1959), 'The simple economics of basic scientific research', *Journal of Political Economy*, **67**: 297–306.
Nooteboom, B. (1994), 'Innovation and diffusion in small firms: theory and evidence', *Small Business Economics*, **6**: 327–47.
Nooteboom, B. (2000), *Learning and Innovation in Organizations and Economies*, Oxford: Oxford University Press.
OECD (1996), *The Knowledge-based Economy*, Paris: OECD.
OECD (2005), 'Knowledge-based economy', in Glossary of statistical terms, http://stats.oecd.org/glossary/detail.asp?ID=6864.
Owen, G. (2004), 'Where are the big gorillas? High technology entrepreneurship in the UK and the role of public policy', Discussion paper, Interdisciplinary Institute of Management, London School of Economics.
Parker, R. (2001), 'The myth of the entrepreneurial economy: employment and innovation in small firms', *Work, Employment & Society*, **15**(2): 373–84.
Pavitt, K. (2001), 'Can the large Penrosian firm cope with the dynamics of technology?', SPRU electronic working paper 68.
Penrose, E.T. (1995), *The Theory of the Growth of the Firm*, 3rd edn, Oxford: Oxford University Press.
Piore, M.J. and Sabel, C.F. (1984), *The Second Industrial Divide: Possibilities for Prosperity*, New York: Basic Books.
Pisano, G. (2006), 'Can science be a business? Lessons from biotech', *Harvard Business Review*, **84**(10): 114–25.
Powell, W.W., Koput, K. and Smith-Doerr, L. (1996), 'Interorganizational collaboration and the locus of innovation: networks of learning in biotechnology', *Administrative Science Quarterly*, **41**(1): 116–45.
Prahalad, C.K. (2004), *The Fortune at the Bottom of the Pyramid: Eradicating Poverty through Profit*, Upper Saddle River, NJ: Wharton School Publishing.
Rogers, E.M. and Larsen, J.K. (1984), *Silicon Valley Fever: Growth of High Technology Culture*, New York: Basic Books.
Romanelli, E. and Tushman, M.L. (1994), 'Organizational transformation as punctuated equilibrium: an empirical test,' *Academy of Management Journal*, **37**: 1141–66
Roper, S., Du, J. and Love, J.H. (2006), 'The innovation value chain', mimeo, Birmingham: Aston Business School, Aston University.
Rosenberg, Nathan (1992), 'Economic experiments', *Industrial and Corporate Change*, **1**: 181–203.
Rosenberg, N. and Birdzell, L. (1986), *How the West Grew Rich*, New York: Basic Books.
Rosenbloom, J.L. (2007), 'The geography of innovation commercialization in the United States during the 1990s', *Economic Development Quarterly*, **21**: 3–16.
Rothaermel, F.T. and Thursby, M. (2005), 'Incubator firm failure or graduation?: The role of university linkages', *Research Policy*, **34**(7): 1076–90.
Rothwell, R. and Dodgson, M. (1994), 'Innovation and size of firm', in M. Dodgson (ed.), *Handbook of Industrial Innovation*, Aldershot, UK and Brookfield, USA: Edward Elgar, pp. 310–24.
Saxenian, A. (1994), *Regional Advantage: Culture and Competition in Silicon Valley and Route 128*, Cambridge, MA: Harvard University Press.

Schumpeter, Joseph A. (1942), *Capitalism, Socialism, and Democracy*, New York: Harper and Brothers.

Schutjens, V.A.J.M. and Wever, E. (2000), 'Determinants of new firm success', *Papers in Regional Science*, 79(2): 135–59.

Servan-Schreiber, J.J. (1968), *The American Challenge*, London: Hamish Hamilton.

Shahid Y. and Kaora, N. (2007), *How Universities Promote Economic Growth*, Washington, DC: World Bank.

Shane, S.A. (2000), 'Prior knowledge and the discovery of entrepreneurial opportunities', *Organization Science*, 11(4): 448–72.

Shane, S. and Eckhardt, J. (2003), 'The individual–opportunity nexus', in Zoltan Acs and David Audretsch (eds), *Handbook of Entrepreneurship Research: An Interdisciplinary Survey and Introduction*, Boston: Kluwer Academic Publishers, pp. 161–91.

Siegel, R., Siegel, E. and Macmillan, I.C. (1993), 'Characteristics distinguishing high growth ventures', *Journal of Business Venturing*, 8: 169–80.

Smith, D.K. and Alexander, R.C. (1988), *Fumbling the Future: How Xerox invented, then ignored, the first personal computer*, New York: William Morrow.

Stam, E. and Garnsey, E. (2006), 'New firms evolving in the knowledge economy: problems and solutions around turning points', in W. Dolfsma and L. Soete (eds), *Understanding the Dynamics of a Knowledge Economy*, Cheltenham, UK and Northampton, MA, USA: Edward Elgar, pp. 102–28.

Stam, E., Gibcus, P., Telussa, J. and Garnsey, E. (2006), 'Dynamic capabilities and new firm growth', paper presented at the Cass Business School Workshop on Scientific and Managerial Knowledge, 7 December.

Stam, E., Suddle, K., Hessels, J. and Van Stel, A. (2007), 'High growth entrepreneurs, public policies and economic growth', *Ekonomiaz, Basque Journal of Economics*, Special issue on entrepreneurship, 62(2): 124–49.

Storey, D.J. (1997), *Understanding the Small Business Sector*, London: International Thomson Business Press.

Stuart, R.W. and Abetti, P.A. (1990), 'Impact of entrepreneurial and management experience on early performance', *Journal of Business Venturing*, 5: 151–62.

Tapon, F., Thong, M. and Bartell, M. (2001), 'Drug discovery and development in four Canadian biotech companies', *R&D Management*, 31(1): 77–90.

Teece, D.J., Pisano, G. and Shuen, A. (1997), 'Dynamic capabalities and strategic fit', *Strategic Management Journal*, 18: 510–33.

Tübke, A. and Empson, T. (2002), 'Companies as incubators', *Entrepreneurship and Innovation*, 3(4): 257–64.

Van Ark, B. (2000), 'Measuring productivity in the "new economy": towards a European perspective', *De Economist*, 148(1), 87–105.

Van Stel, A.J., Carree, M.A. and Thurik A.R. (2005), 'The effect of entrepreneurial activity on national economic growth', *Small Business Economics*, 24: 311–21.

Vivarelli, M. and Audretsch, D. (1998), 'The link between the entry decision and post-entry performance: evidence from Italy', *Industrial and Corporate Change*, 7: 485–500.

Vohora, A., Wright, M. and Lockett, A. (2004), 'Critical junctures in the development of university high-tech spin-out companies', *Research Policy*, 33: 147–75.

Westhead, P. and Cowling, M. (1995), 'Employment change in independent owner-managed high technology firms in Great Britain', *Small Business Economics*, 7: 111–40.

Witt, U. (1998), 'Imagination and leadership: the neglected dimension of an evolutionary theory of the firm', *Journal of Economic Behavior and Organization*, **35**: 161–77.

Witt, U. (2000), 'Changing cognitive frames – changing organizational forms: an entrepreneurial theory of organizational development', *Industrial and Corporate Change*, **9**(4): 733–55.

Wong, P., Ho, Y. and Autio, E. (2005), 'Entrepreneurship, innovation and economic growth: evidence from GEM data', *Small Business Economics*, **24**(3): 335–50.

Yu, T.F.L. (2001), 'Toward a capabilities perspective of the small firm', *International Journal of Management Reviews*, **3**(3): 185–97.

Zucker, L., Darby, M. and Armstrong, J. (1994), 'Geographically localized knowledge: spillovers or markets', *Economic Inquiry*, **36**(1): 65–86.

APPENDIX

Table 7A.1 Characteristics of the samples of studies on employment growth in new firms

Authors	Time period	Industries	Number of firms	Region
Cooper et al. (1994)	1985–87 (3 years)	Representative for new firm population	1053	USA
Vivarelli and Audretsch (1998)	1985–93 (<9 years; mean age 3 years)	All	100	Emilia (Italy)
Brüderl and Preisendörfer (1998)	1985/86–90 (4 years)	All except crafts, agriculture, physicians, architects and lawyers	1710	Munich and Upper Bavaria (Germany)
Almus and Nerlinger (1999)	1992/96–98	Manufacturing industries (both 'high-tech industries' (R&D intensity above 3.5%) and 'non-high-tech industries' (R&D intensity below 3.5%)	8739	Germany
Dahlqvist et al. (2000)	1994–97 (3 years)	All except agriculture, forestry, hunting, fishery and real estate	6377	Sweden
Schutjens and Wever (2000)	1994–97 (3 years)	All except agriculture and mining	563	Netherlands
Bosma et al. (2004)	1994–97 (3 years)	All except agriculture and mining	758	Netherlands
Colombo & Grilli (2005)	1980 (or later)–2004 (max. 13 years)	High-tech sectors (computers, electronic components, telecommunication equipment, optical, medical and electronic instruments, biotechnology, pharmaceuticals, advanced materials, robotics, and process automation equipment, multimedia content, software, internet services, and telecommunication services)	506	Italy
Stam et al. (2006)	1994–2004 (10 years)	All except agriculture and mining	354	Netherlands

PART III

Connecting for Innovation

8. Sustaining breakthrough innovation in large established firms: learning traps and counteracting strategies

Simone Ferriani, Elizabeth Garnsey and David Probert

It must be considered that there is nothing more difficult to carry out, nor more doubtful of success, nor more dangerous to handle, than to initiate a new order of things. For the reformer has enemies in all those who profit from the old order of things, and only lukewarm defenders in all those who would profit by the new order, this lukewarmness arising partly from fear of their adversaries . . . and partly from the incredulity of mankind, who do not truly believe in anything new until they have had actual experience of it. (Niccolò Machiavelli)

1. INTRODUCTION

Several studies show that breakthrough innovations are often likely to originate with entrants rather than with incumbents. Most established firms have too many obligations and too much to lose to justify the obvious risks of chasing radical possibilities that might or might not be market hits. Nevertheless, some room must be found for breakthroughs or the market leader risks being leap-frogged and deposed by upstart market newcomers. Given this imperative, why do large firms not come up with breakthroughs more regularly? Researchers in innovation management have proffered a variety of arguments to explain this tendency. It has been suggested that large established firms underinvest in radical innovation (Henderson 1993), fall into competency traps (Levitt and March 1988), are hampered by their core rigidities (Leonard-Barton 1992) or remain committed only to their main customers (Christensen 1997). As a result, they may not survive the new cycle of creative destruction (Schumpeter 1934).

Although the idea that small firms are the principal drivers of innovation was originally endorsed by Schumpeter (1934), in his later works he espoused a theory much more supportive of large established firms. Schumpeter argued that the increasingly scientific base of economic activities had caused innovation to become more and more costly, as a result of indivisibilities and significant economies of scale and scope (Schumpeter 1942). Because of such conditions and the presence of barriers to entry and weak appropriability conditions, large firms and *ex ante* monopolistic power were expected to be more conducive to innovation than fully competitive markets populated by small entrepreneurial firms (Freeman and Soete 1997).

While the early Schumpeterian view of entrepreneurial innovation is nowadays widely popular among management scholars, emerging evidence has been accumulated in more recent years that seems to recast the relationship between incumbency and innovation in a different light. Methe et al. (1997), for instance, found that industry incumbents and diversifying entrants could be credited with many major innovations in the telecommunications and medical device industries. Chandy and Tellis (2000) traced the origins of a broad range of breakthrough innovations in office products and consumer durables across 150 years and found that after the Second World War, incumbents actually introduced the majority (75 per cent) of these products. Similarly, Klepper and Simons (2000) found that nearly all dominant US manufacturers of televisions previously were dominant producers of radios and that they took the lead in television product and process innovations.

The above findings suggest that the prevailing emphasis on new entrants as engines of breakthrough innovations should perhaps be mitigated. They also raise the interesting issue of identifying the distinctive traits and practices that allow established firms to maintain an innovative lead in the face of the multiple hurdles associated with corporate inertia and resistance to change. Although a few scholars have started to look at this question (Ahuja and Lampert 2001; Fleming 2001), the majority of research interested in the relationship between incumbency and innovative capability is typically concerned with cases where established firms react to technological innovations brought about by other firms, rather than with incumbents' proactive search for technological breakthrough. Examples of this approach include studies on the capabilities of the established firm to adapt to the new technology and redirect its innovative efforts accordingly (Tripsas 1997; Rosenbloom 2000); and, more generally, on the impact of such inventions rather than their creation (for instance, Achilladelis et al. 1990). In other words, the bulk of these studies focus on issues that arise subsequent to the breakthrough event. As a result, little is known about the incumbents' internal organizational practices and process that are conducive to the innovation. What organizational conditions are required to sustain the genesis of breakthrough innovations?

In addition, limited attention has been placed on the critical transition point when the emerging innovation realizes commercial importance (Adner and Levinthal 2002). Many examples exist of major innovations that prove unviable in the marketplace. Witness the many variants of pen-based computing and personal digital assistants (PDAs) that have been commercial failures (McGahan et al. 1997). The subsequent challenge (even before any commercialization strategy is decided) is then to identify market domains that are more likely to accept the new technology. What impact do these market interactions have on firms' further exploration and refinement? These are important topics that have received insufficient attention.

In this chapter we review the main obstacles that encumber large firms' attempts to generate breakthrough innovations and, drawing on theoretical insights and case evidence, we also suggest mechanisms through which large firms may cope with these hurdles. We believe that addressing such issues is important for several reasons. First, breakthroughs are related to the creation of private wealth and the generation of streams of Schumpeterian rents for their inventors, while also enhancing social welfare (Trajtenberg 1990). Identifying strategies that can help such corporations to improve their record of breakthrough innovations can potentially create significant private and social value (Ahuja and Lampert 2001). Second, a better understanding of the drivers of breakthrough innovations has great strategic importance as superior performance often depends on being consistently innovative (Nelson 1991). Understanding how incumbents nurture and sustain their innovative pre-eminence in an industry is therefore a crucial step in the direction of explaining durable sources of superior performance and reducing competitive threats from new entrants.

The chapter is divided into three sections. We start by providing an overview of the main factors that cause large established firms to oppose innovation projects with high potential but also high risk. Next we conceptualize the generation of breakthrough innovation as a two stage search–selection process. We use this stylized process as a template to illustrate ways through which large firms may counteract the innovation traps and sustain radical change. We focus especially on the selection phase under the assumption the large firms typically generate enormous variation as a byproduct of their ongoing operations. The challenge is then to identify selection mechanisms that are best suited for sifting most promising opportunities from the larger pool of possibilities. We conclude by summarizing the contribution of the work to theory and practice.

2. LEARNING TRAPS AND INNOVATION BARRIERS IN LARGE ESTABLISHED FIRMS

Innovation is the generation and/or acceptance of ideas, processes, products, or services that the relevant adopting unit perceives as new. It can be new to either the firm or the firm's customers. Depending on their 'newness', innovations can be incremental (continuous) or breakthrough (discontinuous). Incremental innovations are critical to sustaining and enhancing shares in mainstream markets (Baden-Fuller and Pitt 1996) and focus on improving existing products and services to satisfy ever-changing customer demands (Bessant 2003). Incremental innovations usually emphasize cost or feature improvements on existing products or services. In contrast, breakthrough innovation concerns the development of new businesses or product lines based on new ideas or technologies (Morone 1993) or substantial cost reductions that may significantly alter the consumption patterns of a market (Yamanouchi 1989; Garcia and Calantone 2002).

Over the years researchers have proffered a variety of terminologies to address this distinction and characterize innovations that represent a major departure from existing practices in products or processes development. So, for instance, innovations have been described as revolutionary, radical, disruptive, or discontinuous. While it is beyond the scope of this chapter to go through a taxonomical analysis, we are well aware that there are some important conceptual differences underlying the semantics.

As Table 8.1 illustrates, the conceptual emphasis in the literature that has first introduced these nomenclature changes depends on such aspects as the cost/performance improvements, the relationship with the existing

Table 8.1   Review of Technological Innovation Types

Innovation type	Emphasis	Authors
Radical technology/ radical innovation	Depth of impact on industries substantial cost/ performance improvements	Foster (1986); Utterback (1994)
Revolutionary innovation or competence-altering innovation	Requires change in firm capabilities	Abernathy and Clark (1985); Utterback (1994); Tushman and Anderson (1986)
Discontinuous innovation	Break with preceding technologies	Utterback (1994)
Disruptive technology	New-tech attributes, new competences enable new entrants to take market share from incumbent firms	Christensen (1997)

Source: Adapted from Maine and Garnsey (2006).

capabilities, the receptiveness of the market, and so on. Yet we also note that these terms overlap to a considerable extent since they all share at least two important characteristics. First, they all embody a fundamental distinction between predictable and original innovations. Second, they are all premised on the general assumption that innovations that are unique, original and unexpected are far more valuable from a competitive standpoint than innovations that are predictable, incremental or mundane. Hence, in the rest of the chapter, we will use the term 'breakthrough innovation' as an overarching notion to represent any creative and original action by individuals or project teams that enables firms to capture at least temporary monopoly profits or that results in a significant increase in market share (Mascitelli 2000).

The ability to create new and valuable breakthroughs offers firms an unambiguous competitive advantage. However, because breakthrough innovations represent a major rupture with established ways of conveying value in a given market, they are much less likely to emerge from established firms inside an industry than from new entrants. Many really novel technologies have been the work of outsiders who were not operating within the conventions shared by members of established organizations. It was not at Kodak, for instance, that the photocopier was developed. Kodak was not an office-products company and the photocopier idea was not launched on the market until Xerox was established. But Xerox in its turn proved blind to the importance of an emerging technology; as at IBM, there was a failure to recognize the importance of the micro-computer. A user-friendly micro-computer with graphical interface and networking potential was pioneered within Xerox's own research laboratories in Palo Alto in 1973, but a product based on these ideas was only taken to market after Apple Computers was founded by 'computer kids' – Steve Jobs and Steve Wozniack. Similarly, the biotechnology industry in the USA was to a large extent the result of bioscientists starting their own enterprise, rather than the fruit of the laboratories of big pharmaceutical companies. Schumpeter believed towards the end of his life that 'invention factories' in well-resourced large companies would become the main source of innovation under capitalism, but he did not deal with the gap between invention (R&D) and innovation which has plagued large firms since his time (Florida and Kenney 1990). The example of the biotech and micro-computer industries, among others, is much closer to his earlier perspective on entrepreneurial innovation (Garnsey 2004).

The opportunities for enterprising new companies in the generation and diffusion of innovations result in part from the ambivalence of established companies towards emergent technologies. On the one hand there is recognition in the corporate sector that innovation may be essential for survival.

Indeed, most established companies have specialist R&D units specifically to promote technological innovation. On the other, it does not make business sense for the established company to develop radical new technologies which may threaten to destabilize the markets that it currently controls. That is why large corporations are usually better at incremental innovation, at modifications and refinements to their products and processes, than at radical innovation (Garnsey and Wright 1990). Researchers in technology management have highlighted a number of reasons why this is often the case (Todd 1999). Below we discuss each of them in some detail.

2.1 Organizational Inertia

Organizational theorists have emphasized the roles that organizational inertia (Hannan and Freeman 1984) and structured routines (Nelson and Winter 1982) play in constraining the actions and limiting the success of incumbent firms. Organizational inertia limits the abilities of incumbent firms because the structures and systems that facilitate survival in stable and predictable environments become liabilities in environments undergoing rapid change (Amburgey et al. 1993; Hannan and Freeman 1984). A similar and related literature argues that organizations develop highly structured routines in order to reduce the costs associated with information acquisition and coordination (Nelson and Winter 1982).

Establishing efficient routines is crucial for organizational learning and performance (Nelson and Winter 1982), yet an increasing reliance on efficient routines may prevent firms from sensing valuable opportunities that would drag them beyond those practices. Indeed, organizational evolution and learning literatures suggest that as organizational routines become entrenched, organizations tend to exploit existing knowledge and capabilities, possibly crowding out more exploratory activities (e.g. March 1991). Routines represent a truce for intra-organizational conflict which, as noted by Perrow (1986), binds the firm in a network of practices that are difficult to alter. This is especially the case for successful organizations, where the growing reliance on well-established routines makes it extremely hard to shift individuals' attention towards new ideas and practices. Indeed, as companies grow, they develop structures and systems to handle the increased complexity of the work. These structures and systems are interlinked so that proposed changes become more difficult, more costly, and require more time to implement, especially if they are more than small, incremental modifications (Garnsey and Wright 1990).

Additionally, an inability to 'recognize the value of new information, assimilate it, and apply it to commercial ends' (Cohen and Levinthal 1990: 128) has also been argued as a reason why incumbent firms have difficulties reacting to discontinuous innovations. Because the 'absorptive capacity' of firms is built incrementally upon prior and related knowledge, and because radical innovations generally require knowledge that exists outside of the firm, incumbent firms are unable to recognize and fully embrace new technological paradigms.

2.2   Cultural Resistance and Cognitive Barriers

As successful organizations become bigger and older, they tend to develop shared expectations about how things are to be done, leading to forms of cultural resistance to change (Dougherty 1992). As noted by Tushman and O'Reilly (1996: 19): 'Cultural inertia, because it is so ephemeral and difficult to attack directly, is a key reason managers often fail to successfully introduce revolutionary change even when they know that it is needed.' This problem is exacerbated by the tendency of organizational departments to develop rather impermeable cognitive orientations and 'thought worlds' (Dougherty 1992) that partition information and insights, thereby inhibiting the kind of collective action that is necessary to innovation. Very often such orientations reflect the identity of the departments as organization members create their identities on the basis of what they know how to do well (Orr 1990). Thus, deviating from existing knowledge domains poses threats to the identity of an organization as well, as Cook and Yanow (1993) documented for one the world's most famous flute building companies.

Well-known illustrations of these phenomena include Tripsas and Gavetti's (2000) analysis of how two prevalent beliefs among Polaroid senior managers hindered the company's developing capabilities in digital cameras, in spite of its leading-edge digital-imaging capabilities. On the one hand, the belief in the primacy of technology led the company to invest aggressively in R&D on digital imaging. On the other hand, the belief of Polaroid managers that their company could not make money on hardware (cameras) but only on consumables (film) severely impeded the exploitation of its digital technologies. Another representative example of the dynamics of incumbents' innovative underperformance due to cognitive impediments is Kasper Instruments' failure in the face of Canon's entry into this market. When Canon, the entrant firm, introduced its innovative proximity aligner in 1973, Kasper Instruments, the incumbent, asked its own engineers to evaluate the competitor's piece of equipment. The team of engineers 'overlooked' the highly innovative features in Canon's proximity aligner because they were 'blinded' to them by their former organizational experience. In both the Polaroid and Kasper cases, the incumbent had the technical resources to develop the new technology, but failed to do so in the light of cultural resistance and scepticism from the established power structures. It is difficult to unlearn received wisdom that has become irrelevant as the fundamentals of the industry shift (Hamel and Prahalad 1994).

2.3   Lack of Incentives and Aversion to Risk

While there are clear benefits to proactive change, only a small minority of farsighted firms initiate discontinuous change before a performance decline. Part of this stems from the risks of proactive change. Because radical new technologies may destabilize established companies' core competences (Tushman and Anderson 1986), large corporations are comfortable with sustaining technologies that build on their existing strengths. They excel at knowing their market, and having a mechanism in place to develop existing technology. Yet such an innovation trajectory reduces technical variation and stunts a firm's learning potential. The greater the threat of cannibalizing existing products, the more intense is the resistance to new ideas. For instance, IBM remained focused too long on mainframes at the expense of personal computers because PCs required a different distribution system, had lower margins, and was not as after-sales service intensive as the mainframe business. As noted by Henderson (1993: 251), 'in some circumstances extensive experience with a technology may be a substantial disadvantage. Large established firms have an advantage over entrants in the pursuit of incremental innovation because incremental innovation builds upon their existing knowledge and capabilities, but these assets can simultaneously reduce substantially the effectiveness of their attempts to exploit radical innovation.'

Moreover, radical technologies often look financially unattractive or excessively risky to incumbents. There are in fact many potentially disruptive technologies that fail (e.g. the Iridium global satellite phone system; see Finkelstein and Sanford 2000). Plus, these high-risk initiatives can only be developed over a considerable period of time, sometimes 10 to 20 years in different industries. The time-horizon to build up knowledge about the potential markets and the further development of the technologies involved is much longer than in the case of incremental innovations. Consequently, there is also more 'patient money' to be invested and it is very likely that in a corporate world where shareholders pay attention on a quarterly cycle to return on investments, the need for quick results will get in the way of the patience needed for breakthroughs (Rice et al. 1998). When companies lack the patience to shepherd radical innovations through early markets, the main benefits of the innovation may be lost or captured by others.[1] As a result, it is not uncommon for hundreds of new ideas to get killed before any one innovation successfully hits the market (Freeman and Soete 1997).

In addition, contribution to earnings from these new products may not significantly influence the firm's stock price or enhance its multiple. The potential market is small and its cost structure too high to serve the radical technology. At the same time the market may be new, so it is difficult to determine who the most likely customers are and what they actually need. New insights must be continuously incorporated and the premises themselves reconsidered. This is what makes decision making about breakthrough innovations so difficult. The successes and failures only become clear well after the smoke of battle has cleared.

As a result of these factors, there is very limited incentive to invest in breakthrough innovations and it is unusual for major companies to reinvent themselves to develop whole new business models. New and more flexible companies take advantage of these technologies because they have nothing to lose by applying such knowledge (Garnsey and Wright 1990) and for them a disruptive technology may be the only chance to gain a foothold. The tendency for major innovations to be brought by mobile scientists from the science base also demonstrates that there are not sufficient incentives internal to an industry to innovate; these may have to be brought in from outside the industry, just as an ecosystem is altered by the arrival of new species external to that ecological system.

2.4 Overcommitment to Current Customers

Market orientation endorses the classic marketing principle that urges firms to stay close to their customers and put their customers at the top of the organizational chart. Several studies have provided empirical support for the positive link between market orientation and firm performance. Yet in recent years scholars in innovation management have raised increasing doubts about the unquestioning focus that firms may place on their markets (Danneels 2004). They caution that an overemphasis on customers could lead to trivial innovations and myopic research and development (R&D), which might lower the firm's innovative competence (Danneels 2002).

Because customers are inherently shortsighted, market-oriented firms may risk losing the foresight of innovating creatively in their attempt to serve customers' existing needs (Hamel and Prahalad 1994). As a result, established firms are sometimes blinded by their customers and therefore miss the boat on emerging technological opportunities. An interesting example of this phenomenon is given by Christensen and Bower in their account of the computer disk drive industry. Computer disk drive manufacturers were highly responsive in pursuing technological opportunities that yielded benefits to their immediate customers; however, these firms ignored other opportunities that were not critical to the needs of these same customers. As Christensen and Bower (1996: 198) stated: 'a primary reason why such firms lose their positions of industry leadership when faced with certain types of technological change is because they listen too carefully to their customers'. The heart of the argument of Christensen and Bower is that major innovations introduce a very different package of attributes from the one that mainstream customers historically value. The attributes of a radical technology that will in future be valued by the consumer are very difficult for companies to predict with accuracy. This is one reason that Christensen (1997) highlights for the fact that companies often fail to spot the true future market potential for breakthrough technologies, as they listen closely to the current wants of their customer base. As such, a customer-oriented firm may risk itself in the 'tyranny of the served market': the firm pursues innovations that directly address existing customers' unsatisfied needs and that promise the best return, but it is unlikely to invest substantially in market-based innovations that have an unknown future, resulting in poor prospects for generating radical innovations (Danneels 2004).

3.   COUNTERING THE INNOVATION BARRIER: SEARCH–SELECTION PROCESSES AND BREAKTHROUGH GENERATION

The above studies argue strongly that incumbent firms' dominant practices and routines supporting success in one market and/or technological domain may represent significant barriers to developing breakthrough innovations. Despite the myriad examples of incumbent firm failures and entrant firm successes, however, there are more and more counter-examples that have been illustrated in the literature (Cattani 2006; Fleming 2001; Macher and Richman 2004; Methe et al. 1997; Stefik and Stefik 2004; Yamanouchi 1989). Because these counter-examples exist, the research question importantly becomes how certain incumbent firms who enjoy market and technological leadership through periods of incremental innovation can simultaneously respond to the challenges of change.

Based on extant research, we suggest that two fundamental enabling conditions can be identified for an established firm wishing to engage in breakthrough innovation – first, it must create an environment conducive to idea generation; second, it must have the fortitude and risk tolerance to persevere and allow the most promising ideas to have a fair chance to succeed. The first is the upstream creative challenge of developing the ability to 'see differently'. Since radical concepts often spring from the imagination of individuals or teams, the challenge is to create an organizational context where creativity may flourish. The second is the downstream implementation challenge of successfully applying and marketing the unique concept, which requires the ability to implement the concept by matching it to the actual market needs. Without the ability to see differently, the firm is unable to change the rules of the game and without the ability to implement it, the firm will join the ranks of companies that failed to capitalize on their pioneering inventions such as Xerox with personal computers and EMI with scanner technology.

Although the history of breakthrough innovations is rife with elements of serendipity and unpredictability, there are also organizational elements that incumbent firms may act upon in order to foster the emergence of these discoveries (challenge 1) and facilitate their subsequent pursuit (challenge 2). The following section develops a general characterization of the organizational practices conducive to breakthrough innovation. We characterize the generation of breakthrough innovation as a two-stage search–selection process. The first stage is aimed at increasing the likelihood of coming across new opportunities, by prompting experimentation and search. The selection phase is aimed at sifting and supporting the implementation of the most valuable opportunities. In considering each of these approaches we are interested in understanding the underlying basis of selection and the potential of the selection approach for changing the organization's trajectory. How likely is the approach to foster a breakthrough?

3.1 Search Phase

The problem of innovativeness is often attributed to an inability to combine a variety of components (Kanter 1988). The solutions to this problem are then mechanisms to enhance the level of experimentation or 'slack search' (Cyert and March 1963) on the part of the organization. More precisely, because inventions are either novel combinations of physical components (Basalla 1988) or rearrangements of previously tried combinations (Henderson and Clark 1990), an organization wishing to enhance its innovativeness should encourage experimentation with a wide variety of components. This well-established 'recombinant view' of the creative process (Simonton 1999), suggests that the first step to innovation is to bring together components (competences, technologies, schemas, etc.) that were previously separated. So, for instance, we know from Fleming's research (2001) that inventors from firms that encompass experimentation with greater technological diversity are more likely to put together a previously untried combination because they are allowed a greater space for combination (Fleming and Sorenson 2001). We also know that experimentation with diverse components stimulates critical processes in breakthrough

generation such as analogies, divergent thinking and technological brokering (Hargadon 2003).

Schumpeter's (1934) research is one of the early studies that support this
idea. It suggests that recombination of existing physical and conceptual
material leads to radical innovation. Utterback's (1994) argument that
breakthroughs come from recombination of well-understood technologies
also supports this strategy. Similarly, Ahuja and Lampert's (2001) findings
indicate that experimentation with unfamiliar technologies, emerging technologies and pioneering technologies enhances the chance of creation of
breakthrough inventions. This form of experimentation can be facilitated in
various ways. Fleming (2001), for instance, in his in-depth analysis of the
organizational sources of HP's thermal ink-jet discovery, describes how HP
actively supported juxtaposition and turbulence by recycling engineers
across disciplines, encouraging social networking and favouring physical
proximity of scientists, to promote idea circulation and interaction (planned
and unplanned). In order to encourage experimentation, Nissan Design
International deliberately promotes 'creative abrasion' by hiring people in
contrasting pairs (e.g. balancing nerds with hippies). At Philips Research,
researchers were encouraged to spend some of their time, about half a day
a week, on topics not immediately related to their day-to-day work.
According to Vanhaverbeke and colleagues (Vanhaverbeke et al. 2003), this
officially legitimized bootlegging work, labelled 'Friday-afternoon experiments', created a fertile ground for some major technological opportunities
at Philips. In a similar way, to allow for experimentation and serendipity, at
3M researchers are encouraged to spend up to 15 per cent of their time on
a research project of their choice. This ensures that problem-driven research
does not preclude curiosity-driven search.

The vision of technological foresight would be wonderful to have, but
such insight is likely to be rare and certainly unpredictable *ex ante*. Lacking
such vision, organizations must exploit their current wisdom about the
world and engage in ongoing search as to future possibilities. To the extent
that an incumbent can continuously experiment and recombine disparate
knowledge components, its ability to invent breakthrough may not
decrease over time. Experimentation can counter learning traps and blinders by providing experience in novel areas. Thus, rather than worrying
about externally caused technological obsolescence, firms might stay inventive by focusing on increasing the recombinant mixing and turbulence
among their current set of inventors and technologies. On the other hand,
however, experimentation commonly leads to highly ambiguous results
which, for the reasons already illustrated, tend to be judged by old or inadequate innovation routines. As noted by Sharma (1999: 150), 'the overwhelming volume of new ideas and the need to invest to carefully evaluate

any one of them in itself restrains innovative activity in large firms'. This brings us to the critical issue of adopting appropriate selection criteria for sifting out promising opportunities from the larger pool of emerging possibilities. We discuss this aspect in the next section.

3.2 Selection Phase

Search is just a first step towards change. It provides the raw material for a selection process. The real challenge is to allow this diversity at a subsystem level to provide the basis for a 'cascade' of actions influencing resources allocation and strategy that have the potential to prompt step changes at the organizational level. As suggested by Adner and Levinthal (1995: 1–2), 'the experiments must be complemented by sufficient variety in the feedback mechanisms which inform the internal selection process within an organization . . . the challenge is how an organization can intelligently select the "thousand flowers" growing in its fields'. What makes this phase extremely challenging is that organizations tend to have a singular set of selection criteria since initiatives are typically evaluated as to whether they fit an organization's existing strategic context. This limited granularity of the selection criteria reinforces the general tendency of large organizations to persist in a given trajectory. How, then, can organizations counteract this propensity and induce variegated feedback mechanisms that may enable promising experiments to emerge?

Three types of organizational selection regimes can be distilled from the literature that seem to have proved particularly effective to this purpose. We characterize these selection regimes as follows: individual driven, lead user driven and application domain driven. We discuss each of them separately.

3.2.1 Selection driven by individuals

Individual-driven selection processes are characteristic of large firms that aim to simulate the entrepreneurial conditions usually associated with new entrants (Baden-Fuller 1995; Stopford and Baden-Fuller 1994). The selection process is led by enterprising employees or champions who emerge in an organization and make a decisive contribution to an innovation by actively and enthusiastically promoting its progress through critical stages, particularly those early on in the process (Burgelman and Sayles 1986; Humble and Jones 1989; O'Connor and Rice 2000). Such processes can take several forms: *internal ventures, spin-offs* or *skunk works*. Internal venture and spin-offs usually result from deliberate efforts on the part of the incumbent to take advantage of new promising opportunities that require protection from the counterproductive forces within the mainstream. Skunk works projects arise more informally when product champions encounter organizational forces unsympathetic to their ideas. In this setting projects are managed outside of the organizational context and, indeed, may not even be exposed to a selection process.

The internal venture may or may not constitute a separate division or separate project that is organizationally or geographically separate from the rest of the firm, but it is unique in that it has objectives that are largely independent from, and in some senses counter to, the rest of the firm. For large corporations, stimulating an enterprising unit internally is an especially valuable option when the new opportunity has a lower profit margin than the mainstream business and should serve the unique needs of a new set of customers. The logic is to keep units small and autonomous so that employees feel a sense of ownership and are responsible for their own results. This encourages a culture of autonomy and risk taking that could not exist in a large, centralized organization. For instance, Tushman and O'Reilly (1996) suggest that at HP the $7 billion printer business emerged not because of strategic foresight and planning at HP's headquarters, but rather due to the entrepreneurial flair of a small group of employees who had the freedom to pursue what was believed to be a small market. The same approach allows J&J to enter small niche markets or develop unproven technologies without the burdens of a centralized bureaucratic control system. These companies promote both local autonomy and risk taking and ensure local responsibility and accountability through strong, consistent financial control systems (O'Reilly and Tushman 2004).

As an alternative, large firms may opt for spinning out the venture altogether, to allow for greater autonomy and strategic latitude. A spin-off can also avoid short-term shareholder pressure which cannot tolerate capital costs, long lead times and technology uncertainty. According to Christensen and Bower (1996), in the entire history of the hard-disk-drive industry only three incumbents achieved commercial success with a radical technology. Christensen and Bower (1996) attributed the success of two out of three to their spinning out an independent spin-off organization to pursue the new technology. However, because spin-offs are more autonomous and detached from the main organization than internal ventures, they may forgo synergies with the parent company. So, it has been suggested that when resource complementarities between the new venture and the mainstream business are crucial, and these complementarities require intracompany coordination, a more integrated approach may be advised (Iansiti et al. 2003). On the other hand, if the fit is poor and if the firm is unwilling to leverage the emerging opportunity to stretch its competence base, then a spin-out approach may be preferred.

Whether it is based on internal development or spin-out, the entrepreneurial pursuit of a new opportunity by large firms is often led by corporate venturing entities (Birkinshaw 1997; Rice et al. 2000). These are typically specialized entities established by large firms with the purpose of identifying and nurturing new business opportunities for the corporation, either by incubating internal business ideas or spinning out businesses. A well-known example of this approach is Xerox Technology Ventures (XTV), the venture fund established by Xerox to commercialize internal technologies that might otherwise have languished. The XTV concept is to create an entrepreneurial vehicle, funded by the parent company, that mimics the traditional venture capital model. The new product venture can have the best of both worlds by exploiting new technologies without giving up access to all corporate resources. The parent company benefits by not having to manage areas outside of its core business (Chesbrough 2002).

Unlike formal internal ventures and spin-offs, skunk works projects are not subject to an explicit selection process. Instead they typically rely on the drive and commitment of their supporters in defying the sceptical scrutiny of the organization. Skunk works typically originate from informal bootlegging initiatives. An individual (or group of individuals) low in the corporate hierarchy identifies a problem and a route to a solution, and decides to go ahead with it without the knowledge or permission of people higher in the organization (Peters and Waterman 1982). Sometimes skunk works may also descend from the top, when a division head decides to start a project that he thinks should be kept secret but well funded. At IBM, for instance, small skunk works have a full year of low-overhead operation without any question asked before managers decide whether or not to move their projects into the company's mainstream. Likewise, DuPont's pioneering success in developing and commercializing Lycra Spandex to replace rubber in the leg bands of baby diapers stemmed from the activity of a sequestered team of six people which operated for a few years unencumbered with normal bureaucratic haggling that comes with typical corporate resistance (Gwynne 1997).

While skunk works may prove highly effective in fostering breakthroughs, they are not without problems. Because evaluation of the project is typically haphazard, occurring at a variety of time points (need of resources that cannot be hidden, accidental discovery of the project, turnover of key managers, etc.) and the allocation of resources ranges from subversive hoarding to scrounging, the pursuit of skunk works initiatives is fraught with uncertainty. As noted by Adner and Levinthal (1995: 6): 'because commitment is elicited on a project by project basis, skunk works are unlikely to change firms' strategic contexts'. In addition, the insulation of these groups from the corporate environment reduces access to key resources such as technological or market know-how, distribution channels, key individual knowledge sets, etc.

3.2.2 Lead-user-driven selection

User-driven selection processes have gained popularity thanks to von Hippel and colleagues' (von Hippel 1988, 2005; von Hippel et al. 1999) seminal studies. Lead users are not to be confused with customers *à la* Christensen. As previously discussed, Christensen warns that incumbents who are too tightly tied to their customers may consequently miss the boat on radical technologies. Lead users, however, are not necessarily customers; in fact, most often they are not customers at all. Lead users – companies, organizations or individuals – are progressive users that have a high motivation to obtain a solution to their so far unmet needs (von Hippel 1988). They are well ahead of market trends and have needs that go far beyond those of the average user.

The lead-user approach is designed to collect information about both needs and solutions from the leading edges of the target market and from markets that face similar problems in a more extreme form. As such it may be an effective way to probe and validate breakthrough innovations. For example, an automobile manufacturer that wants to improve its braking system can look at auto racing teams. Furthermore, this company can look at users out of the target market that face similar problems. Braking, for example, plays an important role in aerospace, too, especially the military section. In fact, aerospace is where innovations such as ABS braking were first developed (military and commercial aircraft pilots have a very high incentive to stop their vehicles before running out of runway!). The firm's effort is therefore more in the direction of finding prototype product and service ideas that have already been generated and selected by lead users than it is on selecting in-house-generated ideas (von Hippel et al. 1999). Thus, in essence, the lead-user approach moves the emphasis of the selection process from the firm to its external environment.

This approach can provide useful insights, especially for companies that operate at the front end of their target market, because they depend on ideas from analogue fields to further improve their products. This is also consistent with the study of Chandy and Tellis (1998) finding that companies focusing on future customers, rather than on current customers, have a greater degree of radical product innovation. An advantage of the lead-user approach, in particular, is that lead users have a thorough understanding of the problem to be solved and are usually open minded enough to combine knowledge from different fields. Therefore they can help to identify valuable analogue fields that are in certain aspects ahead of them and from which they expect useful insights for the products they use. Besides, the lead-user method can be applied by any company; no experts or special training is needed. A fundamental challenge, however, is to identify a lead user to follow and to extract the relevant information from; this can be difficult and time-consuming as there are often many more innovating users thinking about a problem than there are manufacturer-based developers, and these users are thinking about and testing many different ideas.

3.2.3 Application-domain-driven selection

Pursuing new avenues for innovation requires a complex coupling between technical development and a technology's market application. Because of the belief that technological revolutions usually occur in the lab, managers sometimes undervalue the importance of applications. Yet the selection forces that operate at the level of the niche domain of application may prove highly consequential in triggering radical step changes in a firm's route to innovation. Compared to the low dimensionality of the basis by which initiatives are internally evaluated, the market environment is composed of numerous niches, each with its own requirement and selection criteria. This requisite variety may induce critical shifts in the functionality by which a technology is evaluated and unleash resources for supporting its development and commercialization. As noted by Adner and Levinthal (2002: 63): 'beneath the revolutionary emergence of new technologies is often a process of shifting application domains and rapid subsequent growth in the new domain'.

Levinthal (1998) and Adner and Levinthal (2002) refer to the discontinuities that may arise from matching an existing technological know-how to different domains of application with the notion of technological speciation. The term technological speciation denotes an innovation that shares a prior lineage with another technology but has radical new features that result from its exposure to new selection forces in the market, with different niche structure of the resource space. Even where the advance of science and technology is steady and incremental, when a new domain of application is opened up for such knowledge (or for a combination of known technologies) a major discontinuity may ensue. For example, video recording involved a combination of known image-storing and signal-processing technologies applied to the consumer video market. EMI's medical scanning instrumentation was a combination of known X-ray technology and information-processing technologies, combined in a new application for the medical market.

The implication of these observations is that there are probably many technological developments that could take off, *in the appropriate application domain* (Adner and Levinthal 2002). Thus the speciation approach shifts attention from the selection of the proper technology to the selection of the appropriate domain of application and thereby to its matching with the technology. This approach points to the importance of 'probing' (Lynn et al. 1996) multiple domains of application and learning from gradual feedbacks that accumulate as the existing technology is adapted to the selected niche. The challenge is then for managers to anticipate the structure of the possible selection environments in which they can develop a given technological initiative and actively look for ways to move the speciation process forward. By expanding a firm's search space, and fostering engagement in new markets, firms learn about consumer preferences and requirements (Adner and Levinthal 2002). As a result of this feedback-driven interaction with the environment, new technologies may develop and eventually come to commercial fruition. In a recent study by Cattani (2006), for instance, it is shown how Corning developed fibre optics by leveraging its long-standing expertise in specialty glass into the telecommunication industry. The breakthrough was a result of the adaptation of this existing know-how to the performance demands and requirements of the new domain. Similarly, in an in-depth study of Acorn computers and its spin-off ARM, Garnsey et al. (2008) provide evidence of speciation mechanisms behind the pioneering development and commercialization of ARM microprocessors.

This approach is not without limitations, though. Multiple domains may generate noisy rejection signals. For instance, an innovation rejected in one niche may find acceptance in another or at a later stage of development. As noted by Adner and Levinthal (2002: 62): 'The open ended nature of this feedback can make abandoning projects quite difficult.' Because firms need to be able to probe opportunity paths in a timely and efficient manner, adopting efficient resource allocation processes is a critical component in implementing this approach.

4. DISCUSSION

Breakthrough innovations are important to firms. They enable firms to challenge the existing technological order and shape new paths, and allow them to engage in corporate reinvention, growth and new business development (Burgelman 1983). They represent rare, valuable and inimitable sources of competitive advantage for firms (Barney 1991). Ever since Schumpeter associated the advent of revolutionary technologies with 'waves of creative destruction', there has been debate about the relative role of incumbent large firms and new entrants in sparking these waves of change. Over the past 20 years, most of the analytical writing has been piled against incumbents, based on the general observation that new firms are less likely to be affected by the kind of change preventing forces typically associated with large companies' entrenched wisdom. Entrants have less to lose, and for them a breakthrough innovation may be the only chance to gain a foothold. On the contrary, numerous 'antibodies' are present in large organizations that systematically detect and hamper the progression of fundamentally different ideas that may threat the company's *status quo* (Garnsey and Wright 1990).

As we have illustrated in the first part of the chapter, some of these barriers are a result of firms' path dependence and their consequent resistance to modify competences or organizational practices that have proved successful in the past, while others stem from their aversion to the inevitable risks and uncertainties that beset the early stages of radically new technologies. These barriers are pervasive in large organizations and represent a powerful impediment to breakthrough innovations.

But despite these consolidated arguments, we also have noted that there is an increasing number of counter-examples indicating that at least some large firms are able systematically to reinvent themselves and contribute major innovations (Ahuja and Lampert 2001; Fleming 2001). Accordingly, in the second part of the chapter we have sought to illustrate the key mechanisms and practices that underpin such deviations from the mean. To this end, we have conceptualized the generation of breakthrough innovation as a two-step search–selection process. In the search phase firms need to widen their solution space by encouraging experimentation and stimulating recombination of previously separated components (knowledge sets, technologies, schemas, etc.). These combinatorial thought trials (Simonton 1999) afford greater creative opportunities and more possibilities of creating new innovations.

The generation of variety, however, is obviously not sufficient. In fact, following Adner and Levinthal (1995), we surmise that the lack of variety is not even the key constraining factor in bounding the incumbent's innovative potential. Indeed, variety very often results as a natural by product of large firms' ongoing operations (Kanter 1988). As illustrated by Garud and Nayyar (1994), most established firms have vast storehouses of knowledge that is poorly exploited. The real challenge is then how an organization can select effectively among the 'thousand flowers' growing in its fields (Adner and Levinthal 1995). Experiments and search, in other words, must be integrated by sufficient variety in the feedback mechanisms which guide the internal selection process within an organization.

We have characterized three types of selection regimes that summarize the canonical forms of such processes: individual driven; user driven; and application domain driven. As we have discussed, all of these approaches, in their various declinations, may prove viable in choosing among promising alternatives and thereby supporting radical innovations in large companies. Yet we also emphasize a salient difference in the degree of granularity that is embodied by these selection processes. Individual- and user-driven approaches both rely on single sources of feedback, be it an individual intuition (champion-driven) or a user need or perception (user-driven). The application domain approach, instead, relies on a broader set of niches each with its own requirements and selection criteria. Because the first two approaches allow for a higher degree of focus, they are probably better suited to enhance and capitalize on existing competences. Focus, however, is likely to cause lower sensitivity to the broader range of opportunities available in the environment, which might prevent the organization from adequately assessing the signals that come from the market. As noted by Adner and Levinthal (1995: 17) 'A firm that is relying on a singular view of the environment, whether an internally derived one or one based on a particular set of actors within its environment is likely to get it wrong. Wrong either in the sense of inappropriate product development efforts . . . or wrong in the sense of over-reacting to signals from a particular segment of the environment.' These arguments suggest that the narrower the dimensionality of the selection approach, the more conservative the choice of projects is likely to be and the less the potential of the approach to fostering radical change.

Given the multifaceted nature of the environment, we suggest that the domain-driven approach is the one that allows for higher dimensionality in the selection criteria and for that reason the one that is more likely to be conducive to successful breakthrough innovations. By experimenting with diverse domains and thus relying on variegated feedbacks to drive initiatives, firms may engender radical changes even in the face of modest initial impetus (Adner and Levinthal 2002). High-technology domains, in particular, are subject to a great deal of uncertainty and, as a result, offer some of the richest opportunities for experience-based learning. In such settings the use of variegated feedback from multiple facets of the application environment may serve as superior selection mechanisms to the judgement of corporate headquarters.

5.   CONCLUSIONS

We would like to close by pointing to some avenues for research enquiry that also have practical implications for managers. First, we encourage researchers to look beyond financial impact when examining the relationship between breakthrough innovation and performance. The path to breakthrough innovations is almost invariably punctuated by setbacks and failures. Yet these failures often lead to new resource combinations and trigger learning platforms that may facilitate future value creation. New skills, know-how and valued relationships can be developed that may well be exploited at some point in the future. These are valued outcomes that are likely to emerge as a byproduct of breakthrough pursuit and that should be factored into large firms' innovations strategies (Dess and Lumpkin 2002).

Second, numerous researchers have suggested that the pursuit of innovation requires established companies to strike a fragile balance between engaging in activities that use what they already know, while at the same time exploring new activities and opportunities to rejuvenate themselves (Floyd and Wooldridge 1999; March 1991; Tushman and O'Reilly 1996). Clearly, advocates of the need for a firm to pursue breakthroughs would implicitly favour the need for exploration over the short-term needs for the efficient allocation of a firm's existing resource base. Hence further research could examine what organizational and managerial practices are best suited to promote a balance between these two often opposing forces (O'Reilly and Tushman 2004). And, as noted earlier, such research would be enhanced by having a broad perspective of the dependent variable of organizational performance. Insights into the relevance of such important trade-offs for value creation and competitive advantage have important implications for both scholars and practitioners. In particular, from a practitioner standpoint, explaining the determinants of breakthrough innovations is of significant importance given the economic and societal stakes associated with them (Harhoff et al. 1999).

Additionally, the combined illustration of innovation traps and viable counteracting strategies also has practical utility as it provides managerially actionable guidelines that large firms can follow to break from their inertial constraints. If firms acknowledge the importance of breakthrough innovations to their future, routines that fail to produce such innovations are likely to be revised and challenged. Thus, by describing practices to counteract such pathologies we deepen and enrich the learning trap arguments from a conceptual standpoint (Ahuja and Lampert 2001).

A breakthrough is something new, something unheard of, which rises above the culture and the environment from which it surfaces. Given the many intangible elements that affect a breakthrough, it would be foolish to extrapolate a flawless success formula. None the less large firms exist that do display consistent and systematic ability in spurring breakthroughs and coping with change. Not only should this observation serve as a testament to the viability of 'managing for breakthroughs', but it should also encourage further empirical efforts in this direction. Breakthrough innovations are rapidly in the spotlight *ex post*, and companies are then praised or criticized for their decisions to pursue or overlook them. But the winners rarely have the foresight to anticipate tomorrow's waves of creative destruction. Yet this is the challenge faced by managers (Bessant et al. 1996; Huston 2004). We hope our work will provide fruitful suggestions to address this challenge and allow for a more systematic understanding of the organizational and strategic processes that may help overcome corporate resistance and support breakthrough innovations in large established firms.

NOTE

1. For instance, packet switching, now a key technology for digital data transmission, was not adopted until the early 1990s, despite being first worked on in the late 1960s. Similarly, it was not until after 20 years from Percy Spencer's original invention that the microwave was successfully introduced into the marketplace (Nayak and Ketteringham 1993).

REFERENCES

Abernathy, W. and Clark, K. (1985), 'Innovation: mapping the winds of creative destruction', *Research Policy*, **14**(1): 3–22.

Achilladelis, B., Schwarzkopf, A. and Cines, M. (1990), 'The dynamics of techno-logical innovation: the case of the chemical industry', *Research Policy*, **19**(1): 1–34.

Adner, R. and Levinthal, D. (1995), 'Organizational renewal: variegated feedback and technological change', Working paper of the Sol C. Snider Entrepreneurial Research Center, The Wharton School.

Adner, R. and Levinthal, D.A. (2002), 'The emergence of emerging technologies', *California Management Review*, **45**: 50–66.

Ahuja, G. and Lampert, C.M. (2001), 'Entrepreneurship in the large corporation: a longitudinal study of how established firms create breakthrough inventions', *Strategic Management Journal*, **22**: 521–43.

Amburgey, T., Kelly, D. and Barnett, W. (1993), 'Resetting the clock: the dynamics of organizational change and failure', *Administrative Science Quarterly*, **38**, 51–73.

Baden-Fuller, C. (1995), 'Strategic innovation, corporate entrepreneurship and matching outside-in to inside-out approaches to strategy research', *British Journal of Management*, **6**(6): 3–16.

Baden-Fuller, C. and Pitt, M. (1996), *Strategic Innovation*, London: Routledge.

Barney, J. (1991), 'Firm resources and sustained competitive advantage', *Journal of Management*, **17**(1): 99–120.

Basalla, G. (1988), *The Evolution of Technology*, New York: Cambridge University Press.

Bessant, J. (2003), *High-Involvement Innovation: Building and Sustaining Competitive Advantage Through Continuous Change*, Chichester: Wiley.

Bessant, J., Caffyn, S. and Gilbert, J. (1996), 'Learning to manage innovation', *Technology Analysis and Strategic Management*, **8**: 59–70.

Birkinshaw, J. (1997), 'Entrepreneurship in multinational corporations: the charac-teristics of subsidiary initiatives', *Strategic Management Journal*, **18**: 207–29.

Burgelman, R.A. (1983), 'A process model of internal corporate venturing in the diversified major firm', *Administrative Science Quarterly*, **28**: 223–45.

Burgelman, R.A. and Sayles, L.R. (1986), *Inside Corporate Innovation: Strategy Structure and Managerial Skills*, New York: The Free Press.

Cattani, G. (2006), 'Technological pre-adaptation, speciation, and emergence of new technologies: how Corning invented and developed fiber optics', *Industrial and Corporate Change*, **15**(2): 285–318.

Chandy, R. and Tellis, G.J. (2000), 'The incumbent's curse? Incumbency, size and radical product innovation', *Journal of Marketing*, **64**(3): 1–17.

Chesbrough, H. (2002), 'Graceful exits and foregone opportunities: Xerox's management of its technology spinoff organizations', *Business History Review*, **76**(4): 803–38.

Christensen, C.M. (1997), *The Innovator's Dilemma*, Boston, MA: Harvard Business School Press.

Christensen, C.M. and Bower, J.L. (1996), 'Customer power, strategic investment and the failure of leading firms', *Strategic Management Journal*, **17**: 197–218.

Cohen, W.M. and Levinthal, D.A. (1990), 'Absorptive capacity: a new perspective on learning and innovation', *Administrative Science Quarterly*, **35**: 128–52.

Cook, S.N. and Yanow, D. (1993), 'Culture and organizational learning', *Journal of Management Inquiry*, **2**: 373–90.

Cyert, R.M. and March, J.G. (1963), *A Behavioral Theory of the Firm*, Englewood Cliffs, NJ: Prentice-Hall.

Danneels, E. (2002), 'The dynamics of product innovation and firm competences', *Strategic Management Journal*, **23**(12): 1095–121.

Danneels, E. (2004), 'Disruptive technology reconsidered: a critique and research agenda', *Journal of Product Innovation*, **21**: 246–58.

Dess, G. and Lumpkin, G. (2002), 'The role of entrepreneurial orientation in stimulating effective corporate entrepreneurship', *Academy of Management Executive*, **19**(1): 147–56.

Dougherty, D. (1992), 'Interpretive barriers to successful product innovation in large firms', *Organization Science*, **3**: 179–203.

Finkelstein, S. and Sanford, S.H. (2000), 'Learning from corporate mistakes: the rise and fall of iridium', *Organizational Dynamics*, **29**(2): 138–48.

Fleming, L. (2001), 'Finding the organizational sources of technological breakthroughs: the story of Hewlett Packard's thermal ink-jet', *Industrial and Corporate Change*, **11**(5): 1059–84.

Fleming, L. and Sorenson, O. (2001), 'Science as a map in technological search', *Strategic Management Journal*, **25**: 909–28.

Florida, R. and Kenney, M. (1990), *The Breakthrough Illusion*, New York: Basic Books.

Floyd, S.W. and Wooldridge, B. (1999), 'Knowledge creation and social networks in corporate entrepreneurship: the renewal of organizational capability', *Entrepreneurship Theory and Practice*, **23**(3): 123–43.

Foster, R.N. (1986), *Innovation: The Attacker's Advantage*, New York: Summit Books.

Freeman, C. and Soete, L. (1997), *The Economics of Industrial Innovation*, 3rd edn, Cambridge, MA: The MIT Press.

Garcia, R. and Calantone, R. (2002), 'A critical look at technological innovation typology and innovativeness terminology: a literature review', *The Journal of Product Innovation Management*, **19**: 110–32.

Garnsey, E. (2004), 'Notes on technological innovation', Working paper of the Centre for Technology Management, University of Cambridge.

Garnsey, E. and Wright, S. (1990), 'Technical innovation and organizational opportunity', *International Journal of Technology Management*, **5**(3): 267–91.

Garnsey, E., Lorenzoni, G. and Ferriani, S. (2008), 'Technology speciation through entrepreneurial spin-off: the Acorn–ARM story', *Research Policy*, forthcoming.

Garud, R. and Nayyar, P. (1994), 'Transformative capacity: continual structuring by intertemporal technology transfer', *Strategic Management Journal*, **15**: 365–85.

Gwynne, P. (1997), 'Skunk works, 1990's style', *Research Technology Management*, **40**(4): 18–23.

Hamel, G. and Prahalad, C.K. (1994), *Competing for the Future*, Boston, MA: Harvard Business School Press.

Hannan, M.T. and Freeman, J. (1984), 'Structural inertia and organizational change', *American Sociological Review*, **49**(2): 149–64.

Hargadon, A. (2003), *How Breakthroughs Happen*, Cambridge, MA: Harvard Business School Press.

Harhoff, D., Narin, F., Scherer, M. and Vopel, K. (1999), 'Citation frequency and the value of patented inventions', *Review of Economics and Statistics*, **81**(3): 511–15.

Henderson, R.M. (1993), 'Underinvestment and incompetence as responses to radical innovation: evidence from the photolithographic alignment equipment industry', *RAND Journal of Economics*, **24**: 248–70.

Henderson, R.M. and Clark, K.B. (1990), 'Architectural innovation: the reconfiguration of existing product technologies and the failure of established firms', *Administrative Science Quarterly*, **35**: 9–30.

Humble, J. and Jones, G. (1989), 'Creating a climate for innovation', *Long Range Planning*, **22**(4): 46–51.

Huston, L. (2004), 'Mining the periphery for new products', *Long Range Planning*, **37**(2): 191–6.

Iansiti, M., McFarlan, W. and Westerman, G. (2003), 'Leveraging the incumbent's advantage', *MIT Sloan Management Review*, **44**(4): 58–64.

Kanter, R.M. (1988), 'When a thousand flowers bloom: structural, collective, and social conditions for innovation in organizations', *Research in Organizational Behavior*, **10**: 169–211.

Klepper, S. and Simons, K. (2000), 'Dominance by birthright: entry of prior radio producers and competitive ramifications in the US television receiver industry', *Strategic Management Journal*, **21**: 997–1016.

Leonard-Barton, D. (1992), 'Core capabilities and core rigidities: a paradox in managing new product development', *Strategic Management Journal*, **13**: 111–25.

Levitt, B. and March, J.G. (1988), 'Organizational learning', *Annual Review of Sociology*, **14**: 319–40.

Lynn, G.S., Morone, J.G. and Paulson, A.S. (1996), 'Marketing and discontinuous innovation: the probe and learn process', *California Management Review*, **38**(3): 8–37.

Macher, J. and Richman, B.D. (2004), 'Organizational responses to discontinuous innovation: a case study approach', *International Journal of Innovation Management*, **8**(1): 87–114.

Maine, E. and Garnsey, E. (2006), 'Commercializing generic technology: the case of advanced materials ventures', *Research Policy*, **35**(3): 375–93.

March, J.G. (1991), 'Exploration and exploitation in organizational learning', *Organization Science*, **2**: 71–87.

Mascitelli, R. (2000), 'From experience: harnessing tacit knowledge to achieve breakthrough innovation', *Journal of Product Innovation Management*, **17**(3): 179–93.

McGahan, Anita, Vadasz, Leslie and Yoffie, David (1997), 'Creating value and setting standards: the lessons of consumer electronics for personal digital assistants', in D. Yoffie (ed.), *Competing in the Age of Digital Convergence*, Boston, MA: Harvard Business School Press, pp. 239–54.

Methe, D.T., Swaminathan, A., Mitchell, W. and Toyama, R. (1997), 'The underemphasized role of diversifying entrants and industry incumbents as the sources of major innovations', in H. Thomas and D. O'Neal (eds), *Strategic Discovery: Competing in New Areas*, New York: Wiley, pp. 99–116.

Morone, J.G. (1993), *Winning in High Tech Markets*, Boston, MA: Harvard Business School Press.

Nayak, P. and Ketteringham, J. (1993), *Breakthroughs!*, New York: Rawson Associates.

Nelson, R. (1991), 'Why do firms differ, and how does it matter?', *Strategic Management Journal* (special issue) **12**: 61–74.

Nelson, R.R. and Winter, S.G. (1982), *An Evolutionary Theory of Economic Change*, Boston, MA: Harvard University Press.

O'Connor, G.C. and Rice, M.P. (2000), 'Opportunity recognition and breakthrough innovation in large established firms', *California Management Review*, **43**: 95–117.

O'Reilly III, C.A. and Tushman, M.L. (2004), 'The ambidextrous organization', *Harvard Business Review*, **82**(4): 74–81.

Orr, J. (1990), 'Sharing knowledge celebrating identity: war stories and community memory in a service culture', in D.S. Middleton and D. Edwards (eds), *Collective Remembering: Memory in Society*, Beverley Hills, CA: Sage Publications, pp. 169–89.

Perrow, C. (1986), *Complex Organizations: A Critical Essay*, New York: Random House.

Peters, T.J. and Waterman, R.H. (1982), *In Search of Excellence*, New York: Harper and Row.

Rice, M.P., O'Connor, G.C., Peters, L.S. and Morone, J.E. (1998), 'Managing discontinuous innovation', *Research Technology Management*, **41**(3): 52–8.

Rice, M.P., O'Connor, G.C., Leifer, R., McDermott, C.A. and Standish-Kuon, T. (2000), 'Corporate venture capital models for promoting radical innovation', *Journal of Marketing Theory and Practice*, **8**: 1–11.

Rosenbloom, R.S. (2000), 'Leadership, capabilities, and technological change: the transformation of NCR in the electronic era', *Strategic Management Journal*, **21**(10–11): 1083–104.

Schumpeter, J.A. (1942), *Capitalism, Socialism and Democracy*, New York: Harper.

Schumpeter, Joseph A. (1934), *The Theory of Economic Development*, Cambridge, MA: Harvard University Press.

Sharma, A. (1999), 'Central dilemmas of managing innovation in large firms', *California Management Review*, **41**(3): 146–64.

Simonton, K. (1999), *Origins of Genius: Darwinian Perspectives on Creativity*, New York: Oxford University Press.

Stefik, M. and Stefik, B. (2004), *Breakthrough: Stories and Strategies of Radical Innovation*, Cambridge, MA: MIT Press.

Stopford, J.M. and Baden-Fuller, C. (1994), 'Creating corporate entrepreneurship', *Strategic Management Journal*, **15**: 521–36.

Todd, A. (1999), 'Managing radical change', *Long Range Planning*, **32**(2): 237–44.
Trajtenberg, M. (1990), 'A penny for your quotes: patent citations and the value of innovations', *The RAND Journal of Economics*, **21**(1): 172–87.
Tripsas, M. (1997), 'Unraveling the process of creative destruction: complementary assets and incumbent survival in the typesetter industry', *Strategic Management Journal* (special issue), **18**: 119–42.
Tripsas, M. and Gavetti, G. (2000), 'Capabilities, cognition, and inertia: evidence from digital imaging', *Strategic Management Journal*, **21**(10–11): 1147–62.
Tushman, M.L. and O'Reilly III, C.A. (1996), 'Ambidextrous organizations: managing evolutionary and revolutionary change', *California Management Review*, **38**(4): 8–30.
Tushman, M.L. and Anderson, P. (1986), 'Technological discontinuities and organizational environments', *Administrative Science Quarterly*, **31**: 439–65.
Utterback, J.M. (1994), *Mastering the Dynamics of Innovation*, Boston, MA: Harvard Business School Press.
Vanhaverbeke, W., Berends, H., Kirschbaum, R. and de Brabander, W. (2003), *Knowledge Management Challenges in Corporate Venturing and Technological Capability Building through Radical Innovations*, Proceedings of the 10th International Product Development Management Conference, 10–11 June, Brussels, Belgium.
Von Hippel, E. (1988), *The Sources of Innovation*, New York: Oxford University Press.
Von Hippel, E. (2005), *Democratizing Innovation*, Cambridge, MA: MIT Press.
Von Hippel, E., Thomke, S. and Sonnack, M. (1999), 'Creating breakthroughs at 3M', *Harvard Business Review*, **77**: 47–57.
Yamanouchi, T. (1989), 'Breakthrough: the development of the Canon personal copier', *Long Range Planning*, **22**(5): 11–21.

9. Search strategies for discontinuous innovation

John Bessant and Bettina von Stamm

1. INTRODUCTION

A key aspect of innovation is the search activity that organizations undertake to find trigger signals to start the process. Whether these are 'push' signals – creation of new knowledge or acquisition and internalizing of that generated elsewhere – or 'pull' signals (market demand, competitor behaviour, shifts in the regulatory framework, etc.), the key questions concern how well the organization manages the search process. The concept of search routines was first outlined by Nelson and Winter, building on earlier work by Cyert, March and Simon (Cyert and March 1963; Simon and March 1992; Nelson and Winter 1982). In essence they argue that firms undertake a trial and error approach working within a bounded search space – the selection environment. This search behaviour is also shaped by pathways – technological and other trajectories – that become established over time as promising avenues in which the firm and its competitors carry out their search (Dosi 1982). While search is initially a trial and error process, it becomes 'routinized' – successful strategies are reinforced and embedded in procedures and systems while less successful strategies are abandoned. Zollo and Winter have elaborated the concept, building on work by, *inter alia*, Cohen and Levinthal, and Zahra and George, stressing links with absorptive capacity and dynamic capability (Cohen et al. 1996; Zollo and Winter 2002; Zahra and George 2002).

Importantly, Nelson and Winter draw a distinction between *operational* routines, which maintain the established search patterns, and *strategic* routines, which extend the search in new directions (Costello 1996: 3). These higher-level routines for exploration guide the firms' evolution and are the sources of firm differences. 'Our concept of search obviously is the counterpart of that of mutation in biological evolutionary theory' (Nelson and Winter 1982: 18). Viewed in these terms, interfirm competition is both about efficiency in exploiting existing knowledge sets and about the development of new ways of finding new knowledge sets that can extend the space in which the firm explores.

In the context of innovation systems this has considerable relevance. If knowledge flows are critical to innovation success, then the ways in which firms seek (and, it is hoped, find) new knowledge created in R&D labs, universities, etc. become a significant focus for policy attention – whether the policy actors are firm managers, regional or sectoral associations or national level.

2. THE CHALLENGE OF DISCONTINUITY

Innovation studies have increasingly highlighted the challenge posed by discontinuity. For example, Utterback (1994) looked at disruptions when radically different technologies emerged in sectors such as ice harvesting, where existing incumbents often struggled – in part because new developments occurred outside their 'normal' search space.

By contrast, Tushman and Anderson's work on competence-enhancing and competence-destroying change indicated that incumbents could survive and ride the waves of radical technological change – but much depended on their ability to pick up early warning signals and this in turn required active and extensive search routines (Tushman and Anderson 1987). Henderson and Clark's work underlined the problem in that existing product architectures define a set of knowledge relationships across which exploration takes place – incumbents within this network find it difficult to pick up on cues when the architecture – and hence the knowledge interrelationships – change (Henderson and Clark 1990).

In similar fashion, firms find it difficult to pick up on changes in market demand signals. Christensen (1997) shows that when new market constituencies emerge, they are again at the fringe and not perceived as relevant. The value network involved – suppliers, users, etc. – essentially the new section environment – is different but the incumbent's response is often to reinforce the existing one – even listening hard to the voice of the wrong customers.

A third strand of the problem argument concerns the ways of seeing that firms develop. Tripsas and Gavetti (2000) show how cognitive make-up effectively blinds the incumbent in interpretation of weak incoming signals about new directions of a possibly life-threatening nature. Prahalad (2004) talks of the blinders of dominant logic, while Day and Schoemaker (2006) use the metaphor of peripheral vision to suggest that firms often rely on narrow focus and miss key signals.

There is also the problem of trajectories and directions of search. As Hargadon (2003) and others show, innovation is often triggered not by absolute advances at the frontier of knowledge but rather by application of existing knowledge in a new context. The role of 'recombinant' innovation has been critical in many breakthrough innovations – for example, Hargadon cites the cases of both Ford and Edison as typical of this approach. This places emphasis on the mechanisms through which bridging between 'small worlds' of knowledge can take place – and arguably firms that lack routines of this combinatorial/bridging side will be at a disadvantage.

This also underlines the point identified by many writers – the problem of existing search networks acting as a constraint on innovation. Under certain conditions searching even harder across existing networks is the worst thing to do – 'the ties that bind become the ties that blind' (Burt 1992). In this sense weak ties (Granovetter 1973) become important and strong ties a limitation – a challenge to theories about inter-organizational innovation which often stress the value of close relationships with suppliers and customers (Delbridge 2004; Birkinshaw et al. 2007). For incremental innovation – doing what we do but better – these relationships based on strong ties are valuable, but for radical or recombinant innovation they may be limiting.

All this points to a special but not insignificant case – that of discontinuity – where 'traditional' search routines are inadequate. Even well-developed search behaviour may not stop disruption. When the rules of the game change, old skills may not be relevant and new ones are needed – at least in detecting the shift early enough to compile a response. Even if – as Markides and others have argued – the better strategic response of an incumbent player is to be a 'fast second', there is still a requirement for advanced early warning systems to allow time for the timing decision to be made, and this poses the question of *how* to search differently (Markides 1997).

The converse is also true: where established incumbents have succeeded in riding the waves of discontinuous change in their environments, it has been in large measure due to their having both the scanning capability to pick up early warning and weak signals, and the organizational information processing and decision making to act proactively and ahead of the challenge. This point – made elsewhere in this book by Ferriani and colleagues (Chapter 8) – helps explain why it is not always new entrants that are able to exploit new opportunities created by disruptive technological or market conditions.

Extending search behaviour and especially expanding existing and developing new exploration routines is an important issue in the 'open innovation' debate. The arguments for why firms should open up their processes to enable higher-variety knowledge flows are well made by Chesbrough and others (Chesbrough 2003). But assuming the philosophical shift is accepted – whether on grounds of rising R&D costs, difficulties in matching the variety in complex environments or whatever – the problem remains of *how* such strategies are operationalized. Moving beyond the rhetoric, what do firms need to do in terms of developing new search strategies and then routinizing them? And how do they align these with the existing mainstream and manage the tensions between exploitation and exploration?

3. SEARCH STRATEGIES

Figure 9.1 highlights the challenge. The innovation search challenge in zone 1 is (relatively) straightforward – how to systematically explore a space in which we understand where and how to search. A well-defined 'selection environment' is one in which we know the key technological lines of enquiry, we understand the markets and the competitive dynamics, the regulatory and political framework and constraints, etc. It is a space into which we can deploy controlled variations on what we currently do. *Exploitation* here, as an innovation strategy, has a relatively high chance of success and can be based on established search routines.

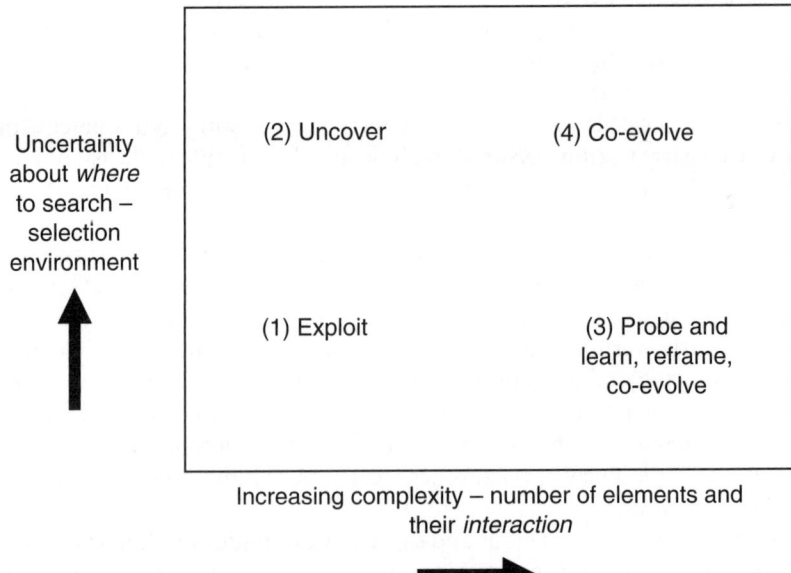

Figure 9.1   Search strategies for innovation

Zone 2 poses more problems because we have to search beyond our current context – but in a (relatively) predictable environment we have a good idea of the directions in which to place our research bets – for example, along particular technological or regulatory pathways, into new but promising market opportunities. *Exploration* is certainly involved here but it is bounded by the envelope of the rules of the overall business game that we (and our competitors) understand. Again this makes use of established routines for extending the search space – for example, following established trajectories in R&D.

In zone 3 the challenge is one of searching in territory where there is some familiarity involving the same set of markets and technologies, competitors, regulators and other players. But it is unpredictable – the sheer number of elements and the ways in which they can interact means that the focus of innovation here may shift and require the ability to reframe and unlearn old 'rules' of the game. Here the challenge becomes one not just of searching but of ways of seeing – the ability to reframe an existing set of parameters can lead to a new business model – a new paradigm – in which different problems and solutions become relevant. For example, the airline business remains one of transportation of people and freight using aircraft and airports. But the low-cost revolution has radically changed the rules of the game – a new business model has emerged. And it isn't simply a matter of low prices – it's a complete reframing of the way different elements operate and interact. Successful low-cost airlines need rapid turnaround capabilities, sophisticated pricing models and adaptive IT to support their deployment, flexible and multi-skilled working arrangements, etc.

The implications for search routines under these circumstances are not just to look more widely but to look in a different way – and it is a powerful explanation of why existing incumbents struggle and new models are usually linked to new players. Arguably search routines here require both extensive peripheral vision – to spot alternative models as they emerge as weak signals – and the ability to consider multiple interpretations of the patterns in the selection environment. Low-cost and traditional carriers both 'saw' the same signals, but the former interpreted them in a very different way. Search routines need to build the organization's capability for dealing with ambiguity – being able to think of multiple parallel futures at the same time.

In zone 4 there is unpredictability and no clear place to start – it's an open space where completely new games can emerge. A dominant design is not established, but emerges gradually as a result of trial and error, feedback and learning. Complexity theory talks about gradual *emergence* of new configurations of elements – the new dominant design – not as the result of planned and targeted search but rather of co-*evolution* with a changing environment. Under these conditions firms need additional and different search capabilities suited to picking up quickly on emergent new phenomena and on 'co-evolving' with them.

For example, the shift in the music industry towards downloading and viral marketing has fundamentally changed the business. But this was not predictable 20 years ago: although all the elements might have been visible – Internet, new compression technologies, rising demand for customized products, market segmentation, social networking, etc. – their precise interaction and the innovations that gave rise to would not have been. They were essentially *emergent* – and the pioneers of the new businesses they enabled were newcomers. They didn't necessarily have particular financial or technological assets as a competitive edge but rather they were better placed to pick up on and shape the emergent new dominant design.

The implications for search routines here are very complex – essentially the message from history is that under emergent conditions firms need to be in there, early, and be able to shape the emergent dominant design. But how to find the game and the players is a big question since the solution isn't obvious. Search routines therefore need to be about 'positioning possibilities' rather than following established trajectories.

Firms can choose to search proactively for such triggers – and thus buy themselves time and at least an entrance ticket to the emerging new game – or they can try to respond quickly once the new patterns begin to emerge.

4.   DEVELOPING SEARCH CAPABILITIES

For successful players the problems of how to search zones 1 and 2 are not insurmountable. They have good road maps for zone 1 and experienced surveyors able to extend that map-making into zone 2. Zone 1 represents line extensions, variations on existing product/service portfolios, entry into new geographical markets, etc. The search behaviour here uses well-defined routines of market and technological R&D. Zone 2 represents a targeted and strategic (re)search approach, investing in directions which – while unknown and therefore risky – have close links with what the company knows about and how to use. Trajectories – both market and technological – are well established (Dosi 1982).

But how does a firm begin to tackle zones 3 and 4? No organization has enough resources to explore all the potential space, so strategies for searching must be developed. Yet these strategies need to deal with the problem of finding a way forward in the fog that characterizes the right-hand side of the diagram. They need to use a mixture of judicious experimentation and a great deal of fast adaptive feedback to emerging situations. It's essentially an approach based on 'probe and learn', developing new routines in Nelson and Winter's trial and error mode.

One way of getting to grips with this challenge is for firms to learn together about managing discontinuous innovation – sharing experiences, trying new things out, reflecting on what has and hasn't worked, and looking at new ideas and models. This is the idea behind the AIM 'Discontinuous Innovation Laboratory', which involves networks of firms in the UK, Germany and Denmark working through such a shared learning process (see http://www.innovation-lab.org/ for details). Activities have involved a mixture of experience-sharing workshops (both in host countries and internationally) coupled with in-depth case research of discontinuous innovation (DI) experiences and experiments in each of the participating firms.

A key focus in this work has been trying to map the exploratory development of search routines for dealing with DI. Although by their nature these are still emergent models, we have identified a set of 12 common mechanisms around which such experimentation is taking place (see Figure 9.2).

It is important to recognize that these are not a replacement for good practice search strategies around R&D or market research – rather they are *additional* and *experimental* approaches to help deal with the DI challenge. Few firms make use of the whole suite, although there is clearly scope for interfirm learning and imitation of apparently successful approaches that broaden a particular firm's repertoire. We look briefly at each of these approaches in the following subsections.

4.1 Sending Out Scouts

This is a widely used strategy that involves sending out people (full or part time) whose role is to search actively for new ideas to trigger the innovation process. (In German they are called *Ideenjäger* – idea hunters – a term that captures the concept well). They could be searching for technological triggers, emerging markets or trends, competitor behaviour, etc., but what they have in common is a remit to seek things out, often in unexpected places. Search is not restricted to the organization's particular industry; the fringes of an industry or even currently entirely unrelated fields can be of interest (Hargadon 2003). Nor is their search confined to products and technologies – it is also about changes in social trends, new business models and even shifts in political situations. In this they are building a capacity for ambiguity, the capability to reframe existing situations on behalf of the firm and to explore relevant connections resulting from a change of worldview.

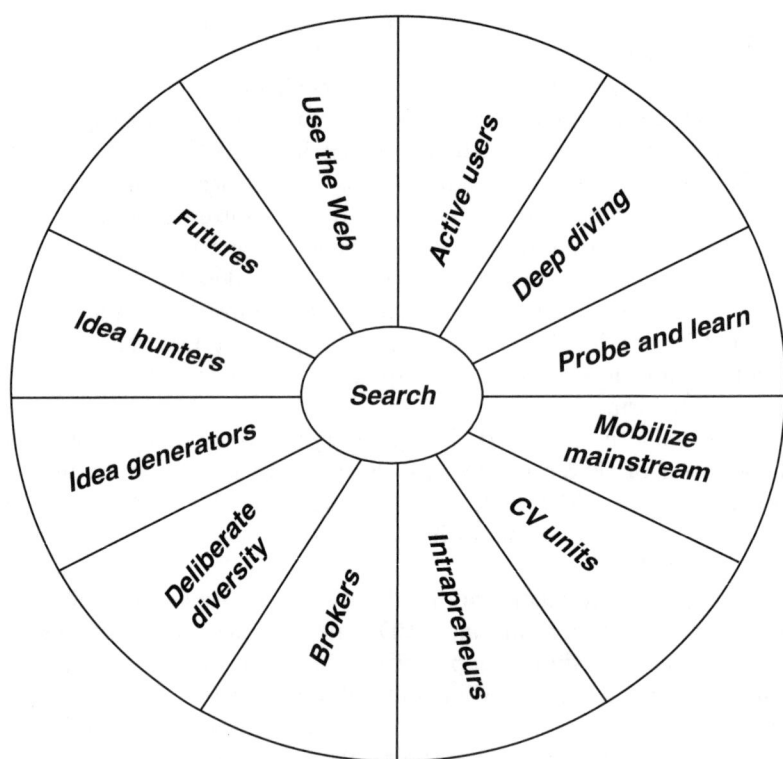

Figure 9.2 Search strategies for discontinuous innovation

A well-known example is that of Procter & Gamble, which has set itself the target of sourcing 50 per cent of its innovations from outside the company as it pursues its 'Connect and develop' open innovation approach (Huston and Sakkab 2006). To help make this happen it employs around 80 'technology entrepreneurs' – essentially scouts who are licensed to roam the world with a wide remit to find and bring back interesting new ideas. The mobile telecoms provider O2 has a trend-scouting group of about ten people who interpret externally identified trends into O2's specific business context. Once a year the group meets with the board to discuss and select ideas. BT has a scouting unit in Silicon Valley which assesses some 3000 technology opportunities per year in California. The four-man operation was established in 1999 to make venture investments in promising telecom start-ups, but after the dotcom bubble burst it shifted its mission towards identifying partners and technologies that BT was interested in. The small team looks at more than 1000 companies per year and then, based on their deep knowledge of the issues facing the R&D operations back in the UK, they target the small number of cases where there is a direct match between BT's needs and the Silicon Valley company's technology. While the number of successful partnerships that result from this activity is small – typically four or five per year – the unit fulfils an invaluable role in keeping BT abreast of the latest developments in its technology domain. The company has now established a similar operation in China. Arvato, a German mobile telecoms content provider, makes extensive use of scouts to try to keep track of emerging trends among the different segments of its user base. Many of these are 'boutique' market and technology research firms that provide an observatory on a high-velocity innovation environment.

4.2 Exploring Multiple Futures

Another source of ideas about possible innovation triggers is to imagine alternative futures. Creating and exploring such futures can be performed through trend extrapolation or 'standard' forecasting techniques, but can also be done using more advanced scenario-based approaches (Schwartz 1996). For example, the concept of 'scenario planning' is often associated with Shell, which has taken a pioneering role in this approach (de Geus 1996). Shell's 'Gamechanger' programme makes extensive use of alternative futures as a way of identifying domains of interest for future business that may lie outside the 'mainstream' of its current activities. Increasingly these rich 'science fiction' views of how the world might develop (and the threats and opportunities that this might pose in terms of discontinuous innovations) are being constructed by using a wide and deliberately diverse set of inputs rather than using the relatively narrow frame of reference that company staff might have. One consequence has been the growth of specialist service companies, which offer help in building and exploring models of alternative futures.

Another related approach is to build 'concept' models and prototypes to explore reactions and provide a focus for various kinds of input that might shape/co-create future products and services. Concept cars are commonly used in the automotive industry not as production models but as stepping stones to help understand and shape what will be products in the future. Similarly Airbus and other aerospace firms have concept aircraft, while Toyota is working on concept projects around housing, transportation and energy systems.

More recently companies have started to see value in developing such scenarios jointly with other organizations and have discovered exciting opportunities for cross-industry collaboration (which often means the creation of an entirely new market).

For example, the chemicals firm BASF went through a multi-stage process, using the mega-trend 'ageing population' as a starting point. The process began with a discussion of experts from a wide variety of professions (e.g. airport, newspaper, medical profession, etc.) about what life for the aged would be and feel like in 2020. Next, internal experts looked at the results of the discussion and related them to BASF's industries. Further steps were taken to verify conclusions internally as well as externally. GSK Consumer Health used a quite similar process to strengthen its medium- to long-term pipeline. Ordnance Survey in the UK uses an annual conference to bring leading thinkers from related industries together to discuss what the future might look like. Novo Nordisk, a major Danish pharmaceuticals business, makes use of a company-wide scenario-based programme to explore radical futures around its core business. Its 'Diabetes 2020' process involved exploring radical alternative scenarios for chronic disease treatment and the roles that a player such as Novo Nordisk could play.

4.3   Using the Internet

A third strategy seeks to use the power of the Internet to access and explore different developments – connecting to multiple sources of information and operating various forms of web-enabled marketplace. At one level the Web offers a vast library – and the mechanisms to make new connections to and among the information it contains. This is, naturally, a widely used approach but it is interesting to look in more detail at how particular firms are developing and shaping this powerful tool.

In its simplest form, the Web is a passive information resource to be searched – an additional space into which the firm might send its scouts. Increasingly there are professional organizations that offer focused search capabilities to help with this hunting – for example, in trying to pick up on emerging 'cool' trends among particular market segments. High-velocity environments such as mobile telecoms, gaming and entertainment depend on picking up early warning signals and often make extensive use of these search approaches across the Web.

Developments in communications technology also make it possible to provide links across extranets and intranets to speed up the process of bringing signals to where they are needed. Firms such as Zara and Benetton have sophisticated IT systems giving them early warning of emergent fashion trends that can be used to drive a high-speed flexible response on a global basis.

Such a rich information source can quickly be exploited if it is seen as a two-way or even multi-way information marketplace. One of the first companies to take advantage of this was Ely Lilly, which set up Innocentive as a match-making tool, connecting those with scientific problems with those being able to offer solutions. As Innocentive CEO Darrel Carroll says, 'Lilly hires a large number of extremely talented scientists from around the world, but like every company in its position, it can never hire all the scientists it needs. No company can.'

In similar fashion Procter & Gamble makes use of multiple websites such as http://www.homemadesimple.com for consumer ideas, and http://pg.t2h.yet2.com/t2h/page/homepage for technology ideas. BMW makes use of the Web to enable a 'Virtual Innovation Agency' – a forum where suppliers from outside the normal range of BMW players can offer ideas that BMW may be able to use. These can be both product-related and process-related – for example a recent suggestion was for carbon recycling out of factory waste. Although this carries the risk that many 'cranks' will offer ideas, they may also provide stepping stones to new domains of interest.

A further extension of this is to use websites in a more open-ended fashion, as laboratories in which experiments can be conducted or prototypes tested. For example, a site that is growing in popularity is www. secondlife.com – essentially a role-playing game with over 6 million users. In this alternative world people can create different characters for themselves and interact in an alternative world – in the process creating a powerful laboratory for testing out ideas. Since by definition Second Life is the result of people projecting their aspirations and interests in a different space, it offers significant scope for early warning about or even creating new trends. The potential of 'advergaming' is being explored, for example, by US clothing retailer American Apparel, which opened a virtual store in Second Life in 2006, and by IBM, Vodafone and others, which have a number of interactive facilities at different virtual locations where potential customers can engage with their range of service offerings and ideas. Other social networking sites such as My Space (with 106 million members) have become a powerful channel for finding and developing music and other entertainment ideas, challenging 'traditional' marketing approaches.

Beyond these uses come those that bring users into the equation as 'co-creators' – a theme we discuss in the next section.

4.4 Working with Active Users

An increasingly significant strategy involves seeing users not as passive consumers of innovations created elsewhere, but rather as active players whose ideas and insights can create new markets, products and services. User-led innovation of this kind was first observed in the medical world, where it remains a strong model for innovation (Von Hippel 2005). But recent years have seen it spread across to many sectors – and with the advent of power-ful new tools there is huge scope for engaging users in active 'co-creation' of products and services. For example, the Internet has enabled the 'open source' movement to develop high-quality software as a cooperative process while tools such as rapid prototyping, simulation and computer-aided design help create the spaces where active users can interact with pro-fessional designers. Linux is the best-known example of such open source software, but there is a growing trend towards other products, for example around entertainment, bringing in users as co-creators in what one com-mentator has called a 'perpetual beta' world where there is constant adjust-ment, refinement and development across a shared population.

Active users become particularly important in the DI context because very often the challenge is to find the things that no one has yet noticed, or the markets that don't yet exist. In diffusion theory such users are not even early adopters, but rather active innovators. They are tolerant of failure and prepared to accept that things go wrong, but through mistakes they believe they can achieve something better – hence the growing interest in partici-pating in 'perpetual beta' testing and development of software and other online products. More often than not active users want to get involved because they feel strongly about the product or service in question; they really want to help and improve things (Von Hippel 2005).

For example, when LEGO originally launched the Mindstorms RCX they immediately learned that the most advanced users/hackers cracked the code and developed their own updated versions in a few days. However, they also learned that these advanced users produced variants of the product that were superior to the original. In 2006 LEGO was launching a radical new Mindstorms product – the NXT. This time it took advantage of the enthusiastic users and invited a few of the best to participate directly in the development. Furthermore, the source-code has revealed a 'right to hack' disclaimer in the documentation. As Senior Vice President Mads Nipper explained, 'We came to understand that this is a great way to make the product more exciting. It's a totally different business paradigm.' In recognition of the success of this programme, LEGO stated in January 2006 that it was looking for 100 more citizen developers (see http://mind-storms.lego.com). It now uses the approach across the business and has set up the LEGO Factory website where users can design their own model online and then have the ready-to-assemble set sent out to them (http://factory.lego.com/).

This kind of approach is being explored by the British Broadcasting Corporation in trying to deal with the discontinuous challenges of the new digital media environment. One alternative to trying to predict how new digital media will evolve is to try to engage a rich variety of players in the emerging spaces via a series of 'open innovation' experiments. BBC Backstage is an example, trying to acheive with new media development what the open source community did with software development. The model is deceptively simple – developers are invited to make free use of various elements of the BBC's site (such as live news feeds, weather, TV listings, etc.) to integrate and shape innovative applications. The strap line is 'use our stuff to build your stuff' – and since the site was launched in May 2005 it has already attracted the interest of hundreds of software developers. Ben Metcalf, one of the programme's founders, summed up the approach. 'Top line, we are looking to be seen promoting innovation and creativity on the Internet . . . if someone is doing something really innovative, we would like to . . . see if some of that value can be incorporated into the BBC's core propositions' (see www.bbcbackstage.com).

4.5 'Deep Diving'

Market research has become adept at listening to the 'voice of the customer' via interviews, focus groups, panels, etc., but in parallel there is growing use of anthropological-style techniques to get closer to what people need/want in the context in which they operate. 'Deep dive' is one of many terms used to describe the approach – 'empathic design' and 'ethnographic methods' are others (Kelley et al. 2001; Koen et al. 2001; Goffin and Mitchell 2005). This offers another set of search strategies aimed at extending coverage of demand-side signals that might trigger innovation.

The Danish medical devices firm Coloplast has been experimenting with this on the back of its successful use of health professionals as active 'lead users'. Their approach has been to find new communities, or those less well served – for example, in developing-country contexts, where the nature of the context differs radically but where a deep understanding of how their devices might be used in such a context will be critical for entry to those markets. However, in building this understanding there may be important lessons to transfer back to the mainstream – for example, how to make products that are 'good enough' in quality but available at a fraction of the current price in developed markets.

This echoes work by Unilever in the Indian market, where community-based observations led to the development of a reduced-foam detergent for the Indian market where washing is mainly done by hand. This meant that less water was required to rinse clothes, saving up to two buckets of water per wash. Again this innovation has potential to transfer back to more highly developed markets.

When Smith & Nephew shut down most of its projects as a consequence of a major strategic review, the people whose projects had been closed down were sent into the field to observe how their products were used in the market. The ideas they brought back into the organization led to a number of exciting and radical new developments. So successful was the exercise that the company has established a programme under which anyone in the organization can spend a day in the field.

4.6 Probe and Learn Prototyping

There are two complementary dimensions here – the concept of 'prototyping' as a means of learning and refining an idea, and the concept of pilot-scale testing before moving across to a mainstream market. In both cases the underlying theme is essentially one of 'learning as you go', trying things out, making mistakes but using the experience to get closer to what is needed and will work. In the context of DI, where we may simply not know where we are going, probe and learn offers a way of feeling our way through the fog towards a clearer innovation concept. It draws on a rich tradition in design methods aimed to give shape, form and function to new but poorly articulated ideas (Schrage 2000). Trying things out often puts precise words to rather complex ideas and thereby facilitates involvement and validation by stakeholders – enabling 'co-creation'.

Piloting is a way of exploring the potential of an idea without risking large-scale failure or damage to reputation. It means selecting a small but relevant testing ground: small so that it can be controlled, and relevant so that the insights cannot be refuted as not being meaningful. Importantly when applied to DI it is much more than 'traditional' test marketing – it offers a deliberate learning strategy, and experiments may be designed with the prime intention of getting more information about what and what not to do.

One important aspect of 'probe and learn' strategies is that they allow firms to devise experiments to explore alternative hypotheses – for example, looking for opportunities in new or emerging market spaces. They offer a way of dealing with Christensen's 'innovator's dilemma', where new markets emerge around very different value propositions at the fringes of the mainstream (Christensen 1997). Firms that disrupted established markets did so by fast learning around emergent new areas – essentially deploying probe and learn strategies. So if an existing incumbent wishes to anticipate disruptive threats, it would carry out experiments at the fringes of its existing business.

This approach forms the basis of several challenging 'probe and learn' experiments going on at what C.K. Prahalad calls 'the bottom of the pyramid' (Prahalad 2006). His work looks at how radically different products and services emerge when firms try to create them for the vast market of poor people earning below the $2/day poverty line. By radically rethinking the business model – essentially moving to a low-margin, high-volume premise – major innovation opportunities may be identified that could potentially migrate to the mainstream in the fashion Christensen describes. But exploring these new options requires a high degree of probe and learn experimentation. Cases include new communication systems, low-cost sustainable energy systems, low-cost high-quality health care and microfinance models for purchasing consumer goods and services.

This is a common pattern across the high-velocity mobile telecoms environment where new features are introduced, the reaction is gauged and then the feature is either made part of the base offering or rejected. Arvato uses multiple probe and learn experiments around content, user segments and financial models, while O2 reports similar activities around handsets and service packages. In the medical field Novo Nordisk is making extensive use of probe and learn approaches in trying to understand the possible evolution of new diabetes-related services and care pathways that may represent an important new direction for that has hitherto been a 'traditional' pharmaceutical firm with emphasis on drugs and delivery systems. Much of this work is going on in 'laboratories' where very different conditions apply – for example, in Africa where the need is for holistic solutions involving education, infrastructure (clinics and treatment centres) and prevention methods – but all delivered from a very low cost base.

Similar work is going on in the UK with the National Health Service Institute for Innovation and Improvement. One major project forms part of the Design Council RED initiative and has been prototyping new options for dealing with chronic diseases such as diabetes, heart conditions and Alzheimer's disease. As with the Novo Nordisk experiments, the aim is to learn by doing and also by engaging with the multiple stakeholders who will be part of whatever new system co-evolves.

Bang & Olufsen has revitalized its prototyping department and made it refer directly to the innovation hub of the company. The prototyping department is engaged in new ideas as early as possible and experience to date show that this strongly supports the process. After a period with disappointing results in applying electronics in toys, LEGO made a change in its development approach towards more intensive use of prototypes. Prototypes were created within days – often within hours – after the ideas matured. The result was a much more precise dialogue both within the organization and with the main customers. Eventually, this led to simpler technology – and more success in terms of sales.

4.7 Mobilize the Mainstream

A significant issue in DI is that the organization is often already stretched and lacks resources for new and different search activities. One way is to make better or different use of existing resources – to mobilize mainstream players in new or additional roles. For example, it could refocus the core tasks of groups such as procurement, sales or finance staff to pick up peripheral information about trends in the wider world.

The scope within the mainstream is considerable – not only all a company's own staff but the network of other organizations with which it works. In a recent IBM survey of 750 CEOs around the world, 76 per cent ranked business partner and customer collaboration as top sources for new ideas, while internal R&D ranked only eighth. The study also indicated that 'outperformers' – in terms of revenue growth – used external sources 30 per cent more than underperformers. As one CEO put it, 'We have at our disposal today a lot more capability and innovation in the marketplace of competitive dynamic suppliers than if we were to try to create on our own', while another stated simply, 'If you think you have all of the answers internally, you are wrong.'

Firms such as BMW in Germany and Rolls Royce in the UK are using suppliers to create and suggest new ideas to fill gaps in their roadmap, for example through the BMW Virtual Innovation Agency. At Bang & Olufsen a number of inspiration clubs have been formed. Each club has appointed a chair, with the role of facilitator and driver. Although the efficiency of the clubs varies, the set-up ensures that ideas from the whole organization are identified and elaborated.

Reckitt Benckiser has a network of 'internal correspondents' who feed what is happening in their market into a central team whose role is to take this information and combine it with information gathered from other sources. They produce a bulletin that is published every six to eight weeks and sent to select senior management. Larry Wendling, Vice President for Corporate Research at 3M, talks about their 'secret weapon' – the rich formal and informal networking that links the thousands of R&D and market-facing people across the organization.

A variant of the 'mobilizing mainstream' approach is the use of multiple stakeholders – people who are players in the game but who may not always share the same values. Working with them – and using their objections and concerns as a stimulus for new innovation direction – can open up space for experimentation. For example, in recent years there has been a recognition within Novo Nordisk that its investment in corporate social responsibility (CSR) – necessary to secure a continuing licence to operate and helpful in brand building – is also a powerful *innovation* resource. Not only does it offer an opportunity to pick up on weak signals about future directions in the marketplace, but it also puts the company in a position to help shape the emergent new game (Birkinshaw et al. 2007).

4.8 Corporate Venturing

A widely used approach has been the setting up of special units with the remit – and more importantly the budget – to explore new diversification options. Loosely termed 'corporate venture' (CV) units, they actually cover a spectrum ranging from simple venture capital funds (for internal and externally generated ideas) through to active search and implementation teams, acquisition and spin-out specialists, etc.

The purpose of corporate venturing is to provide some ring-fenced funds to invest in new directions for the business. Such models vary from being tightly controlled (by the parent organization) to being fully autonomous. Although they are a popular option, the experience of many firms is that CVs are not always successful; indeed, many fail. Part of the problem is that expectations are often very high, but it isn't always easy to spot the significant new directions in which to invest (Buckland et al. 2003).

There is growing sophistication in the design and operation of such approaches, reflecting the trial and error learning towards effective routines. For example, Unilever has not one but three vehicles by which to draw the benefits from corporate venturing. These are introduced briefly below.

1. A fund that is 40 per cent owned by Unilever, with the rest being owned by banks and other funders. The purpose of the fund is to buy companies from entrepreneurs and see whether they can be scaled up and made more successful.
2. Unilever Technical Venture (UTV), based in San Francisco, which is wholly owned by Unilever. However, investment decisions do not involve the parent organization. The fund's purpose is to invest in early-stage technology start-ups and take a minority stake. There does not have to be a direct fit with what Unilever currently does; the main purpose is to see things around the periphery of their core operations.
3. Unilever Venture, which has a twofold purpose: to invest in technical spin-outs, and to invest in businesses that are close to the core.

Nokia Venturing Organization is focused on corporate venturing activities that include identifying and developing new businesses, or as the organization puts it, 'the renewal of Nokia'. Nokia Venture Partners invests exclusively in mobile and IP-related start-up businesses. It has a very interesting third group called Innovent that directly supports and nurtures nascent innovators with the hope of growing future opportunities for Nokia.

SAP has set up a venture unit called SAP Inspire to fund start-ups with interesting technologies. The mission of the group is to 'be a world-class corporate venturing group that will contribute, through business and technical innovation, to SAP's long-term growth and leadership'. It does so by:

- seeking entrepreneurial talent within SAP and providing an environment where ideas are evaluated on an open and objective basis;
- actively soliciting and cultivating ideas from the SAP community as well as effectively managing the innovation process from idea generation to commercialization;
- looking for growth opportunities that are beyond the existing portfolio but within SAP's overall vision and strategy.

Novozymes is developing a 'Radical Innovation Hub' with a similar remit – to experiment and challenge with new business models and directions. It includes insourcing, acquisitions and idea scouting as part of this mandate. The team currently involves around 30 people worldwide and is trying to grow this by recruiting entrepreneurs into the company (see below).

4.9 Corporate Entrepreneuring/Intrapreneuring

The other side of the CV coin, corporate entrepreneuring, includes various ways of mobilizing high-involvement innovation across the organization. Sometimes called 'intrapreneurship', it tries to build on ideas generated within and across the organization to move it into new areas. Perhaps the classic example is 3M, which has achieved an enviable record of 'breakthrough' innovations that create new categories and markets, and that owe their genesis to the company's commitment to intrapreneurship. Creating the culture to enable this is not simple – it requires a commitment of resources (for example, the well-known case of 3M, which gives its scientists 15 per cent to work on projects of their own choosing) but also a set of mechanisms to take bright ideas forward, including various internal development grants and an increasingly difficult venture funding process. At its heart is a strong incentive scheme for those willing to put their drive and ideas into innovation projects – if these succeed, the originators get to run the businesses they create (Gundling 2000).

Intrapreneurs offer a powerful route to new ideas, but they also provide an implementation pathway to make sure those ideas are taken forward. Many intrapreneurship programmes stress the importance of informal networking, 'bootlegging' and other mechanisms to take ideas forward below the radar screen of formal corporate systems. For example, Novozymes is in the process of building an internal network of entrepreneurs. Besides identifying internal people, it is also recruiting people with entrepreneurial spirit from the outside – partly using the criterion of seeking out those who have built up their own businesses. The organization is aware that such people might be very different and that this could introduce the risk that they would leave after a short period of time. But it decided that the process would still be worthwhile if such people stayed with the organization for two years; this would give them at least that time for inspiration and learning. Within its Radical Innovation Hub there is also scope for migration in and out, so that new ideas can move across from the group to the mainstream.

Within BMW there is a strong commitment to 'bootlegging' – encouraging people to try things out without necessarily asking for permission or establishing a formal project (Augsdorfer 1996). This approach – in BMW these are called 'U-boat projects' – means that people deploy their natural entrepreneurial abilities and often come up with creative solutions. Importantly they also learn ways of getting the attention of the mainstream and managing changes in attitudes. A good example was the Series 3 Estate version, which the mainstream company thought was not wanted and would conflict with the image of BMW as a high-quality, high-performance and somewhat 'sporty' car. A small group of staff worked on a U-boat project, even using parts cannibalized from other cars to make a prototype – and the model has gone on to be a great success and opened up new market space.

4.10   Use Brokers and Bridges

Increasingly organizations are looking outside their 'normal' knowledge zones as they begin to pursue 'open innovation' strategies. But sending out scouts or mobilizing the Internet can result simply in a vast increase in the amount of information coming into the firm – without necessarily making new or helpful connections. There is a clear message that networking – whether internally across different knowledge groups, or externally – is one of the big management challenges in the twenty-first century. Organizations are making growing use of social networking tools and techniques to map their networks and spot where and how bridges might be built – and this is a source of a growing professional service sector activity. Firms such as IDEO specialize in being experts in nothing except the innovation process itself – their key skill lies in making and facilitating connections. A number of new brokers today use the Internet to facilitate innovation. We have already mentioned Innocentive and CommuniSpace in the 'Using the Web' strategy above. Other Web-based brokers are companies such as YET2.com, which provides bridging capabilities for (external) inventors with ideas or concepts to corporate development units.

Examples include P&G, whose 'Connect and Develop' mechanisms including internal websites, gatekeepers and communities of practice through which people can meet and find out about projects and expertise (see http://www.scienceinthebox.com/en_UK/pdf/C_DbrochureFINAL. pdf for more information). The Danish pump-maker Grundfos is trying to use the wide experience across the company to mobilize knowledge via a series of mechanisms including a DI-linked intranet for radical new ideas. Arvato makes extensive use of intermediaries to connect to different and often hard-to-reach market segments and communities. The UK engineering services company Arup has done extensive work on mapping its social networks inside and outside the business to better exploit the connectivity. It has a map of the Arup 'brain' which indicates where connections are made and could be made, and who could engineer such links. And a number of organizations – Coloplast, B&O, Grundfos for example – hold regular events and conferences where the prime purpose is to bring people together and enable networking and potential brokering across what are otherwise diverse organizations.

4.11   Deliberate Diversity

Another strategy in discontinuous innovation is to try to create diversity of vision by hiring different skills and experience sets, or by creating heterogeneous groups and teams within the firm. For example, IDEO hires people from backgrounds as diverse as medicine, engineering, anthropology and physics to create a team with capabilities for developing groundbreaking new ideas (Kelley et al. 2001).

A variation on this theme is to collaborate with 'strange' partners to learn new perspectives – instead of focusing on strategic alliances, firms are exploring 'strategic dalliances' (Philips et al. 2006). Firms are increasingly trying to extend and build new networks with such partners – but this raises questions about how to find them, form relationships with them and make those networks perform effectively (Birkinshaw et al. 2007). At the limit there is considerable scope for learning across sectors and out of industry – for example, some of the apparently radical innovation in health care began life in automobile factories.

For example, when seeking some radical departures from its existing 'playing ground', Danish medical company Coloplast hired an astrophysics PhD to help it think about products of the future. He asked 'stupid questions' as he did not know – nor was he expected – to understand specifics about the industry. Automotive components maker Webasto had an extensive discussion on the topic of '*Querdenker*' (people who think against the grain). It realized that it had stopped recruiting such people – one reason was that they can be quite demanding on resources. It also noted that it sometimes used consultants or other external people to take on the role of a '*Querdenker*'.

4.12 Idea Generators

Last – but by no means least – is the strategy of using creativity tools and techniques to increase the flow of radical ideas. A long tradition of research and the development of powerful tools and techniques has shown that organizations can 'get out of the box' and develop new insights and ideas (Rickards 1997). This can be done internally or through the use of external intermediaries.

Internally, organizations have increasingly been recognizing the potential of using their workforce not simply as pairs of hands but as idea sources, but much of this has been targeted at 'continuous improvement' – essentially 'do what we do but better' (Bessant 2003). However, there is plenty of scope for mobilizing this resource towards more radical options and helping with discontinuous search. Firms such as Siemens, Unilever, Microsoft and others are increasingly reassessing their employee involvement activities in this more radical light.

Rather than using internal resources to do the searching and scanning for potential discontinuities, an increasing number of firms are using an external agency. Such external agents are not necessarily required to produce detailed concepts or ideas, but rather act as early warning systems for weak signals about changing trends (Kingdon 2002). For example, trend agencies provide companies with more general insights on socio-demographic development etc. But the interpretation for the company-specific context does generally still need to be undertaken by someone more familiar with the organizational context. However, an increasing number of design and innovation consultancies offer exploration of future scenarios and investigation of company-specific implications.

5. CONCLUSIONS

A key theme in twenty-first-century innovation is the growing importance of knowledge *flows* across increasingly wide networks. Under 'open innovation' regimes boundaries between firms, sectors and regions become blurred and permeable, and an extensive range of new and profitable connections are made. To take a neurological analogy, significant increases in cognitive capacity emerge as a result of new connections being made.

This raises the question of *how* new connections are found and made – and this 'rewiring' process poses challenges for both users and producers of knowledge. How do firms search for new connections and how do generators – universities, research institutes, etc. – develop different and productive relationships with them? The question assumes a particular significance under what can be termed 'discontinuous' conditions where established knowledge sets and the links that deliver them may be insufficient or redundant. The search problem here is a particularly acute case of the general open innovation challenge.

This chapter has outlined some of the emerging new search strategies being deployed by organizations trying to develop capabilities in dealing with turbulent and unpredictable environments. Many of these take the form of experiments and have yet to become formalized into routines, and there is considerable scope for elaboration and for developing firm-specific competitive advantage through them. They are also not necessarily new (although many of the more creative uses of the Internet represent significant novelty in search opportunities and spaces unavailable in previous eras), but they are being used in a much more systematic and integrated fashion.

The focus in this chapter has been on firms as users of new knowledge and how they seek it out. But there is an implied challenge in this for the 'supply side' of knowledge production – the universities and research institutes that generate and hold large stocks of new knowledge. Just as user firms are finding that their established search routines are not effective under discontinuous conditions and are being prompted to experiment with new search behaviours, so knowledge creators will increasingly need to review the ways in which they connect to users and form effective networks.

There is also a challenge for policy agents – those responsible at institutional, region or national level for building effective innovation systems in which knowledge flow can take place to generate social and economic value. Rewiring – or extending the wiring – of these systems by experimenting with novel forms of connection (and how these can be facilitated) should be an important agenda item for the future.

REFERENCES

Augsdorfer, P. (1996), *Forbidden Fruit*, Aldershot: Avebury.
Bessant, J. (2003), *High Involvement Innovation*, Chichester: John Wiley & Sons.
Birkinshaw, J., Bessant, J. and Delbridge, R. (2007), 'Finding, forming, and performing: creating networks for discontinuous innovation', *California Management Review*, **49**(3): 67–83.

Buckland, W., Hatcher, A. and Birkinshaw, J. (2003), *Inventuring: Why Big Companies Must Think Small*, London: McGraw-Hill Business.

Burt, R. (1992), *Structural Holes: The Social Structure of Competition*, Cambridge, MA: Harvard University Press.

Chesbrough, H. (2003), *Open Innovation: The New Imperative for Creating and Profiting from Technology*, Boston, MA: Harvard Business School Press.

Christensen, C. (1997), *The Innovator's Dilemma*, Cambridge, MA: Harvard Business School Press.

Cohen, M., Burkhart, R., Dosi, G., Marengo, L., Warglien, M. and Winter, S. (1996), 'Routines and other recurring patterns of organization', *Industrial and Corporate Change*, **5**(3): 653–98.

Costcllo, N. (1996), 'Learning and routines in high tech SMEs', *Journal of Economics Issues*, **30**: 591–7.

Cyert, R. and March, J. (1963), *A Behavioral Theory of the Firm*, Engelwood Cliffs, NJ: Prentice-Hall.

Day, G. and Schoemaker, P. (2006), *Peripheral Vision: Detecting the Weak Signals that will Make or Break your Company*, Boston, MA: Harvard Business School Press.

De Geus, A. (1996), *The Living Company*, Boston, MA: Harvard Business School Press.

Delbridge, R. (2004), 'How motorsport companies collaborate and share knowledge', London: AIM and Government Motorsport Unit.

Dosi, G. (1982), 'Technological paradigms and technological trajectories', *Research Policy*, **11**: 147–62.

Goffin, K. and Mitchell, R. (2005), *Innovation Management*, London: Pearson.

Granovetter, M. (1973), 'The strength of weak ties', *American Journal of Sociology*, **78**: 1360–80.

Gundling, E. (2000), *The 3M Way to Innovation: Balancing People and Profit*, New York: Kodansha International.

Hargadon, A. (2003), *How Breakthroughs Happen*, Boston, MA: Harvard Business School Press.

Henderson, R. and Clark, K. (1990), 'Architectural innovation: the reconfiguration of existing product technologies and the failure of established firms', *Administrative Science Quarterly*, **35**: 9–30.

Huston, L. and Sakkab, N. (2006), 'Connect and develop: inside Procter & Gamble's new model for innovation', *Harvard Business Review* (March): 58–66.

Kelley, T., Littman, J. and Peters, T. (2001), *The Art of Innovation: Lessons in Creativity from Ideo, America's Leading Design Firm*, New York: Currency.

Kingdon, M.E. (2002), *Sticky Wisdom: How to Start a Creative Revolution at Work*, London: Capstone.

Koen, P.A., Ajamian, G., Burkart, R., Clamen, A., Davidson, J., D'Amoe, R., Elkins, C., Herald, K., Incorvia, M., Johnson, A., Karol, R., Seibert, R., Slavejkov, A. and Wagner, K. (2001), 'New concept development model: providing clarity and a common language to the "fuzzy front end" of innovation', *Research Technology Management*, **44**(2): 46–55.

Markides, C. (1997), 'Strategic innovation', *Sloan Management Review*, Spring: 9–24.

Nelson, R. and Winter, S. (1982), *An Evolutionary Theory of Economic Change*, Cambridge, MA: Harvard University Press.

Philips, W., Lamming, R., Bessant, J. and Noke, H. (2006), 'Discontinuous innovation and supply relationships: strategic dalliances', *R&D Management*, **36**(4): 481–91.

Prahalad, C. (2004), 'The blinders of dominant logic', *Long Range Planning*, **37**(2): 171–9.

Prahalad, C.K. (2006), *The Fortune at the Bottom of the Pyramid: Eradicating poverty through profit*, Upper Saddle River, NJ: Wharton School Publishing.

Rickards, T. (1997), *Creativity and Problem Solving at Work*, Aldershot: Gower.

Schrage, M. (2000), *Serious Play: How the World's Best Companies Simulate to Innovate*, Boston, MA: Harvard Business School Press.

Schwartz, P. (1996), *The Art of the Long View: Planning for the Future in an Uncertain World*, New York: Doubleday.

Simon, H. and March, J. (1992), *Innovations*, Oxford: Basil Blackwell.

Tripsas, M. and Gavetti, G. (2000), 'Capabilities, cognition and inertia: evidence from digital imaging', *Strategic Management Journal*, **21**: 1147–61.

Tushman, M. and Anderson, P. (1987), 'Technological discontinuities and organizational environments', *Administrative Science Quarterly*, **31**(3): 439–65.

Utterback, J. (1994), *Mastering the Dynamics of Innovation*, Boston, MA: Harvard Business School Press.

Von Hippel, E. (2005), *The Democratization of Innovation*, Cambridge, MA: MIT Press.

Zahra, S.A. and George, G. (2002), 'Absorptive capacity: a review, reconceptualization and extension', *Academy of Management Review*, **27**: 185–94.

Zollo, M. and Winter, S.G. (2002), 'Deliberate learning and the evolution of dynamic capabilities', *Organization Science*, **13**(3): 339–51.

10. Accelerating diffusion among slow adopters

Richard Adams and John Bessant

INTRODUCTION

While a systematic approach to product innovation has long been recognized as critical for firms' continued growth, even survival, many UK organizations overlook or choose not to pursue the opportunities offered from also innovating in their business processes. Emerging from fields of research and practice in recent years are sets of innovative practices that have been demonstrated to deliver benefits to those adopting organizations that deeply embed them. Included among these are world class manufacturing (WCM) techniques (Schonberger 1986), which have a track record of demonstrable success. Through their effective adoption and implementation, WCM techniques offer the promise of significant performance improvements, particularly in terms of exploitative innovation and efficiency gains (Stoneman and Kwon 1996; Benner and Tushman 2003; Montes and Jover 2004).

However, in spite of this promise, the adoption of WCM techniques has been slower than might be expected. This apparent failure, or reluctance, to adopt and embed has been cited as one reason for the UK's performance gap relative to important international competitors (Porter and Ketels 2003). Beginning in the 1990s, a series of studies looking to firm-level discrepancies for answers to international variations in productivity performance has consistently found UK firms to be slower at adopting process and practice innovations than their international counterparts (see, *inter alia*, Hanson and Voss 1995; Voss 1995; Waterson et al. 1999; Rigby 2001; Clegg et al. 2002; Lucking 2004; Wood et al. 2004). Accordingly, the potential benefits of WCM adoption for the UK economy have not been fully realized.

Consequently, the UK government has long been interested in devising policy to encourage and facilitate domestic organizations to take on these modern management techniques. Such policy has frequently been premised on the notion of managers as rational decision makers assuming that, if they are made aware of the potential benefits of the practices, then adoption will surely follow. The 'long tail' of slow-to-adopt organizations would appear to suggest that such policies have not been widely successful. To promote accelerated adoption of such practices, it is therefore important for government to understand and clarify the dynamics and mechanisms that facilitate the successful diffusion, adoption and implementation of WCM.

Following West and Farr's (1990: 6) definition of innovation as 'the intentional introduction and application within a[n] . . . organization of ideas, processes, products or procedures, new to the relevant unit of adoption, designed to significantly benefit the individual, the group, organization or wider society', innovation diffusion theory can be used as a framework to consider the reasons behind the slow spread of WCM and its implications for policy.

The current chapter considers the issue of accelerating adoption among a population of UK organizations, and is structured as follows: first, we discuss WCM and the nature of the long tail in the context of innovation diffusion theory. Next, we propose a typology of government- or institutionally sponsored modes and mechanisms that captures the evolutionary history of interventions in support of adoption and diffusion. The typology comprises five elements, including: *laissez-faire*, self-help, broadcast, agent-assist and peer-assist modes. In the penultimate section of the chapter we discuss reasons for the success of a recent intervention in the UK, the Industry Forum initiative, within the context of this typology. Finally, the chapter closes with some reflections for policy implications.

DIFFUSING WCM

We define WCM as combining a customer-focused, quality-oriented philosophy emphasizing waste minimization and continuous improvement with a set of practical techniques associated with organizing and managing manufacturing operations. Adopting and implementing WCM techniques can be a complex affair for many organizations; Eveland and Tornatzky (1990) would describe it as a 'lumpy' innovation. Characteristically, their impact is wide and enduring, throughout the socio-technical context of an organization. Further, they are characterized by their high knowledge and abstract content, making it hard for potential users first to trial and later to implement, frequently needing to be modified or adapted for local conditions.

By aggregating adoptions over time, the pattern of spread of an innovation through populations can be presented graphically as a diffusion curve. Numerous studies note an empirical regularity in the tendency of diffusion curves to present sigmoidally, that is, S-shaped (Mahajan and Peterson 1985); see Figure 10.1. The curve naturally depicts three separate regimes of diffusion activity. The first is characterized by the curve's shallow gradient

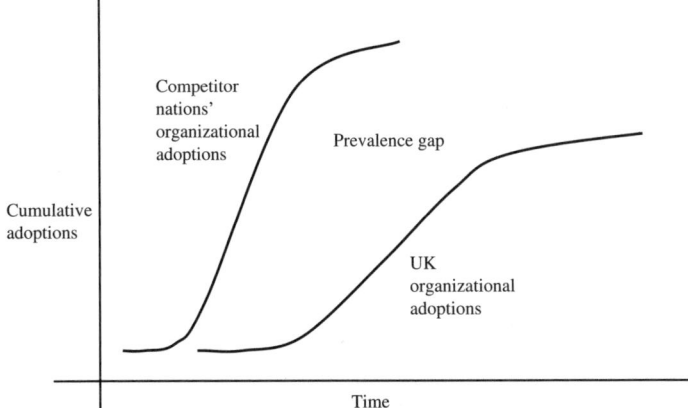

Figure 10.1 The WCM prevalence gap: UK versus international competitors

at the left-hand side, marking an innovation's first introduction and gradual acceptance. This is followed by a rapid steepening of the curve as the innovation is taken on by the majority. Finally, to the right, the curve begins to flatten, indicating a slowing down in the rate of adoption to the point where either the market is saturated or prevalence, the point at which all potential adopters who are going to adopt have adopted (Valente 1995), is reached. In the case of the UK, the prevalence of WCM is lower than in many competitor nations (see above). Figure 10.1 illustrates this prevalence gap: UK firms, in general, have started to adopt later, in fewer numbers and less rapidly than international competitors.

In previous studies, scholars have identified different categories of adopter along the diffusion curve, ranging from Moore's (1991) *early adopters* and *mainstream adopters* to Rogers's (2003) *innovators, early adopters, early majority, late majority* and *laggards*. Each of these typologies is underpinned by an assumption of diffusion under market conditions. However, another category can be postulated, one consisting of slower adopters who are reluctant to adopt except by means of the facilitating hand of external support, typically by government or their agents (Bessant and Rush 1993). However, later or slower adoption is not simply the obverse of early adoption and, to facilitate and accelerate adoption among this community, promoters of WCM must take special account of the needs, context and characteristics of this group. Diffusion theory provides a framework to facilitate this (see Figure 10.2).

Diffusion theory is a diverse and distributed body of work consisting of several competing explanatory models which can be broadly organized into

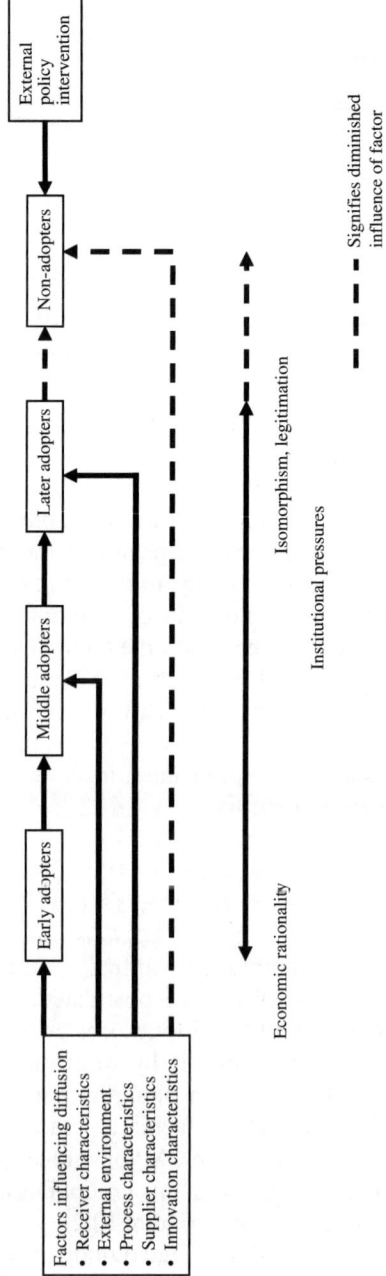

Figure 10.2   Factors influencing the adoption of innovations

two categories, the communications model (Rogers 2003), and the economically rational model (Mansfield 1968). In the former, diffusion is likened to the spread of a virus through a population, in which the rate of adoption is a function of the relative proportions of the population that are 'infected' or 'uninfected' by communications with prior users. In the latter, diffusion is principally related to a cost–benefit analysis by potential adopters. Within these perspectives, five overarching factors have been identified as important in understanding the dynamics of adoption and diffusion: adopter characteristics; innovation characteristics; supplier characteristics; process characteristics; and environmental context.

Adopter Characteristics

A range of organizational characteristics has been associated with adoption, and the suggestion is that organizational heterogeneity on these dimensions can explain adoption variance. That is, different firms with different goals and abilities are likely to want to adopt new technologies at different times (Geroski 1999). A number of these factors lie beyond the influence of government, such as organizational size (Camison-Zornoza et al. 2004) and organizational slack (Nohria and Gulati 1996), and so are not considered here. However, important from the policy perspective is the concept of organizations' need and readiness for and receptivity to change (Cunningham et al. 2002; Eby et al. 2000; Jansen 2000).

Many organizations simply fail to recognize that there is any need for change (Bloom and Van Reenen 2005). That is, there is variance in the extent to which organizations are open or receptive to change and value it as a legitimate objective (Siegel and Kaemmerer 1978). The implication for interventions of this is that, first, owners and managers must be convinced of a need for change and, second, as the corollary of low receptivity to change appears to be an unsupportive organizational climate for change (Holahan et al. 2004), significant efforts must be made to endow employees with the spirit of change, and develop the structures and norms to allow for and facilitate implementation. Similarly, adopters require the capability to absorb (Cohen and Levinthal 1990) and deploy the innovation through their organizations by means of championing and management support (Sohal 1996).

Innovation Characteristics

Innovations differ one from another, but research has also demonstrated that the same innovation can mean different things across organizations (Krackhardt 1997); that is, for some organizations it is more difficult to implement a particular innovation than for others. Rogers (2003) suggested that organizations will find innovations more readily adoptable where they are perceived to have a greater degree of compatibility, observability, relative advantage and trialability, and lower levels of complexity.

The greater the adopter–innovation compatibility, in terms of existing values, past experiences and perceived needs, the greater the likelihood of adoption (Thio 1971). Arguably, as a reframing of the 'dominant design', WCM requires a major adaptation on the part of users.

Observability of an innovation is the degree to which its results are visible to others. The principal component of WCM is ideational, and the more ambiguous the knowledge, the more difficult it is to transfer (Simonin 1997), and structural mechanisms (e.g. training, internal consulting and assistance) can affect the degree of knowledge transfer. Seeing proven results in other similar organizations can help overcome reluctance.

Relative advantage is the degree to which an innovation is perceived as being better than the idea it supersedes. In the absence of a physical artefact it is arguably difficult to see how organizations might conceive the relative advantage for process innovations. In spite of accumulated evidence indicating the benefits of adoption, not every case has been successful, particularly where adoption has been only nominal rather than embedded within organizational practice (Benner and Tushman 2003). Indeed, WCM may be perceived by potential adopters as simply another management fad (Abrahamson 1991). Promoters of the innovation must be sensitive to this history and adequately communicate the benefits to potential adopters.

Trialability is the degree to which an innovation may be experimented with. Organizational concepts require 'interpretative viability' in order to be successful (Benders and Van Bijsterveld 2000); that is, concepts must lend themselves to different interpretations and be modifiable according to contextual requirements. Early adoptions of WCM were flawed by the belief in a single model's universal applicability and failed to take account of contextuality and contingency. To be successful, WCM demands flexibility in its implementation. This is particularly significant in the case of socio-technical innovations such as WCM, whose adoption impact can ripple throughout the firm. Early adopters of robotic arms, for instance, experienced unforeseen effects: the necessity of running three-shift systems to cover the cost of the equipment (affecting task organization); the requirement for technicians to be on hand to cover breakdowns; and improved precision engineering of components to accommodate the repeated accuracy and low error tolerances of welding arms.

Complexity refers to the extent to which the innovation is perceived as difficult to understand or absorb. In the case of WCM it represents a very different view and its Japanese origins, labelling many core practices with unfamiliar language such as *kaizen* and *kanban*, may mean that many firms dismiss it as too complex and different. This makes it unlikely that they will accept a new practice that renders obsolete the taken-for-granted assumptions that have become organizationally embedded as a result of persistent performance of previous practices (the previous operating paradigm).

Supplier Characteristics

Supplier or promoter characteristics are important in helping improve rates of adoption, but, relative to the other characteristics discussed here, have been neglected in the literature (Greenhalgh et al. 2004). Many organizations will defer adoption until knowledge barriers relating to the innovation have been lowered. This leaves open the question of how and by how much these knowledge barriers might, should or could be reduced by promoting agents (Attewell 1992). Adopting organizations see similar organizations as more relevant and easier to learn from, and so diffusion can be amplified where there is similarity of social, organizational or strategic characteristics between the promoter and receiver (Brass et al. 2004), and organizations that are very different may not be as likely to imitate, transfer or learn practices from one another as organizations that are similar (Lane and Lubatkin 1998).

Consequently, knowledge is more likely to be transferred between people and organizations with similar knowledge, training and background characteristics (Reagans and McEvily 2003). Rogers (2003) calls this 'homophilitic learning', the tendency for people to listen to and interact more with their own kind (either by preference or prejudice) than with others.

The lesson for policy agents is, then, to design interventions with clear homophilitic credibility, derived from interaction with similar others, in order to: (1) facilitate transmission of tacit knowledge; (2) simplify coordination; and (3) avoid or lessen potential conflicts (Borgatti and Foster 2003). In particular, HR relations should be positive and supportive; the two systems should have a common language, meanings and value systems, and should enable and facilitate networking and collaboration among organizations; and the consequences of innovation should be jointly evaluated (Greenhalgh et al. 2004).

Process Characteristics

One of the fundamentals of the classical communications model is that later adopters lag behind earlier adopters because they learn later about the existence of the innovation. Knowledge and technical know-how are important barriers to diffusion, and supply-side institutions have to innovate in the development of novel mechanisms for reducing the knowledge and learning burden on end users (Attewell 1992). Several targeted communications and risk reduction activities of promoting institutions have been associated with innovation adoption (Frambach and Schilliwaert 2002). In the case of slower adopters, these activities need to take on different characteristics if supplying to innovative organizations predisposed to adopt. Communications strategies need, for example, to emphasize the benefits of the innovation and address the attributional characteristics that are detrimental to adoption. Risks may be reduced by designing opportunities for trial, or at least making it possible to gain experience of the innovation in action or by reducing financial risk of adoption. This suggests a process that brings supplier and receiver closely together. With the notion of 'user innovation' Douthwaite et al. (2001) suggest an alternative to the insinuation of adopter passivity in the communications model; they emphasize co-production in the process of innovation implementation in which both supplier and receiver are fully engaged.

Environmental Context

The external environment of an organization is generally held to comprise forces beyond the control of the organization, and includes influences of a social, political, regulatory, competitive and technical nature. Organizations in highly variable environments require higher levels of innovation (Burns and Stalker 1961; Lawrence and Lorsch 1967), and pressures from these external sources for adherence to conformity can also drive adoption (Westphal et al. 1997).

Firms feel the pressure of these forces more or less keenly, and the institutional perspective (DiMaggio and Powell 1983) suggests that those who respond early to these external pressures do so for reasons of technical or economic rationality, while those who respond later do so for reasons of external or social legitimacy. That is, if the first to adopt an innovation improves their performance as a consequence, followers-on will try to institutionalize the same practices in the same way (Montes and Jover 2004), although the benefits sought and derived from later adoption might incline more towards social than economic approval (Staw and Epstein 2000).

In the UK, early adoptions of WCM tended to be limited to the automotive and electronics sectors, particularly among those organizations with affiliations of some sort with Japanese and North American firms. Consequently, isomorphic pressures for adoption in other sectors were, initially, weak. Also, isomorphic pressures tend to normalize towards configurations implemented by early adopters. As some early adopters reported mixed successes, for example having only nominally adopted the practices or having failed to embed some critical components (Newell and Swan 1993), isomorphic pressures were weak.

In the absence of some external stimulus and in stable conditions, or because of reports of failed implementations, potential adopters may deliberately choose to delay adoption (Womack et al. 1990). For reasons of economy, avoidance of the disruption of change, organizational reluctance or market conditions, non-adoption may be a rational response (Beck and Walgenbach 2005; Bloom and Van Reenen 2005). So, despite isomorphic pressures towards social legitimacy, it appears that contextual factors may be more significant and reinforce the status quo. And, while firms that adopt early can create some advantage over others' slower adoption, promoters must find original modes and mechanisms to facilitate diffusion of WCM that address the poor functioning of the market, manage beliefs and attitudes of potential adopters, and configure the innovation appropriately for their specific needs and context.

POLICY RESPONSES

Since the second half of the twentieth century, at least, there have been a number of policy initiatives directed towards shaping the rate and direction of innovative activity through sustained promotional activities. Over time, there has been a recognizable evolution in the nature of these interventions, due both to the recognition of limitations of previous initiatives and as a response to the increasingly complex nature of the practices being diffused. Within this evolution, four distinct modes of intervention can be identified: *broadcast*, *self-help*, *agent-assist* and *peer-assist* (see Table 10.1), the alternative to which is to leave diffusion to take place under market conditions. Many of the earliest initiatives were only loosely targeted, a 'scattergun' approach, in terms of both the ideas being promoted and the potential adopters targeted, predicated on a pipeline model down which new knowledge flowed, without significant modification, to passive adopters. It was not until the 1970s that schemes aimed at diffusing specific technologies or techniques to specific user groups began to emerge.

Broadcast

'Broadcast' refers to knowledge deployment in its simplest form to potential recipients. Emphasis is on a one-way flow of information from a source to a passive recipient. The *'broadcast'* mode represents the type of policy geared to improving the flow of information to potential adopters – and can

Table 10.1　Outline typology of communication modes to accelerate adoption

Mode	Deployment options within mode	Policy examples	Comments
Laissez-faire Leave it to the market (no intervention)	–	–	Evidence of 'market failure'. Despite WCM being available for 25 years, there is still a long tail of non-adopters
Broadcast mode Raise level of awareness by different ways of letting target population know	Non-specific – general broadcast awareness campaign	'Manufacturing into the 90s', DTI/CBI 'Fit for the future' campaign	Evidence suggests this raises general level of awareness but does not deal with specific concerns of individual firms; nor does it help with configuration and adaptation issues (Hobday et al. 2004)
Self-help mode	Users self-audit against a checklist to identify gaps against 'ideal' model	Innovation Management Toolkit, Innovation Your Move, MINT	Provides a toolkit, but degree of external facilitation varies (and tends to be limited)
Agent-assisted mode Use of agents to engage with the adopter to help frame and explore the promising practice	'Missionary' work to take the message to isolated potential adopters. Consultancy support for: articulation of need; exploration of options; and configuration and implementation	Innovation counsellor's model – Business Links (UK), IRAP (Canada), TEKES (Finland) etc. Support for professional consultants to undertake this under broad regulatory framework offered by policy agent, e.g. Enterprise	Evidence suggests such assistance can help deal with changing perceptions of key innovation attributes (Northcott and Bessant 1986; Rush et al. 1994)

Table 10.1   (continued)

Mode	Deployment options within mode	Policy examples	Comments
		initiative, Inside UK Enterprise	
Peer-assisted mode Use of approaches that encourage learning from and with others	Vicarious experience sharing via learning by visiting and questioning 'people like us' Sectoral firm-to-firm programmes incorporating, e.g., 'Masterclass' models or peer-to-peer learning Regional programmes (clusters and best-practice clubs) Topic clubs, e.g. CIRCA, JIT clubs Supply chain learning	SMMT Industry Forum and other forums established in aerospace, textiles, oil and gas, etc. Supply chain learning initiatives	Evidence suggests peer-assisted learning not only deals with perceived attribute problems but also with aspects of the social system such as homophilitic learning (Dyer and Nobeoka 2000; Bessant et al. 2003)

make use of a number of different mechanisms. Early UK government policies designed to promote innovation were often simply letters and circulars from the relevant ministries. In the last 25 years, these have grown more sophisticated. For example, the 'Managing in the 90s' programme and its successor 'Fit for the future' (operated by the DTI and the Confederation of British Industry) made use of public presentations, road shows, breakfast briefings, television and radio, video, CD and DVD, and extensive web presence to promote innovation to a manufacturing audience (DTI/CBI 2001). Much of this was targeted, both in terms of the technologies/techniques being promoted and in the sectors or firm types being addressed.

Self-help

This mode articulates a transition between *broadcast* and *agent-assisted* modes. The knowledge being transferred is more targeted and the mechanisms typically consisted of workbooks, self-assessments, score-cards and training materials (Voss et al. 1994). The value of the approach lay in organizations being able to self-diagnose, enabling them to identify gaps in their own capabilities and areas in need of development. Best value appeared to derive from instances where the process was facilitated by external agents, an addition that became more fully developed in the *agent-assist* mode.

Agent-assist

This mode tends to be predicated on the concerns potential adopters have with configuration and compatibility. They may have a general awareness of the potential of the innovation on offer, resulting from broadcast efforts, but do not see its relevance or applicability to them. In terms of innovation characteristics there is little *perceived* relative advantage, there may be high perceived complexity, there may be major concerns about compatibility, and slow adopters may be unconvinced or unwilling to undertake trials. Agent-assist mechanisms have become an increasingly important element in many national and regional innovation policy frameworks and have had demonstrable influence in capability development and in accelerating take-up of new techniques (Bessant and Rush 1995).

Peer-assist

Understanding the potential value of working within the social system underpins the third approach, which emphasizes firms learning from and with each other. Such learning mechanisms are based on the principle that adoption can be accelerated by engaging with 'people like us' (e.g. similar industry, similar problems, etc.), embracing the idea of homophily. As with earlier diffusion studies (such as those on the adoption of hybrid corn varieties in US agriculture, e.g. Ryan and Gross 1943, 1950; Griliches 1957), the potential of people in a similar context to act as catalysts and accelerators of adoption is significant. Applied at the firm level it suggests that potential adopters will be persuaded by both broadcast messages and agent-led intervention communicated by individuals and organizations perceived to be from similar contexts. This has relevance for several dimensions of the diffusion problem – for example, peer-assist mechanisms can help convert a generic set of attributes into something of specific relevance and interest for the potential adopting group. Relative advantage perceptions can be shaped, issues and concerns about compatibility can be addressed, and the influence of observability can be enhanced. This fits with the institutional perspective outlined earlier in which mimetic isomorphic pressures become increasingly significant in later stages of the diffusion curve. That is, peer-assist mechanisms define and mobilize the population of peers who can create these pressures.

In the UK, a major sectoral initiative congruent with the peer-assist mode has successfully addressed the issue of diffusing complex innovations such as WCM to slower adopters. The 'Industry Forum' (IF) initiative originated as a sector-level activity in the automotive components field, and is briefly reviewed below.

THE UK INDUSTRY FORUM INITIATIVE

Industry Forum was a DTI-sponsored initiative to help whole industry sectors introduce best practice based upon the MasterClass techniques successfully used by the Society of Motor Manufacturers and Traders (SMMT) to transfer practices within the automotive supply chain. Its success in the automotive sector led to more widespread promotion as a policy option, and in the 2000 White Paper on Competitiveness, provision was made to launch up to 13 other programmes in different sectors, all with the aim of promoting WCM (DTI/CBI 2001; Bateman and David 2002). Subsequently, the forum concept was adapted to deliver the same benefits across a range of UK industrial sectors, including: aerospace, ceramics, construction, furniture, tourism and hospitality, red meat industries, shipbuilding. Each sectoral forum is led by a sector trade association or representative organization, and at its core the MasterClass process improvement programme addressing the implementation of lean principles and a culture of continuous improvement among client members. Within this guiding framework, programmes are targeted at the particular needs and abilities of client companies.

MasterClass programmes, for which organizations are required to pay (though subsidy was provided in some sectors), are led by 'Master Engineers', industry-trained individuals with backgrounds and expertise in process engineering. In a structured 15-day programme staged over a period of approximately six months, Master Engineers work with employees on process improvements in a 'learn-by-doing' fashion, the focus of which is to eliminate waste and maximize the potential of available resources. Participants are drawn from all disciplines and levels within host companies, working under the guidance of Master Engineers to undertake process improvement on the shop floor. The structure of the programme is, however, flexible enough to accommodate individual contingencies.

The MasterClass process follows a systematic and consistent pattern (see Figure 10.3). An improvement team made up of company personnel is selected and this team works with the Master Engineers to identify and

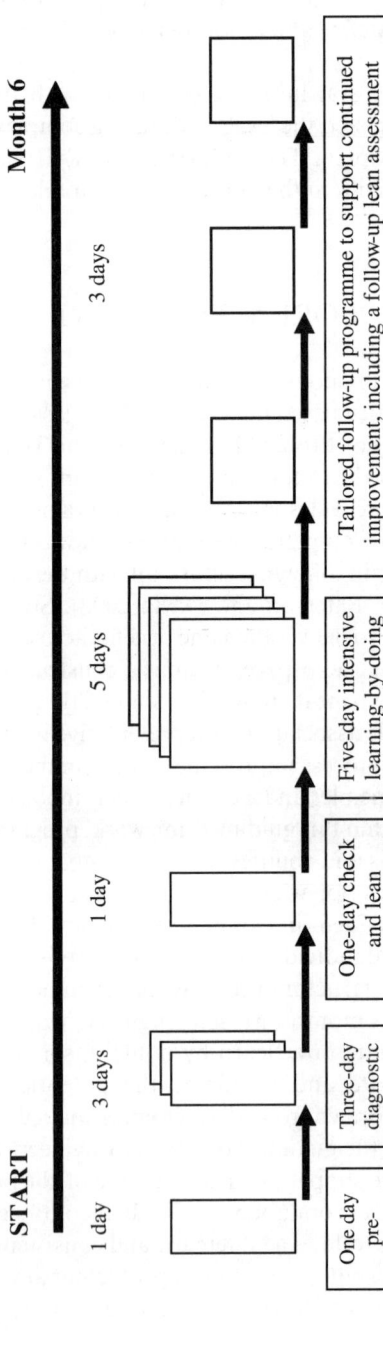

START

Month 6

1 day | 3 days | 1 day | 5 days | 3 days

One day pre-diagnostic

Three-day diagnostic

One-day check and lean assessment

Five-day intensive learning-by-doing workshop

Tailored follow-up programme to support continued improvement, including a follow-up lean assessment

Source: SMMT (2006).

Figure 10.3 MasterClass programme

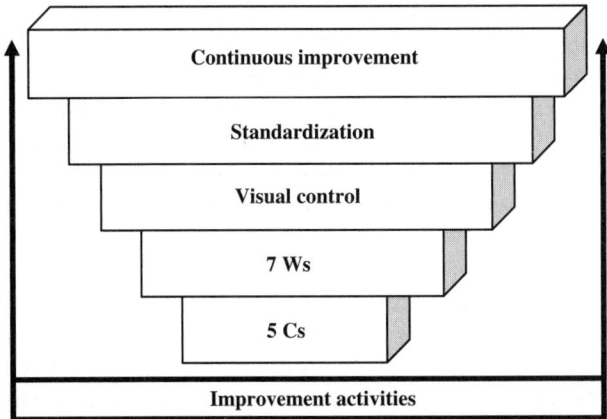

Source: SMMT (2006).

Figure 10.4 *The MasterClass lean toolkit*

implement improvement. Each MasterClass activity involves a pre-diagnostic day to agree the target area for improvements; a three-day diagnostic phase to gather data and systematically analyse the performance problems identified; a five-day implementation workshop and three separate follow-up days. The implementation workshop brings the building blocks of five controls (5C), seven wastes (7W), visual control, standardization and the philosophy and practice of continuous improvement (see Figure 10.4), as appropriate, into play to make process improvements. Improvement teams are trained in the use of specific tools such as Pareto analysis, flowcharting and one-piece flow. During subsequent follow-up days, the Master Engineers continue to support the team in process improvement with the aim of encouraging the development of a culture of continuous improvement within the target area and the wider site.

For example, a team of individuals drawn from across the participating organization addresses a live organizational problem over the duration of the intervention to demonstrate the efficacy of the approach. Successes and benefits percolate rapidly through the organization because of the involvement of individuals across multiple layers of the organization, and these provide a convincing motivation for senior management to continue to pursue the initiative. Such multi-level involvement also helps identify early likely organization-wide impacts.

As Figure 10.4 demonstrates, the MasterClass operates on the basis of multiple mechanisms to engage individuals and enable adoption. During the pre-diagnostic and diagnostic sessions, the Master Engineer works closely with employees identifying areas for improvement. Typically, this will involve a process of benchmarking to define and measure improvements. During the workshop stage, five fundamental philosophies and practices are drawn upon and targeted towards the identified areas for improvement. Depending on a particular company's requirements, capacity and capability, not all of the tools may be implemented. The subsequent, tailored follow-up programmes, spread over several months, are critical to the sustained implementation of the practices and to encourage continued culture change within the organization.

One important feature of the process is the collection of company data to develop the starting-point benchmarks. This diagnostic element of the programme appears to be significant in that it fully engages employees, the within-company change agents, with the nature and size of issues facing the company. That is, it is not 'merely' an administrative task, a filed report; rather, change agents are confronted with the magnitude and significance of the task they are facing (WEAF 2005a). It also serves as the basis for highlighting limitations and weaknesses of existing working practices.

Although such data are widely available outside peer-assist schemes, the exposure to benchmarking in the company of peers exerts a strong isomorphic pressure for change which underpins and reinforces the adoption decision. Exposure to others' experiences, together with phases of facilitated learning-by-doing, enables the development of local configuration to suit particular contexts and deals with many of the perceived compatibility questions raised by WCM. These mechanisms include a high level of people-based support, for example through the loan of engineers and other experienced personnel as transfer agents.

The success of Industry Forum in general and the MasterClass concept in particular to support the transfer and sustained implementation of innovative management practices can be attributed to several factors:

- Contextualizing abstract knowledge to give it local meaning. This issue is addressed particularly in the diagnostic phase, giving attention to company needs and directly addressing any misgivings owners, managers and employees may have about the benefits and advantages (i.e. innovation characteristics) of adoption.
- Implementation facilitated by the Master Engineer directly addresses the critical issue of transferor credibility. Previous interventions have had less success because of the lack of credibility. Master Engineers' backgrounds in industry lend personal and technological credibility.
- Locating forums within representative trade organizations has helped establish credibility at the sectoral level. Executive and management boards of these representative organizations frequently consist of members drawn from community practitioners who, by

means of their endorsement and (often) utilization of the initiative, establish sectoral-level compatibility.

- Widespread engagement of individuals from all levels of the business ensures that the activities are transferred throughout, rather than being imposed by dictat. This recognizes the social process of learning, places internal change agents at the core of the programme's design, and is particularly important in sustaining implementation of culture change within the organizations (Chedzey and Kennedy 2003).
- Recognizes innovation as a social process. The MasterClass set-up not only facilitates the transfer of the technology of WCM, but also provides means and mechanisms of support for the individual change agents in support of the implementation of WCM.
- Learning-by-doing. MasterClasses consist of a mix of theoretical and classroom activities with learning-by-doing. Because the learning was relevant to individuals' jobs and organizations' identified problems, participants were able to identify, address and resolve issues directly meaningful to their own situation.
- Resource commitment. Compared to previous initiatives, resource commitment for the MasterClass is high (though sometimes subsidized, thus helping to overcome the barrier of perceived high costs). MasterClasses include considerable group and organizational support.
- Recognizing the socio-technical nature of WCM, Brown et al. (2003) stress the importance of rapport and mutually supportive working relationships between the Master Engineers and MasterClass participants. The use of a wide range of learning methods helped improve commitment towards learning. These methods included: participation in production process improvement reviews and implementation; Master Engineer workshops; group discussions; assignments; portfolio building; discussions with tutor; use of computer-mediated communications for discussions, document transfer and tutor feedback.

DISCUSSION

The arguments regarding whether or not intervention is justified or legitimate are complex and beyond the scope of this chapter but, where diffusion deviates from the 'welfare optimal path', then policy intervention is desirable (Stoneman and Diederen 1994). Thus, much innovation policy is predicated on concepts such as 'market failure', where intervention is justified on the basis of bridging gaps, and evaluated in terms of 'additionality', where change takes place that would otherwise not have happened (Dodgson and Bessant 1996).

Interventions such as have been described are purposefully designed to engage organizations reluctant to adopt. Reasons for reluctance can be myriad: from fatalistic scepticism in the face of mounting international competitive pressures to simple lack of desire to grow or change because historically the business has performed satisfactorily. However, today a firm's survival is increasingly dependent on its ability to innovate, not radically but incrementally: focusing on 'doing what we do but doing it better'. WCM offers one route to this objective.

Directly or indirectly governments become involved in innovation through the development and application of different policy tools directed towards shaping the supply side, demand side or the environment. WCM techniques originally began to diffuse in the UK as a result of demand-pull factors, inward investors (notably in the automobile industry) cascading best-practice expectations down through the domestic supply chain (Wickens 1987; DTI 1994; Stoneman and Diederen 1994). However, demand-pull proved to be insufficient on its own to generate widespread diffusion, and consequently the UK government has provided a series of supply-push initiatives.

The success of recent interventions has depended on more than simply the transmission of technical facts, upon which earlier 'broadcast mode' initiatives relied. Evidence in the academic (e.g. Haque 2003; Tilson 2001; Venables 2004; Bateman 2005; Crute et al. 2003) and grey literature (e.g. Bevis 2004; Chedzey and Kennedy 2003; WEAF 2005b, 2006) points to the success of the Industry Forum initiative, but fails adequately to identify the reasons. This chapter has considered the question of adoption of process innovations with particular reference to the insights that diffusion theory can offer to help policy agents accelerate the process within populations of reluctant adopters.

In their achievements, Industry Forums have succeeded in targeting populations of UK organizations that the innovation literature might disparagingly refer to as 'laggards', at least with regard to their slow and reluctant adoption of certain innovative management practices. Indeed, it is questionable whether or not some or all of these organizations would ever have adopted WCM without the influence of a policy of intervention. The success of the initiative lies in a programme that intimately links together the processes of knowledge transfer with tangible organizational problems identified in partnership between the Master Engineers and the host company.

The transfer of 'know-what' and 'know-how', particularly of often complex, abstract, ideational knowledge such as WCM, cannot simply be presumed to be a mechanical process since the new knowledge may not fit with the cultural, taken-for-granted assumptions about existing practices in recipient organizations (Newell et al. 2003). For others, its adoption may represent too great a demand of management time, with apparent increases in bureaucracy; and a lag between adoption and appropriating benefits may be too long for the relationship to be apparent (Hill and Wilkinson 1995).

The MasterClass programme recasts slow adopters as part of the innovation process, not treating them as passive recipients of new ideas but engaging them fully as partners in the process of shaping and configuring the innovation. Von Hippel (1986) argued that 'lead users', exogenous stakeholders at the leading edge of important marketplace trends, can be important to developing and shaping innovation: however, slow adopters are no lead users; indeed the configurations of practice that the slow adopters ultimately implement are modifications, through the auspices of the relationship between Master Engineers and internal change agents, of those of lead users – the innovators and early adopters. That is, MasterClass owes much of its success to the fact that slower adopters have deliberately been engaged as active, contributory participants in the adoption process.

The role of government is often overlooked in classic diffusion studies, previously the diffusion of process improvement programmes on the Japanese model throughout UK industry has been attributed to knowledge spillovers (DTI 2006). But this kind of activity is increasingly viewed as important by policy actors, whether at the level of a supply chain 'owner' (AFFA 2000; Bessant et al. 2003), a trade or sectoral business association (Bateman and David 2002), or a regional or national government (Best 2001; Cooke and Morgan 1991).

There is growing evidence to support the use of peer-assist modes of intervention. For example, in South Africa the domestic automotive components sector was confronted by significant performance gaps resulting from global pressures newly felt in its post-apartheid economy. Catching up to the 'world class' frontier became an urgent priority, and central to this was the need to adopt WCM rapidly and widely. One approach was the formation of a series of 'benchmarking clubs' in key regions where the sector was a significant element in the local economy – around Durban, along the Eastern Cape seaboard and in the areas between Pretoria and Johannesburg. These clubs operated in a similar fashion to IF, using a mixture of benchmarking to develop shared motivation for change allied to extensive interfirm support for experimenting with and learning about WCM and particularly how it could be adapted and configured to suit very different educational, social and cultural conditions (Morris et al. 2006).

A number of writers have identified the need for successful technology transfer policies to take a multidimensional approach, reflecting the principle that one size does not fit all. Policies aimed at promoting diffusion need to be multidimensional and embed an understanding of influences on adoption. This chapter has argued that developing such an understanding opens up some interesting new research questions but also new directions for policy agents to take in their search for ways of accelerating the adoption of innovative management practices.

ACKNOWLEDGEMENT

The authors gratefully acknowledge the contribution of Philip Sowden for helpful comments on previous versions of this chapter.

REFERENCES

Abrahamson, E. (1991), 'Managerial fads and fashions – the diffusion and rejection of innovations', *Academy of Management Review*, **16**(3): 586–612.

AFFA (2000), 'Supply chain learning: chain reversal and shared learning for global competitiveness', Department of Agriculture, Fisheries and Forestry, Canberra, Australia.

Attewell, P. (1992), 'Technology diffusion and organizational learning: the case of business computing', *Organization Science*, **3**(1):1–19.

Bateman, N. (2005), 'Sustainability: the elusive element of process improvement', *International Journal of Operations and Production Management*, **25**(3): 261–76.

Bateman, N. and David, A. (2002), 'Process improvement programmes: a model for assessing sustainability', *International Journal of Operations and Production Management*, **22**(5): 515–26.

Beck, N. and Walgenbach, P. (2005), 'Technical efficiency or adaptation to institutionalized expectations? The adoption of ISO 9000 standards in the German mechanical engineering industry', *Organization Studies*, **26**(6): 841–66.

Benders, J. and Van Bijsterveld, M. (2000), 'Leaning on lean: the reception of a management fashion in Germany', *New Technology Work and Employment*, **15**(1): 50–64.

Benner, M.J. and Tushman, M.L. (2003), 'Exploitation, exploration, and process management: the productivity dilemma revisited', *Academy of Management Review*, **28**(2): 238–56.

Bessant, J. and Rush, H. (1993), 'Government support of manufacturing innovation: two country-level case studies', *IEEE Transactions on Engineering Management*, **40**(1): 79–91.

Bessant, J. and Rush, H. (1995), 'Building bridges for innovation: the role of consultants in technology transfer', *Research Policy*, **24**: 97–114.

Bessant, J., Kaplinsky, R. and Lamming, R. (2003), 'Putting supply chain learning into practice', *International Journal of Operations and Production Management*, **23**(2): 167–84.

Best, M. (2001), *The New Competitive Advantage*, Oxford: Oxford University Press.

Bevis, K. (2004), 'Learning at work (automotive college)', *Manufacturing Engineer*, 83(1): 30–33.

Bloom, N. and Van Reenen, J. (2005), *Measuring and Explaining Management Practices across Firms and Countries*, London: Advanced Institute of Management Research.

Borgatti, S.P. and Foster, P.C. (2003), 'The network paradigm in organizational research: a review and typology', *Journal of Management*, 29(6): 991–1013.

Brass, D.J., Galaskiewicz, J., Greve, H.R. and Tsai, W. (2004), 'Taking stock of networks and organizations: a multilevel perspective', *Academy of Management Journal*, 47(6): 795–817.

Brown, A., Rhodes, E. and Carter, R. (2003), 'Supporting learning in advanced supply systems in the automotive and aerospace industries', in H. Rainbird, A. Fuller and A. Munro (eds), *Workplace Learning in Context*, London: Routledge, pp. 166–82.

Burns, T.R. and Stalker, G.M. (1961), *The Management of Innovation*, London: Tavistock Publications.

Camison-Zornoza, C., Lapiedra-Alcami, R., Segarra-Cipres, M. and Boronat-Navarro, M. (2004), 'A meta-analysis of innovation and organizational size', *Organization Studies*, 25(3): 331–61.

Chedzey, M. and Kennedy, N. (2003), Case study on Poeton (Gloucester) Ltd. http://www.weaf.co.uk/competitiveness/case_studies/Poeton%20Industries1.pdf, accessed November 2006.

Clegg, C.W., Wall, T.D., Pepper, K., Stride, C., Woods, D., Morrison, D., Cordery, J., Couchman, P., Badham, R., Kuenzler, C., Grote, G., Ide, W., Takahashi, M. and Kogi, K. (2002), 'An international survey of the use and effectiveness of modern manufacturing practices', *Human Factors and Ergonomics in Manufacturing*, 12(2): 171–91.

Cohen, W.M. and Levinthal, D.A. (1990), 'Absorptive capacity: a new perspective on learning and innovation', *Administrative Science Quarterly*, 35(1): 128–52.

Cooke, P. and Morgan, K. (1991), 'The intelligent region: industrial and institutional restructuring in Emilia-Romagna', *Regional Industrial Research Report No. 7*, Cardiff: University of Wales.

Crute, V., Ward, Y., Brown, S. and Graves, S. (2003), 'Implementing lean in aerospace – challenging the assumptions and understanding the challenges', *Technovation*, 23(12): 917–28.

Cunningham, C.E., Woodward, C.A., Shannon, H.S. and MacIntosh, J. (2002), 'Readiness for organizational change: a longitudinal study of workplace, psychological and behavioural correlates', *Journal of Occupational and Organizational Psychology*, 75(4): 377–92.

DiMaggio, P.J. and Powell, W.W. (1983), 'The iron cage revisited: institutional isomporphism and collective rationality in organizational fields', *American Sociological Review*, 48: 147–60.

Dodgson, M. and Bessant, J. (1996), *Effective Innovation Policy*, London: International Thomson Business Press.

Douthwaite, B., Keatinge, J.D.H. and Park, J.R. (2001), 'Why promising technologies fail: the neglected role of user innovation during adoption', *Research Policy*, 30(5): 819–36.

DTI (1994), *Inside UK Enterprise*, London: IFS Publications on behalf of the UK Department of Trade and Industry.

DTI (2006), *Sector Competitiveness: Analysing the UK Leisure Boatbuilding Industry*, London: Department of Trade and Industry.

DTI/CBI (2001), *FiT for the Future: Supply Chain Learning – A Resource for Management*, London: DTI/CBI.

Dyer, J.H. and Nobeoka, K. (2000), 'Creating and managing a high-performance knowledge-sharing network: the Toyota case', *Strategic Management Journal*, **21**: 345–67.

Eby, L.T., Adams, D.M., Russell, J.E.A. and Gaby, S.H. (2000), 'Perceptions of organizational readiness for change: factors related to employees' reactions to the implementation of team-based selling', *Human Relations*, **53**(3): 419–42.

Eveland, J.D. and Tornatzky, L.G. (1990), 'The deployment of technology', in L. Tornatzky and M. Fleischer (eds), *The Processes of Technological Innovation*, Lexington, MA: Lexington Books, pp. 117–48.

Frambach, R.T. and Schillewaert, N. (2002), 'Organizational innovation adoption – a multi-level framework of determinants and opportunities for future research', *Journal of Business Research*, **55**(2): 163–76.

Geroski, P.A. (1999), 'Models of technology diffusion', *Research Policy*, **29**(4/5): 603–25.

Greenhalgh, T., Robert, G., MacFarlane, F., Bate, P. and Kyriakidou, O. (2004), 'Diffusion of innovations in service organizations: systematic review and recommendations', *Milbank Quarterly*, **82**(4): 581–629.

Griliches, Z. (1957), 'Hybrid corn: an exploration in the economics of technological change', *Econometrica*, **25**(4): 501–22.

Hanson, P. and Voss, C. (1995), 'Benchmarking best practice in European manufacturing sites', *Business Process Re-engineering and Management Journal*, **1**(1): 60–74.

Haque, B. (2003), 'Lean engineering in the aerospace industry', *Proceedings of the Institution of Mechanical Engineers – Part B – Enginneering Manufacture*, **217**(10): 1409–20.

Hill, S. and Wilkinson, A. (1995), 'In search of TQM', *Employee Relations*, **17**(3): 8–25.

Hobday, M., Rush, H. and Bessant, J. (2004), 'Approaching the innovation frontier in Korea: the transition phase to leadership', *Research Policy*, **33**(10): 1433–57.

Holahan, P.J., Aronson, Z.H., Jurkat, M.P. and Schoorman, F.D. (2004), 'Implementing computer technology: a multiorganizational test of Klein and Sorra's model', *Journal of Engineering and Technology Management*, **21**(1–2): 31–50.

Jansen, K.J. (2000), 'The emerging dynamics of change: resistance, readiness, and momentum', *Human Resource Planning*, **23**(2): 53–5.

Krackhardt, D. (1997), 'Organizational viscosity and the diffusion of controversial innovations', *Journal of Mathematical Sociology*, **22**(2): 177–99.

Lane, P.J. and Lubatkin, M. (1998), 'Relative absorptive capacity and interorganizational learning', *Strategic Management Journal*, **19**(5): 461–77.

Lawrence, P.R. and Lorsch, J.W. (1967), *Organization and Environment*, Cambridge, MA: Harvard University Press.

Lucking, B. (2004), 'International comparisons of the third Community Innovation Survey (CIS 3)', London: DTI.

Mahajan, V. and Peterson, R.A. (1985), *Models for Innovation Diffusion*, Newbury Park, CA: Sage Publications.

Mansfield, E. (1968), *Industrial Research and Technological Innovation*, New York: Norton.

Montes, F.J.L. and Jover, A.J.V. (2004), 'Total quality management, institutional isomorphism and performance: the case of financial services', *Service Industries Journal*, **24**(5): 103–19.

Moore, G.A. (1991), *Crossing the Chasm*, New York: Harper Business.

Morris, M., Bessant, J. and Barnes, J. (2006), 'Using learning networks to enable industrial development – case studies from South Africa', *International Journal of Operations and Production Management*, **26**(5–6): 532–57.

Newell, S., Edelman, L., Scarborough, H., Swan, J. and Bresnen, M. (2003), ' "Best practice" development and transfer within the NHS: the importance of process as well as product knowledge', *Health Services Management Research*, **16**(1): 1–12.

Newell, S.M. and Swan, J.A. (1993), 'The potential role of a Canadian professional association in the dissemination of knowledge', *British Journal of Canadian Studies*, **8**: 241–59.

Nohria, N. and Gulati, R. (1996), 'Is slack good or bad for innovation?', *Academy of Management Journal*, **39**(5): 1245–64.

Northcott, J. and Bessant, J. (1986), *Promoting Innovation: Consultancy support for microelectronics applications*, London: Policy Studies Institute.

Porter, M.E. and Ketels, C.H.M. (2003), 'UK competitiveness: moving to the next stage', DTI Economics Paper No. 3, URN 03/899.

Reagans, R. and McEvily, B. (2003), 'Network structure and knowledge transfer: the effects of cohesion and range', *Administrative Science Quarterly*, **48**(2): 240–67.

Rigby, D. (2001), 'Management tools and techniques: a survey', *California Management Review*, **43**(2): 139–60.

Rogers, E.M. (2003), *Diffusion of Innovations*, New York: Free Press.

Rush, H., Bessant, J. and Hoffman, K. (1994), 'Implementing flexible manufacturing: a role for government', *Proceedings of the SPRINT conference on Technology Transfer Practice in Europe*, 20–28 April, Hanover, Germany: European Commission – SPRINT/TII.

Ryan, B. and Gross, N.C. (1943), 'The diffusion of hybrid seed corn in two Iowa communities', *Rural Sociology*, **8**(1): 15–24.

Ryan, B. and Gross, N.C. (1950), 'Acceptance and diffusion of hybrid seed corn in two Iowa communities', Research Bulletin 372, Iowa Agriculture Experiment Station, Ames, Iowa, pp. 665–708.

Schonberger, R.J. (1986), *World Class Manufacturing: The Lessons of Simplicity Applied*, New York: The Free Press.

Siegel, S.M. and Kaemmerer, W.F. (1978), 'Measuring the perceived support for innovation in organizations', *Journal of Applied Psychology*, **63**(5): 553–62.

Simonin, B.L. (1997), 'The importance of collaborative know-how: an empirical test of the learning organization', *Academy of Management Journal*, **40**(5): 1150–74.

SMMT (2006), MasterClass, http://www.industryforum.co.uk/brochures/master. pdf, accessed December 2006.

Sohal, A.S. (1996), 'Assessing AMT implementations: an empirical field study', *Technovation*, **16**(8): 377–85.

Staw, B.M. and Epstein, L.D. (2000), 'What bandwagons bring: effects of popular management techniques on corporate performance, reputation and CEO pay', *Administrative Science Quarterly*, **45**(3): 523–56.

Stoneman, P. and Diederen, P. (1994), 'Technology diffusion and public policy', *Economic Journal*, **104**(425): 918–30.

Stoneman, P. and Kwon, M.J. (1996), 'Technology adoption and firm profitability', *Economic Journal*, **106**(437): 952–62.

Thio, A.O. (1971), 'A reconsideration of the concept of adopter–innovation compatibility in diffusion research', *Sociological Quarterly*, **12**(1): 56–68.

Tilson, B. (2001), 'Success and sustainability in automotive supply chain improvement programmes: a case study of collaboration in the Mayflower cluster', *International Journal of Innovation Management*, **5**(4): 427–56.

Valente, T.W. (1995), *Network Models of the Diffusion of Innovations (The Hampton Press Communication Series. Quantitative Methods in Communication)*, Cresskill, NJ: Hampton Press.

Venables, M. (2004), 'Striving for process excellence', *Manufacturing Engineer*, **83**(6): 40–43.

Von Hippel, E. (1986), 'Lead users: a source of novel product concepts', *Management Science*, **32**(7): 791–805.

Voss, C.A. (1995), 'Operations management – from Taylor to Toyota and beyond?', *British Journal of Management*, **6**(6): 17–29.

Voss, C.A., Chiesa, V. and Coughlan, P. (1994), 'Developing and testing benchmarking and self-assessment frameworks in manufacturing', *International Journal of Operations and Production Management*, **14**(3): 83–100.

Waterson, P.E., Clegg, C.W., Bolden, R., Pepper, K., Warr, P.B. and Wall, T.D. (1999), 'The use and effectiveness of modern manufacturing practices: a survey of UK industry', *International Journal of Production Research*, **37**(10): 2271–92.

WEAF (2005a), Adams Precision case study, http://www.weaf.co.uk/competitiveness/documents/AdamsCaseStudy2.pdf, West of England Aerospace Forum, accessed November 2006.

WEAF (2005b), Traxsys Case Study, http://www.weaf.co.uk/competitiveness/documents/TraxsysCaseStudy.pdf: West of England Aerospace Forum, accessed September 2006.

WEAF (2006), Ultra Electronics case study, http://www.weaf.co.uk/competitiveness/documents/10UltraElectronicsLeancasestudy1.pdf, West of England Aerospace Forum, accessed November 2006.

West, M.A. and Farr, J.L. (1990), 'Innovation at work', in M.A. West and J.L. Farr (eds), *Innovation and Creativity at Work: Psychological and Organizational Strategies*, Chichester, UK: John Wiley & Sons, pp. 3–13.

Westphal, J.D., Gulati, R. and Shortell, S.M. (1997), 'Customization or conformity? An institutional and network perspective on the content and consequences of TQM adoption', *Administrative Science Quarterly*, **42**(2): 366–94.

Wickens, P. (1987), *The Road to Nissan: Flexibility, Quality, Teamwork*, London: Macmillan.

Womack, J.P., Jones, D. and Roos, D. (1990), *The Machine that Changed the World*, London: Macmillan.

Wood, S.J., Stride, C.B., Wall, T.D. and Clegg, C.W. (2004), 'Revisiting the use and effectiveness of modern management practices', *Human Factors and Ergonomics in Manufacturing*, **14**(4): 415–32.

11. Understanding and overcoming resistance to innovation

Sue Morton and Neil Burns

The importance of innovation to the continued success of organizations has been covered extensively in earlier chapters. The purpose of this chapter is to provide a review of the literature surrounding resistance to innovation within organizations, to look at the potential influence of organizational culture and climate, and to identify possible areas of interest where intervention may help overcome organizational barriers to innovation and foster productivity improvement. It will also focus on what elements of the innovation process can be measured and review some of the instruments that are currently available for the task.

If we are to understand and overcome resistance to innovation, we need to make explicit the various forms in which it is manifest. The literature on innovation, including its causes and consequences, has developed over the last half-century and extends across many academic fields. Recent decades have seen the creation of new journals and associations, an ever-increasing number of publications and the emergence of numerous cross-discipline research centres, all with a focus on innovation. While it would not be possible to summarize within this chapter all that has been written about innovation, the introduction to this book and Fagerberg's (2005) overview provide a useful guide to the literature. Elsewhere in this book Ferriani et al. (Chapter 8) include a helpful section characterizing incremental (continuous) and breakthrough (discontinuous) innovation, together with a tabular review of technological innovation types to further clarify the distinction between extant terminologies. For the purposes of the current chapter, innovation is defined as:

- the process of bringing new and improved products and processes to market;
- developing, adopting and adapting manufacturing processes to enhance productivity and product quality;
- developing, adopting and adapting business practices to enhance the performance of the firm.

PSYCHOLOGICAL MODELS OF INNOVATION

It has been well documented that early innovators identify a problem and then attempt to solve it creatively (Beckenbach and Daskalakis 2003). Newell and Simon's seminal work has the perception of a problem as the starting point, followed by a series of actions to achieve the desired result (Newell and Simon 1972): the problem-solving process, the methodology for which can be thought of as a continuum, with concrete and positive steps at one end, through to suggested ways of thinking, a set of heuristics, at the other (Martin 2000). This process is based on the person's knowledge about the domain, the declarative knowledge, combined with knowing how to step through to a plan of action, the person's procedural knowledge. Finke's *genplore* model (Finke et al. 1992) extends this rather simplistic view to an iterative process of idea generation and exploration, where the initial phase *gen*erates new ideas and the next phase ex*plore*s these new ideas, frequently moderating them and leading back to another phase of idea generation. In this way, multiple feedback cycles result in new declarative and procedural knowledge that could lead to a new product (Beckenbach and Daskalakis 2003). Subsequent models have focused on personal qualities, divided into knowledge and skills (Weisberg 1999), personality features (Csikszentmihalyi 1999) and motivation (Amabile 1998) constrained, nevertheless, by available knowledge. The creative problem-solving process is thus a cognitive activity, key aspects of which include:

- individual qualities – knowledge and skills in combination with personal features and motivations of the problem solver. Knowledge defines how to use the available heuristics while knowledge and skills combine with strong motivation for fulfilment that, in turn, can come from intrinsic views or from extrinsic information. Further prerequisites for creative activity include curiosity, patience and a willingness to take risks (Csikszentmihalyi 1997);
- process analysis – the process of verifying all sources of information;
- environment – space must be provided for the process to take place. Here the degree of incomplete knowledge and open-mindedness of the experts in the field are both important factors, determined by education and expertise, prior experience, knowledge and the assessment of potential constraints. (Beckenbach and Daskalakis 2003)

Cognitive psychology describes how people use simplified mental representations (schemata, cognitive maps and mental models) to cut down the amount of complex information processing necessary in human activity (Hodgkinson 2003). Stemming from Bartlett's work in the early 1930s

(Bartlett 1932), modern schema theory incorporates many of his ideas on schemata that filter people's perception of reality according to their knowledge and past experience, with both positive and negative effects. On the one hand this enables them to solve problems quickly, while on the other it constrains their creativity and ability to recognize important information external to their own mental model.

Hewett (2005) provides a useful overview on how insight, creativity and innovation occur, including some of the psychological aspects. Based on a review of the psychological literature, Hewett proposes a set of user requirements that are general enough to be applied across several different domains. He describes how insight, creativity and innovation can be fostered in working environments, if only by avoiding conditions that are known to disrupt or to work against creativity. Innovation is unlikely to occur, for example, where an individual or group does not have a strong involvement with the problem to be solved and belief in the importance of the work (Gruber 1989; Csikszentmihalyi and Sawyer 1995).

Schilling (2005) describes cognitive insight and how an 'atypical association', resulting from either directed research or a random event, can link mental representations and create a new understanding that could lead to innovation. Through the application of graph theory, she demonstrates how shortcuts in an individual's network of mental representations can cause reorientation of understanding and lead to a range of other connections. This lends weight to the importance of bringing in new information from a variety of sources. Dahlander and Gann (Chapter 3 in this volume) discuss how firms benefit from the social landscape in which they are embedded, and the influence of network structures on organizational behaviour and performance. Interaction with customers and suppliers, and through professional and social networks, brings about contact with people and ideas that may challenge the accepted way of thinking about things (see for example the research on the importance of 'weak ties' to innovation Granovetter 1973, 1982; Bryson and Daniels 1998). What is unclear, and thus worthy of further investigation, is how this might fit with the concept of a psychological contract. Is it a separate element in what comprises the *innovation formula*, or is it part of the *predisposition to be creative*, or the *context in which they operate* that makes some people open to using, and not rejecting, a new piece of information that doesn't fit their current mental map?

The psychological contract sets the dynamics for the relationship between an employer and an employee (Rousseau 1995). Distinguishable from the formal written contract of employment, it represents the mutual beliefs, perceptions and informal obligations inherent in the relationship, and defines the detailed practicality of the work to be done (Conway and

Briner 2005; Wikipedia 2006). How then does the psychological contract fit
with the proposition that innovation is a function of the relationship of
person with organization, the context in which they operate, and their predisposition to be creative?

Flood et al. (2001) propose a model of causes and consequences that
relates the antecedents and consequences of the psychological contract.
They hypothesize that the motivation for parting with value-creating
knowledge comes from perceived fairness of organizations' reward and
recognition practices, which creates an obligation to contribute. While this
is not supported directly, perception of met expectations did affect obligation to contribute. Flood found that employers who retain their knowledge
workers provide interesting and challenging work; allow workers to build a
portable portfolio of skills; and ensure that merit and equity are observed
in the administration of rewards. Such motivation is important, particularly in high-technology organizations that need to ensure knowledge
employees are motivated to voluntarily share the tacit knowledge and skills
upon which the innovation process is built. Sharkie's model (2005) shows
the relationship between perceptions that an individual has about the organization in which he or she works. These perceptions are based on the psychological contract; the resultant perception of trust in the organization, in
management or in fellow workers; and the consequent willingness, or otherwise, of the individual to engage in conversations and share knowledge
with others. Sharkie developed six propositions about perceptions of the
psychological contract and the concomitant effects on the level of trust and
on the willingness to converse and to share knowledge. He discusses the
need for trust in order to share knowledge, and how this trust depends on
the employee's perceived psychological contract in a climate where long-
term employment can no longer be relied on. In contemporary society the
psychological contract is a constantly changing set of expectations that,
although unwritten, can be a significant determinant of behaviour in organizations. Perceptions of violation can have lasting effects on trust, with a
concomitant impact upon innovation (see Robinson and Rousseau 1994;
Patterson et al. 2005; Miranda and Kavan 2005).

RESISTANCE TO INNOVATION: BARRIERS AND ENABLERS

In its recognition of a number of levels, the resistance literature is similar
to much that has been written on other aspects of innovation. Resistance
to innovation manifests itself in the form of different barriers. Again,
Ferriani et al.'s chapter (Chapter 8 in this volume) proves helpful in its

consideration of barriers to innovation in large established firms. Loewe and Dominiquini (2006) identify six such barriers in terms of having: a short-term focus; lack of time, resources and/or staff; expectation from leadership of an earlier payoff than is realistic; management incentives not structured to reward innovation; a lack of a systematic innovation process; and a belief that innovation is inherently risky. The lack of senior management support, in combination with a lack of knowledge or capability for learning about markets, is also likely to have significant negative impact on innovation performance (see Adams et al. 1998; Dougherty et al. 2000). Bond and Houston (2003) further extend theory and suggest three sets of barriers to the matching of successful technologies with market opportunities: technology and market; strategic and structural; and social and cultural barriers. They provide a framework of important barriers to innovation, identify the challenges that each pose to the firm and propose example approaches to overcoming such challenges.

Barriers have also been identified as occurring at the level of the individual, the workgroup/team and the organization (see for example King 1990; Kratzer et al. 2005). Zwick (2002) reviews employee resistance to innovation in relation to how future job security may affect whether they are cooperative when it comes to implementing change, which highlights the importance of making sure that motivational factors and goals are all working in the same direction for an innovation to succeed. Other perhaps more significant barriers to innovation relate to the riskiness of the innovation, the costs involved in its implementation, shortage of capital, and how easy the innovation is to copy, together with the associated regulation and standards. Resistance to innovation and change may be attributable to any one factor or to the combination of a number of different factors. However, while earlier chapters have identified a clear distinction between creativity and innovation, the distinction becomes more blurred in the literature on barriers and enablers to innovation. It is therefore helpful to also include reference to some of the literature on creativity.

Research on individual innovation reviews a range of blocks to creativity that challenge a person's beliefs and values, self-image and the perceptual ability to recognize opportunities and threats (King 1990). Perceptions of future job security can affect cooperation when implementing change (Zwick 2002), rigid management structures can have significant negative impact on innovation (Amabile et al. 1996), with high-care atmospheres favouring knowledge creation and transfer (Kratzer et al. 2005; Zarraga and Bonache 2005). Indeed, extant research supports the notion that 'creative cognitions occur when individuals are free from pressure, feel safe, and experience relatively positive effect' (West et al. 2005: 140). Important factors for innovation at the level of the team or working group have been suggested to include leadership and cohesiveness, together with group longevity, composition and structure.

At the level of organizational innovation, resistance can be based on selective perception and the social systems factors of vested interests, rejection of outsiders, misunderstandings, incompatibility of innovation with organization structure, and lack of top-level support. A major source of resistance is regarded by many as being at the level of middle management, where vested interests and issues of motivation may be rife (Barnes et al. 2001; Terziovski et al. 2003). Further issues with the potential to inhibit innovation include project-based working patterns, lack of technology, and lack of time, resources and staff. Indeed, this last point was identified by Loewe and Dominiquini, as mentioned earlier. In terms of the propensity to be innovative, an organization's culture may also have a detrimental effect. In a mature organization, the mechanisms that initially enabled success often inhibit the firm's innovation capability (Dougherty and Cohen 1995; Leifer et al. 2000).

McLaughlin et al. (2005) describe an intervention aimed at creating a culture for radical innovation within a small mature company. An archetypal culture for radical innovation is significantly different to an archetypal culture for incremental innovation (Greenwood and Hinings 1993). It requires very careful handling to be able to promote both radical innovation ('do different') and incremental innovation ('do better') within the same organization (see Nadler and Tushman 1980; Tushman and O'Reilly 1996; Smith and Tushman 2005). A lean organization may not allow for much 'idea time' (Anderson and West 1998), while programmes intended to improve quality and productivity frequently have a negative effect on radical innovation (Benner and Tushman 2003; Benner 2002; Naveh and Erez 2004). Lemon and Sahota (2004) talk about suggested interventions in the context of cultural archetypes and innovative capacity in organizations, albeit focusing on knowledge management. They model organizational culture as a 'bundle of knowledge repositories with storing and information processing capabilities'. Naveh and Erez (2004) describe an intervention aimed at quality improvement, and consider the effect on culture, innovation and productivity. Cameron and Quinn's (1999) seminal work comprises a comprehensive set of tools and procedures for the diagnosis and change of organizational culture.

MEASURING INNOVATION

We have shown that innovation, resistance to innovation and the barriers to and enablers of innovation can take many different forms; thus it is only to be expected that the measurement of innovation is not trivial. Traditional indicators of innovation incorporate measures that look at inputs to the innovation process: R&D expenditure, for example, and outputs such as patents. Given that only a proportion of innovating firms conduct formal R&D, and hence are able to distinguish between it and other expenditure, such indicators have significant problems, while the use of patents varies greatly from firm to firm and between different industries. That is not to say that analysis of such measures is invalid. Indeed, patent data analysis has realized a range of important achievements (see Scherer 1982; Patel and Pavitt 1998; Jaffe et al. 1997). Nevertheless, more recently innovation surveys have tended to take either an object-based approach that focuses on the technology – the objective output of the innovation process – or a subject-based approach focusing on the innovation agent (Archibugi and Pianta 1996). A key illustration of the former is that of the Science Policy Research Unit (SPRU) at the University of Sussex. SPRU compiled a database covering more than 4000 major innovations drawn from all sectors of the economy and encompassing a period of around 40 years to the early 1980s. Results of work using the SPRU database have shown 'the existence of quite different types of innovative activity across different types of industry'; 'early empirical insight into the complexity of what is now called the system of innovation'; and have 'emphasized the inter-sectoral flow of innovations' (Smith 2005: 162). More meaningful indicators have also been used by the Community Innovation Surveys (CIS), which were developed as a follow-up to the *Oslo Manual*, the Organization for Economic Cooperation and Development (OECD) initiative of the early 1990s. The CIS measure:

- expenditure on activities related to the innovation of new products (R&D, training, design, market exploration, equipment acquisition and tooling-up . . .);
- outputs of incrementally and radically changed products, and sales flowing from these products;
- sources of information relevant to innovation;
- technological collaboration;
- perceptions of obstacles to innovation, and factors promoting innovation. (Smith 2005: 163)

Data collection generally took place at the level of the firm and aggregated to national or European level. Resultant literature from CIS data can be categorized in terms of descriptive overviews; European Commission sponsored analytical studies; and studies of innovation that are either econometrically or statistically based (Smith 2005). Tether (2002), Tether and Swann (2003) and Laursen and Salter (2006) exemplify research work that is based on the UK data; the work of Brusoni et al. (2005) uses data from The Netherlands; Inzelt (2002) looks at Hungary; and Lehtoranta (2005) at Finland, while Hinloopen (2003) uses CIS to analyse the differences between the innovation performances of organizations across Europe. The CIS has been undertaken since the mid-1990s within the European Union and the results used to inform innovation policy.

It may be possible to use this type of large-scale instrument to measure innovation on a smaller scale, although limitations exist in the ability to measure innovation within service-based industries (Smith 2005), and organizational innovation due to its firm-specific nature, which makes it 'difficult to summarise in aggregate, sector or economy-wide statistics' (OECD, *Oslo Manual*, paragraph 120, quoted in Tether 2001). It is also unclear as to how well the CIS instrument might measure process innovation, given the conflation of product and process innovation with that of service innovation in the second survey, and the markedly different patterns of innovation in service to those of innovation in process (Tether 2001).

In the context of this chapter, however, it is of more interest to focus on the organization and its capacity to innovate, rather than on the innovations generated by the organization. A number of studies use various indicators to measure organizational level of innovation, its 'innovativeness', although the term is used in different ways. For instance, Romijn and Albu (2002) use 'innovativeness' to describe the magnitude of novelty in a product or process innovation. Our focus here, however, is on firm or organizational innovativeness, which has been defined as the propensity for a firm to innovate or develop new products (Garcia and Calantone 2002, after Ettlie et al. 1984), or the propensity for a firm to adopt innovations (Garcia and Calantone 2002, after Damanpour 1991; Rogers 1995). Persaud (2005) looks at ten indicators rated on a scale of 1 (decreased substantially) to 5 (increased substantially), commonly used in innovation management research to measure the innovativeness of firms.

Freel (2005) also talks about firm innovativeness, although here the measurements focus simply on whether firms are novel innovators, incremental innovators or non-innovators: introducing product or process new to the industry, introducing product or process new to the firm, or doing neither, respectively. Hollenstein (1996) provides a composite measure of firm innovativeness using input, output and market-oriented indicators that have the potential for aggregation to national or international level, a measure more typically reliant on traditional R&D or patent indicators. Tang (1999) also provides a much more thorough approach to assessing organizational innovativeness using nine scales of measurement and 46 indicators. Further, while most studies just take one point in time, Subramanian (1996) believes that measures of innovativeness should include a temporal aspect. The same can also be said for the measurement of organizational culture and climate.

ORGANIZATIONAL CULTURE AND CLIMATE

Organizational culture can be described as that which comprises the attitudes, experiences, beliefs and values of an organization. It lends itself relatively easily to explanation and has been defined as 'the specific collection of values, norms, beliefs, and attitudes shared by people and groups in an organization and that control the way they interact with each other and with stakeholders outside the organization' (Hill and Jones 2004: 404). In contrast, the concept of organizational climate proves hard to define. While there are several approaches to the concept of climate, two have received substantial support in the literature. The *cognitive schema* approach views the concept of climate as an individual perception, while the second approach emphasizes the importance of *shared perceptions* (Anderson and West 1998; Mathisen and Einarsen 2004). In a review of instruments for measuring climate, Mathisen and Einarsen (2004) assess the effectiveness of five such instruments.

INSTRUMENTS FOR MEASURING ORGANIZATIONAL CLIMATE

The KEYS instrument (Amabile et al. 1996) looks at the intrinsic motivation of individuals to be creative and assesses perceived barriers and enablers to creativity. The questionnaire is completed by individuals and records their perceptions of environmental factors on the different levels of individual, supervisory, group and organizational. The main dimensions measured are challenging work; freedom; resources; workgroup support; supervisory encouragement and organizational support. Management practices are also measured in terms of organizational impediments and workload pressure. KEYS also includes a measure of perceived productivity and creativity. Amabile et al. found that workgroup encouragement is particularly important, as is having some work pressure: considering work to be challenging but not overpressured. Organizational impediments such as internal strife, conservatism and rigid management structures may have a significant negative impact (Amabile et al. 1996). The KEYS instrument is well respected and has been used by other researchers (see for example Bommer and Jalajas 2002; McLaughlin et al. 2005), although it could benefit from improvement in factor structure (Mathisen and Einarsen 2004).

The Creative Climate Questionnaire (CCQ) was developed in Swedish (Ekvall 1996), translated into English and refined as the Situational Outlook Questionnaire (SOQ) (Isaksen and Kaufmann 1990; Isaksen et al. 1999). The dimensions of the CCQ cover challenge, freedom, idea support, dynamism/liveliness, playfulness/humour, debates, trust, conflict, risk taking and idea time. The SOQ uses the same dimensions, but dynamism was eliminated as a separate dimension, while challenge was expanded to challenge/involvement and trust to trust/openness. However, there are some doubts about the psychometric quality of these instruments (Mathisen and Einarsen 2004).

Anderson and West's (1998) Team Climate Inventory (TCI) is administered at the individual level and the responses aggregated to team level. It is based on the four factors identified by West (1990) of vision, participative safety (i.e. team participation and safety), task orientation (climate for excellence and constructive controversy) and support for innovation. This last item concerns the extent to which time, practical support, cooperation and resources are given to team members to implement new ideas and proposals, and incorporates four items taken from Siegel and Kaemmerer (1978). During the initial TCI validation (Anderson and West 1998) there emerged a possible fifth factor of interaction frequency. This five-factor version has been further tested by Kivimaki et al. (1997) and is preferred, especially in situations of high job complexity. Kivimaki et al. also proposed a mini-version of the TCI (Kivimaki and Elovainio 1999). The TCI in its various guises has received good reports for validity and psychometric quality (Mathisen and Einarsen 2004).

The Siegel Scale of Support for Innovation (SSSI) (Siegel and Kaemmerer 1978) was developed in schools and uses the dimensions of leadership, ownership, norms for diversity, continuous development and consistency. According to Scott and Bruce (1994), Siegel and Kaemmerer used three subscales of *support for creativity*, *tolerance of differences* and *personal commitment*, but did not distinguish between innovative and non-innovative organizations. Furthermore, there are some doubts about its applicability in work organizations (Mathisen and Einarsen 2004). Scott and Bruce (1994) also conducted an interesting study looking at leadership, workgroup and individual attributes, and how they relate to psychological climate for innovation and ultimately innovative behaviour. Their findings revealed that psychological climate is not very important in determining innovative behaviour, although it does act as a mediator between leader–member exchange and innovative behaviour. Developed more recently, another instrument with a team focus is the Team Factor Inventory (TFI) (Rickards et al. 2001). This assesses seven factors using a large database of information, although no evaluation is provided. A further instrument that may be of use is the Business and Organizational Climate Index (BOCI). This was developed in the early 1970s and assesses 17 dimensions (Payne and Pheysey 1971). It has been used to classify organizations into eight climate types and has high levels of inter-rate reliability (Sparrow and Gaston 1996).

INSTRUMENTS FOR MEASURING ORGANIZATIONAL CULTURE

The literature on organizational culture provides a number of instruments and models for its assessment, a summary of which is given in Table 11.1.

While the TCI is used at the individual level and then aggregated to team level, KEYS measures perceptions on the four different levels of group, organization, individual and supervisory. The OCAI displays current and desired positions for organizational culture, reflecting the temporal aspect desirous by Subramanian (1996) and with the potential for further application after a period of time.

ORGANIZATIONAL LEARNING AND INNOVATION

The ability of an organization to innovate and improve is related to its ability to learn (Montes et al. 2005; Reissner 2005). Successful organizations reflect on the reasons for successes and failures, which is a fundamental and critical component of learning. After a major new product, service, or business launch, organizations frequently produce substantial documents about what worked and what didn't, and spread the learning throughout the organization. Many organizations set up electronic databases, intranet sites and active networks to facilitate learning. For much of the literature concerning this subject, the focus is on knowledge management and how organizations can utilize the knowledge of their own employees and learn from past projects (Hargadon 2002; Brockman and Morgan 2003; Wang and Ahmed 2004). The importance of knowledge and its role in learning and innovation can vary greatly, with firms in different societies depending differentially on different knowledge types (Nonaka and Takeuchi 1995). These differences can, in turn, influence the different approaches of organizations to product innovation and learning strategies (Lam 1996, 1997, 1998).

Another important aspect of organizational learning is the ability to learn from other organizations and industries. Problem solution in one

Table 11.1   Summary of instruments for measuring organizational culture

Author(s)	Assessment	Description
Nadler and Tushman (1980)	Balance and equilibrium between four system characteristics	Organization Individual Task Informal organization
Hofstede et al. (1990)	Modelled on six dimensions	Process versus results oriented Job versus employee oriented Professional versus parochial Open systems versus closed systems Tightly versus loosely controlled Pragmatic versus normative Creates multi-layered model of organizational culture
Harrison and Stokes (1992)	Four dimensions of organization orientation	Power, role, achievement and support
Goffee and Jones (1998)	Modelled on dimensions of sociability and solidarity	Four types of culture: networked, communal, mercenary and fragmented
Reigle (2001)	Organizational Culture Assessment (OCA) – five elements	Language; artefacts and symbols; patterns of behaviour; espoused values and beliefs; and underlying assumptions Based on the organic–mechanistic dimensions of Burns and Stalker (1961) Demonstrates validity and reliability.
Cameron and Quinn (1999)	Organizational Culture Assessment Instrument (OCAI) – based on competing values	Flexibility and discretion; control and stability; internal focus and integration; external focus and differentiation Results in one of four culture types – clan, adhocracy, market or hierarchy Assesses how things are and identifies how people would like to see them change. Good evidence of reliability and validity

organization may come from the ideas of another organization, albeit dependent upon connections being made across the boundaries between them. Such ideas may be new and creative as they change and combine to meet different users' needs, leading to new concepts or objects crafted from extant but previously unconnected ideas. This process is influenced by the *receptivity* or *absorptive capacity* (see Mangematin and Nesta 1999; Cohen and Levinthal 1990; Caloghirou et al. 2004) of the organization towards new knowledge, and relies on both good linkages with external knowledge sources and a pluralist and participative culture with the organization (Vickers and Cordey-Hayes 1999). Although beyond the remit of the current chapter, elsewhere in this book Perkmann and Walsh (Chapter 12) discuss two dimensions of external knowledge sourcing with reference to organizations' strategies for accessing and using knowledge from universities, while Sharifi et al. (Chapter 15) focus on innovation theories and, in particular, upon Chesborough's idea of 'open innovation' (Chesbrough 2006) in their discussion of the flow of knowledge to innovation.

Existing technologies may often be adapted and transformed before they become usable in a new field, although the role that individual actions and organizational routines play in this transformation has not received much attention (Hargadon 2002). Organizations are more willing to adopt knowledge-intensive innovations if they have high organizational learning capacity (OLC), which Teo et al. (2006: 264) define as 'an organization's shared assumptions and mechanisms (in terms of processes or culture) that contribute to its capabilities to sustain and improve performance unfettered' and as learning has to be an ongoing process occurring simultaneously with change and upheaval, 'OLC is an important property that should become part of an organization's normal functioning'. Simonin (2004) found that learning from international strategic alliance partners was greater when there was intent to learn, albeit moderated by the firm's own culture towards learning, its size and what structural form the alliance took, and inhibited by tacitness of knowledge and partner protectiveness.

SUMMARY

The review of the literature in this chapter gives an indication of the breadth and complexity of potential resistance to innovation within organizations. In seeking to understand and overcome resistance and thereby to foster productivity improvement, we have examined motivational issues; the barriers and enablers of innovation; the culture and the climate within the organization; and the capacity of the individual, the group and the organization to learn.

While the innovation literature is broad, it cannot be construed as comprehensive in our particular area of interest and there remain many gaps to be addressed. Further research opportunities include:

- investigation of the mechanisms of innovation at the workgroup level (King and Anderson 1990);
- relating psychological contract to innovation and learning;
- determinants of learning from other organizations (Vickers and Cordey-Hayes 1999), particularly in relation to:
  - absorptive capacity (Mangematin and Nesta 1999; Cohen and Levinthal 1990; Caloghirou et al. 2004)
  - participative safety (Anderson and West 1998);
- investigation of the relationship between a *creative* climate and a *learning* climate.

Investigation of some of these opportunities will take place in the short- to medium-term future, in partnership with collaborating organizations from industry. The objective is to learn from and with key players in the innovation system, and through that interaction to improve knowledge about the system and how it operates. The analysis will take place at the organization and production system levels, enabling detailed understanding of industry mechanisms and allowing ideas to be tested in their context of application. The aim is to use some of the innovation and psychometric inventories and measurements, reviewed earlier, to identify and relate the determinants of barriers and enablers to innovation, and to provide advice on how firms can be more effective in, for example, learning from external sources. In association with extant research within the organization, resultant findings will be analysed and used to develop a framework of interventions for maximizing innovative potential.

REFERENCES

Adams, M.E., Day, G.S. and Dougherty, D. (1998), 'Enhancing new product development performance: an organizational learning perspective', *Journal of Product Innovation Management*, 15: 403–22.

Amabile, T.M. (1998), 'How to kill creativity', *Harvard Business Review*, September–October (Reprint 98501): 77–87.

Amabile, T.M., Conti, R., Coon, H., Lazenby, J. and Herron, M. (1996), 'Assessing the work environment for creativity', *Academy of Management Journal*, 39(5): 1154–84.

Anderson, N.R. and West, M.A. (1998), 'Measuring climate for work group innovation: development and validation of the team climate inventory', *Journal of Organizational Behavior*, 19(3): 235–58.

Archibugi, D. and Pianta, M. (1996), 'Measuring technological change through patents and innovation surveys', *Technovation*, 16(9): 451–68.

Barnes, J., Bessant, J., Dunne, N. and Morris, M. (2001), 'Developing manufacturing competitiveness within South African industry: the role of middle management', *Technovation*, 21(5): 293–309.

Bartlett, F.C. (1932), *Remembering*, Cambridge: Cambridge University Press.

Beckenbach, F. and Daskalakis, M. (2003), 'Invention and innovation as creative problem solving activities – a contribution to evolutionary microeconomics', Discussion Papers in Economics, 47/03.

Benner, M.J. (2002), 'Process management and technological innovation: a longitudinal study of the photography and paint industries', *Administrative Science Quarterly*, **47**(4): 676–706.

Benner, M.J. and Tushman, M.L. (2003), 'Exploitation, exploration, and process management: the productivity dilemma revisited', *Academy of Management Review*, **28**(2): 238–56.

Bommer, M. and Jalajas, D. (2002), 'The innovation work environment of high-tech SMEs in the USA and Canada', *R&D Management*, **32**(5): 379–86.

Bond, E.U. and Houston, M.B. (2003), 'Barriers to matching new technologies and market opportunities in established firms', *Journal of Product Innovation Management*, **20**: 120–35.

Brockman, B.K. and Morgan, R.M. (2003), 'The role of existing knowledge in new product innovativeness and performance', *Decision Sciences*, **34**(2): 385–419.

Brusoni, S., Marsili, O. and Salter, A. (2005), 'The role of codified sources of knowledge in innovation: empirical evidence from Dutch manufacturing', *Journal of Evolutionary Economics*, **15**(2): 211–31.

Bryson, J.R. and Daniels, P.W. (1998), 'Business Link, strong ties, and the walls of silence: small and medium-sized enterprises and external business-service expertise', *Environment and Planning C-Government and Policy*, **16**(3): 265–80.

Burns, T. and Stalker, G.M. (1961), *The Management of Innovation*, London: Tavistock Publications.

Caloghirou, Y., Kastelli, I. and Tsakanikas, A. (2004), 'Internal capabilities and external knowledge sources: complements or substitutes for innovative performance?', *Technovation*, **24**(1): 29–39.

Cameron, K.S. and Quinn, R.E. (1999), *Diagnosing and Changing Organizational Culture: Based on the competing values framework*, New York: Addison Wesley.

Chesbrough, H.W. (2006), 'Open innovation: a new paradigm for understanding industrial innovation', in H.W. Chesbrough, W. Vanhaverbeke and J. West (eds), *Open Innovation: Researching a new paradigm*, Oxford: Oxford University Press, pp. 1–14.

Cohen, W.M. and Levinthal, D.A. (1990), 'Absorptive-capacity – a new perspective on learning and innovation', *Administrative Science Quarterly*, **35**(1): 128–52.

Conway, N. and Briner, R.B. (2005), *Understanding Psychological Contracts at Work: A Critical Evaluation of Theory and Research*, Oxford: Oxford University Press.

Csikszentmihalyi, M. (1997), *Finding Flow: The Psychology of Engagement with Everyday Life*, New York: Basic Books.

Csikszentmihalyi, M. (1999), 'The creative person', in R.A. Wilson and F.C. Keil (eds), *The MIT Encyclopedia of the Cognitive Sciences*, Cambridge, MA and London: The MIT Press, pp. 305–6.

Csikszentmihalyi, M. and Sawyer, K. (1995), 'Creative insight: the social dimensions of a solitary moment', in R.J. Sternberg and J.E. Davidson (eds), *The Nature of Insight*, Cambridge, MA: The MIT Press, pp. 329–64.

Damanpour, F. (1991), 'Oganizational innovation: a meta-analysis of effects of determinants and moderators', *The Academy of Management Journal*, **34**(3): 555–90.

Dougherty, D., Borrelli, L., Munir, K. and O'Sullivan, A. (2000), 'Systems of organizational sensemaking for sustained product innovation', *Journal of Engineering and Technology Management*, **17**(3): 321–55.

Dougherty, D. and Cohen, M. (1995), 'Product innovation in mature firms', in E. Bowman and B. Kogut (eds), *Redesigning the Firm*, New York: Oxford University Press, pp. 87–115.

Ekvall, G. (1996), 'Organizational climate for creativity and innovation', *European Journal of Work and Organizational Psychology*, **5**(1): 105–23.

Ettlie, J.E., Bridges, W.P. and O'Keefe, R.D. (1984), 'Organization strategy and structural differences for radical versus incremental innovation', *Management Science*, **30**(6): 682–95.

Fagerberg, J. (2005), 'Innovation: a guide to the literature', in J. Fagerberg, D. Mowery and R. Nelson (eds), *The Oxford Handbook of Innovation*, Oxford: Oxford University Press, pp. 1–26.

Finke, A., Ward, T. and Smith, S. (1992), *Creative Cognition. Theory, Research and Applications*, Cambridge, MA and London: The MIT Press.

Flood, P.C., Turner, T., Ramamoorthy, N. and Pearson, J. (2001), 'Causes and consequences of psychological contracts among knowledge workers in the high technology and financial services industries', *International Journal of Human Resource Management*, **12**(7): 1152–65.

Freel, M.S. (2005), 'Patterns of innovation and skills in small firms', *Technovation*, **25**(2): 123–34.

Garcia, R. and Calantone, R. (2002), 'A critical look at technological innovation typology and innovativeness terminology: a literature review', *Journal of Product Innovation Management*, **19**(2): 110–32.

Goffee, R. and Jones, G. (1998), *The Character of a Corporation: How Your Company's Culture Can Make or Break Your Business*, London: Harper Business.

Granovetter, M. (1973), 'The strength of weak ties', *The American Journal of Sociology*, **78**(6): 1360–80.

Granovetter, M. (1982), 'The strength of weak ties: a network theory revisited', in P.V. Marsden and N. Lin (eds), *Social Structure and Network Analysis*, London: Sage, pp. 105–30.

Greenwood, R. and Hinings, C.R. (1993), 'Understanding strategic change – the contribution of archetypes', *Academy of Management Journal*, **36**(5): 1052–81.

Gruber, H.E. (1989), 'The evolving systems approach to creative work', in D.B. Wallace and H.E. Gruber (eds), *Creative People at Work*, Oxford: Oxford University Press, pp. 3–24.

Hargadon, A.B. (2002), 'Brokering knowledge: linking learning and innovation', *Research in Organizational Behavior*, **24**: 41–85.

Harrison, R. and Stokes, H. (1992), *Diagnosing Organizational Culture*, San Diego, CA: Pfeiffer.

Hewett, T.T. (2005), 'Informing the design of computer-based environments to support creativity', *International Journal of Human–Computer Studies*, **63**(4–5): 383–409.

Hill, C.W.L. and Jones, G.R. (2004), *Strategic Management*, 6th edn, Boston, MA: Houghton Mifflin.

Hinloopen, J. (2003), 'Innovation performance across Europe', *Economics of Innovation and New Technology*, **12**(2): 145–61.

Hodgkinson, G.P. (2003), 'The interface of cognitive and industrial, work and organizational psychology', *Journal of Occupational and Organizational Psychology*, **76**: 1–25.

Hofstede, G., Neuijen, B., Ohayv, D.D. and Sanders, G. (1990), 'Measuring organizational cultures – a qualitative and quantitative study across 20 cases', *Administrative Science Quarterly*, **32**(2): 286–316.

Hollenstein, H. (1996), 'A composite indicator of a firm's innovativeness. An empirical analysis based on survey data for Swiss manufacturing', *Research Policy*, **25**(4): 633–45.

Inzelt, A. (2002), 'Attempts to survey innovation in the Hungarian service sector', *Science and Public Policy*, **29**(5): 367–83.

Isaksen, S.G. and Kaufmann, G. (1990), 'Adapters and innovators – different perceptions of the psychological climate for creativity', *Studia Psychologica*, **32**(3): 129–42.

Isaksen, S.G., Lauer, K.J. and Ekvall, G. (1999), 'Situational outlook questionnaire: a measure of the climate for creativity and change', *Psychological Reports*, **85**(2): 665–74.

Jaffe, A., Henderson, R. and Trajtenberg, M. (1997), 'University versus corporate patents: a window on the basicness of invention', *Economics of Innovation and New Technology*, **5**(1): 19–50.

King, N. (1990), 'Innovation at work: the research literature', in M.A. West and J.L. Farr (eds), *Innovation and Creativity at Work: Psychological and Organizational Strategies*, Chichester, UK: John Wiley & Sons, pp. 15–59.

King, N. and Anderson, N. (1990), 'Innovation in working groups', in M.A. West and J.L. Farr (eds), *Innovation and Creativity at Work: Psychological and Organizational Strategies*, Chichester, UK: John Wiley & Sons, pp. 81–100.

Kivimaki, M. and Elovainio, M. (1999), 'A short version of the team climate inventory: development and psychometric properties', *Journal of Occupational and Organizational Psychology*, **72**: 241–6.

Kivimaki, M., Kuk, G., Elovainio, M., Thomson, L., Kalliomaki-Levanto, T. and Heikkila, A. (1997), 'The team climate inventory (TCI) – four or five factors? Testing the structure of TCI in samples of low and high complexity jobs', *Journal of Occupational and Organizational Psychology*, **70**: 375–89.

Kratzer, J., Leenders, R.T.A.J. and Van Engelen, J.M.L. (2005), 'Informal contacts and performance in innovation teams', *International Journal of Manpower*, **26**(6): 513–28.

Lam, A. (1996), 'Engineers, management and work organisation: a comparative analysis of engineers' work roles in British and Japanese electronics firms', *Journal of Management Studies*, **33**(2): 183–212.

Lam, A. (1997), 'Embedded firms, embedded knowledge: problems of collaboration and knowledge transfer in global cooperative ventures', *Organization Studies*, **18**(6): 973–96.

Lam, A. (1998), 'Tacit knowledge, organisational learning and innovation: a societal perspective', DRUID Working Paper No. 98-22.

Laursen, K. and Salter, A. (2006), 'Open for innovation: the role of openness in explaining innovation performance among UK manufacturing firms', *Strategic Management Journal*, **27**: 131–50.

Lehtoranta, O. (2005), 'A comparative micro-level analysis of innovative firms in the CIS Surveys and in the VTT's Sfinno Database', *Statistics Finland & VTT Technology Studies*, VTT Working Papers 24.

Leifer, R., McDermott, C.M., O'Connor, G.C., Peters, L.S., Rice, M.P. and Veyzer, R.W. (2000), *Radical Innovation: How Mature Companies Can Outsmart Upstarts*, Boston, MA: Harvard Business School Press.

Lemon, M. and Sahota, P.S. (2004), 'Organizational culture as a knowledge repository for increased innovative capacity', *Technovation*, **24**(6): 483–98.

Loewe, P. and Dominiquini, J. (2006), 'Overcoming the barriers to effective innovation', *Journal of Strategy and Leadership*, **34**(1): 24–31.

Mangematin, V. and Nesta, L. (1999), 'What kind of knowledge can a firm absorb?', *International Journal of Technology Management*, **18**(3–4): 149–72.

Martin, J.N.T. (2000), *Managing Problems Creatively, B822 Book 2, Open University Business School MBA*, Milton Keynes: The Open University.

Mathisen, G.E. and Einarsen, S. (2004), 'A review of instruments assessing creative and innovative environments within organizations', *Creativity Research Journal*, **16**(1): 119–40.

McLaughlin, P., Bessant, J. and Smart, P. (2005), 'Developing an organizational culture that faciliates radical innovation in a mature small to medium sized company: emergent findings', Cranfield University School of Management Working Papers.

Miranda, S.M. and Kavan, C.B. (2005), 'Moments of governance in IS outsourcing: conceptualizing effects of contracts on value capture and creation', *Journal of Information Technology*, **20**(3): 152–69.

Montes, F.J.L., Moreno, A.R. and Morales, V.G. (2005), 'Influence of support leadership and teamwork cohesion on organizational learning, innovation and performance: an empirical examination', *Technovation*, **25**(10): 1159–72.

Nadler, D.A. and Tushman, M.L. (1980), 'A model for diagnosing organizational behavior', *Organizational Dynamics*, **9**(2): 35–51.

Naveh, E. and Erez, M. (2004), 'Innovation and attention to detail in the quality improvement paradigm', *Management Science*, **50**(11): 1576–86.

Newell, A. and Simon, H.A. (1972), *Human Problem Solving*, NJ: Prentice-Hall.

Nonaka, I. and Takeuchi, H. (1995), *The Knowledge Creating Company*, New York: Oxford University Press.

Patel, P. and Pavitt, K. (1998), 'The wide (and increasing) spread of technological competencies in the world's largest firms: a challenge to conventional wisdom', in A.D. Chandler, Ö. Sölvell and P. Hagström (eds), *The Dynamic firm: The Role of Technology, Strategy, Organization and Regions*, Oxford: Oxford University Press, pp. 192–214.

Patterson, M.G., West, M.A., Shackleton, V.J., Dawson, J.F., Lawthom, R., Maitlis, S., Robinson, D.L. and Wallace, A.M. (2005), 'Validating the organizational climate measure: links to managerial practices, productivity and innovation', *Journal of Organizational Behavior*, **26**(4): 379–408.

Payne, R. and Pheysey, D. (1971), 'G.C. Stern's organizational climate index: a reconceptualization and application to business organizations', *Organizational Behavior and Human Performance*, **6**: 77–98.

Persaud, A. (2005), 'Enhancing synergistic innovative capability in multinational corporations: an empirical investigation', *Journal of Product Innovation Management*, **22**(5): 412–29.

Reigle, R.F. (2001), 'Measuring organic and mechanistic cultures', *Engineering Management Journal*, **13**(4): 3–8.

Reissner, S.C. (2005), 'Learning and innovation: a narrative analysis', *Journal of Organizational Change Management*, **18**(5): 482–94.

Rickards, T., Chen, M.H. and Moger, S. (2001), 'Development of a self-report instrument for exploring team factor, leadership and performance relationships', *British Journal of Management*, **12**(3): 243–50.

Robinson, S.L. and Rousseau, D.M. (1994), 'Violating the psychological contract: not the exception but the norm', *Journal of Organisational Behaviour*, **15**: 245–59.

Rogers, E.M. (1995), *The Diffusion of Innovations*, New York: Free Press.

Romijn, H.A. and Albu, M.A. (2002), 'Innovation, networking, and proximity: lessons from small high-technology firms in the United Kingdom', *Regional Studies*, **36**(1): 81–6.

Rousseau, D.M. (1995), *Psychological Contracts in Organisations: Understanding Written and Unwritten Agreements*, London: Sage.

Scherer, F. (1982), 'Inter-industry technology flows in the United States', *Research Policy*, **11**(4): 227–45.

Schilling, M.A. (2005), 'A "small-world" network model of cognitive insight', *Creativity Research Journal*, **17**(2–3): 131–54.

Scott, S.G. and Bruce, R.A. (1994), 'Determinants of innovative behavior – a path model of individual innovation in the workplace', *Academy of Management Journal*, **37**(3): 580–607.

Sharkie, R. (2005), 'Precariousness under the new psychological contract: the effect on trust and the willingness to converse and share knowledge', *Knowledge Management Research & Practice*, **3**: 37–44.

Siegel, S.M. and Kaemmerer, W.F. (1978), 'Measuring perceived support for innovation in organizations', *Journal of Applied Psychology*, **63**(5): 553–62.

Simonin, B.L. (2004), 'An empirical investigation of the process of knowledge transfer in international strategic alliances', *Journal of International Business Studies*, **35**(5): 407–27.

Smith, K. (2005), 'Measuring innovation', in J. Fagerberg, D. Mowery and R. Nelson (eds), *The Oxford Handbook of Innovation*, Oxford: Oxford University Press, pp. 148–77.

Smith, W.K. and Tushman, M.L. (2005), 'Managing strategic contradictions: a top management model for managing innovation streams', *Organization Science*, **16**(5): 522–36.

Sparrow, P.R. and Gaston, K. (1996), 'Generic climate maps: a strategic application of climate survey data?', *Journal of Organizational Behaviour*, **17**(6): 679–98.

Subramanian, A. (1996), 'Innovativeness: redefining the concept', *Journal of Engineering and Technology Management*, **13**(3–4): 223–43.

Tang, H.K. (1999), 'An inventory of organizational innovativeness', *Technovation*, **19**(1): 41–51.

Teo, H.H., Wang, X.W., Wei, K.K., Sia, C.L. and Lee, M.K.O. (2006), 'Organizational learning capacity and attitude toward complex technological innovations: an empirical study', *Journal of the American Society for Information Science and Technology*, **57**(2): 264–79.

Terziovski, M., Fitzpatrick, P. and O'Neill, P. (2003), 'Successful predictors of business process reengineering (BPR) in financial services', *International Journal of Production Economics*, **84**(1); 35–50.

Tether, B.S. (2001), 'Identifying innovation, innovators and innovative behaviours: a critical assessment of the Community Innovation Survey (CIS)', CRIC Discussion Paper No. 48, CRIC, University of Manchester and UMIST, Manchester, UK.

Tether, B.S. (2002), 'Who cooperates for innovation and why: an empirical analysis', *Research Policy*, 31(6): 947–67.

Tether, B.S. and Swann, P. (2003), 'The use by industry of the science base for innovation: evidence from the UK's innovation survey', *CRIC Discussion Paper No. 64*, (8 August, Version 1.2).

Tushman, M.L. and O'Reilly, C.A. (1996), 'Ambidextrous organizations: managing evolutionary and revolutionary change', *California Management Review*, 38(4): 8–30.

Vickers, I. and Cordey-Hayes, M. (1999), 'Cleaner production and organizational learning', *Technology Analysis & Strategic Management*, 11(1): 75–94.

Wang, C.L. and Ahmed, P.K. (2004), 'Leveraging knowledge in the innovation and learning process at GKN', *International Journal of Technology Management*, 27(6–7): 674–88.

Weisberg, R.W. (1999), 'Creativity and knowledge', in H.R.J. Sternberg (ed.), *Handbook of Creativity*, New York: Cambridge University Press, pp. 226–50.

West, M. (1990), 'The social psychology of innovation in groups', in M.A. West and J.L. Farr (eds), *Innovation and Creativity at Work: Psychological and Organizational Strategies*, Chichester, UK: John Wiley & Sons, pp. 309–33.

West, M.A., Sacramento, C.A. and Fay, D. (2005), 'Creativity and innovation implementation in work groups: the paradoxical role of demands', in L.L. Thompson and H.-S. Choi (eds), *Creativity and Innovation in Organizational Teams*, Mahwah, NJ: Lawrence Erlbaum Associates, pp. 137–59.

Wikipedia (2006), Psychological contract, http://en.wikipedia.org/w/index.php?title= Psychological_contract&oldid=88473211.

Zarraga, C. and Bonache, J. (2005), 'The impact of team atmosphere on knowledge outcomes in self-managed teams', *Organization Studies*, 26(5): 661–81.

Zwick, T. (2002), 'Employee resistance against innovations', *International Journal of Manpower*, 23(6): 542–52.

PART IV

Wealth from Knowledge

12. How firms source knowledge from universities: partnering versus contracting

Markus Perkmann and Kathryn Walsh

INTRODUCTION

There is an expanding literature stressing the importance of external knowledge sources for firms. Research alliances and technological collaboration (Freeman 1991; Hagedoorn et al. 2000), open innovation (Chesbrough 2003), networked and distributed innovation (Coombs et al. 2003; Powell et al. 1996) are concepts used for describing and theorizing this phenomenon. Empirically speaking, several trends indicate that external sources of innovation are becoming more important in the overall innovation strategies of firms (Fey and Birkinshaw 2005): knowledge required for innovating is more dispersed within the economy, particularly in rapidly changing areas such as biotechnology (Chesbrough 2006; Powell et al. 1996); products include a broader range of different technologies (Iansiti 1997); some industries are moving towards open standards and modular innovation (Baldwin and Clark 1997), and outsourcing strategies have expanded to include innovation-intensive components and systems (Harabi 1998); and finally, the type of parties involved in innovation processes appear to become more disparate, as for instance in the case of user-driven innovation (von Hippel 1987).

Links between firms and universities can be seen as part of this general scenario. The generic economic and social benefits of universities,[1] such as educating cohorts of graduates, generating scientific knowledge and instrumentation,[2] have long been recognized as an important source of industrial innovation (Pavitt 1991; Salter and Martin 2001). Yet recently, universities have made efforts to engage more directly in industrial innovation processes. Various indicators suggest that firms increasingly seek to access and utilize university-generated knowledge: a rising patenting propensity by universities (Nelson 2001), growing university revenues from licensing (Thursby et al. 2001), increasing numbers of university researchers engaging in academic entrepreneurship (Shane 2005), a growing share of industry funding in university income (Hall 2004; Perkmann and Walsh 2007), the diffusion of technology transfer offices, industry collaboration support offices and science parks (Siegel et al. 2003) and a growing number of university spin-out companies (Lockett and Wright 2005).

Against the background of these trends, it appears that actual *relationships* between universities and industry – rather than generic *links* – play a stronger role in generating innovations. Concepts such as 'technology transfer' and 'commercialization' are used to describe the essence of these relationships: universities generate valuable technologies that are subsequently transferred to industrial firms who commercially exploit them. The prevalence of the transfer metaphor is reinforced by the fact that the generation and transfer of university-generated intellectual property rights (IPR) (patenting, licensing) have been widely studied by academic researchers (Phan and Siegel 2006). In conjunction with university spin-off companies, they are also regarded as key indicators for successful university–industry interaction by policy makers. This is for instance illustrated by the importance attributed to the 1980 Bayh–Dole Act in the USA that granted universities intellectual property rights on research results performed by their staff (Mowery et al. 2001). Behind these efforts lies the idea that universities represent engines of growth by supplying the economy with inventions and discoveries (Florida and Cohen 1999).

Against this focus on technology transfer and technology commercialization, recent research has emphasized the multiple ways in which firms interact with university researchers (Arundel and Geuna 2004; Cohen et al. 2002; Meyer-Krahmer and Schmoch 1998). In this context, the transfer of codified knowledge, via patenting and licensing, appears to be only moderately important compared to other modes of interaction (Agrawal and Henderson 2002; Cohen et al. 2002; Schartinger et al. 2002). Moreover, links with universities impact in various ways on innovation conducted in firms. For instance, interaction with universities not only allows firms to access new discoveries and new ideas, but also contributes to completing ongoing development projects (Cohen et al. 2002), 'scanning the research frontier', generating specific instrumentalities and providing hands-on assistance (Faulkner and Senker 1994).

Although extant research has emphasized the observed variety of interactions and their relative importance for firms, less attention has been paid to why firms use different mechanisms, and the opportunities and tensions involved in each of them. In this chapter, we intend to place specific emphasis on two aspects of this wider question: first, what are the types of knowledge and expertise that firms source from universities, and second, what are the mechanisms available for sourcing university-generated knowledge or expertise?

In terms of the first aspect, we contend that the role of universities for industrial innovation is broader than being suppliers of discoveries and inventions. As authors such as Stokes (1997) have shown, university research is often inspired by industrial use and covers many 'applied' areas. Hence many university researchers have knowledge and expertise that enables them to participate in industrial innovation processes in other ways than via basic discoveries.

In terms of the second aspect, the mechanisms used for university–industry interaction, we relate our discussion to the different models that have been proposed in the literature as to how innovation processes are syndicated between different organizations. On one hand, authors such as Powell et al. (1996) argue that collaborating organizations in newly emerging fields generate innovation by engaging in mutually interdependent partnerships. This suggests a model of inter-organizational collaboration that relies on co-exploration and cumulative learning whereby the locus of innovation resides in the interstices between organizations (Powell et al. 1996).

On the other hand, Chesbrough (2003) proposes a model of open innovation whereby organizations generate an excess of ideas, some of which they seek to sell to other organizations. Simultaneously, in many cases it makes sense for organizations to search for ideas they require in the external environment rather then relying on internal sources. Chesbrough argues that different roles and functions within the innovation process become increasingly differentiated. This means that organizations become specialized in particular stages of the innovation process, as for instance funding innovations, generating innovations or commercializing innovations. In this scenario, contracting mechanisms play an important role for syndicating innovation activities between these different types of organizations. Therefore the management and leverage of intellectual property is becoming increasingly relevant for firms, possibly with the assistance of intermediaries acting as brokers between the participants in the innovation process (Arora et al. 2001; Chesbrough 2006; Hoppe and Ozdenoren 2005). Both scenarios hint at different ways in which inter-organizational relationships are governed, as we discuss in more detail below.

In the remainder of this chapter, we explore the characteristics of university–industry links as inter-organizational relationships. Specifically, we analyse what firms gain from engaging with universities, and what type of relationships are involved. To this purpose, we first comment on the two key dimensions mentioned above: the type of knowledge sourced from universities, and the way in which relationships between firms and universities are governed. We then analyse the most common types of university–industry relationships (research partnerships/collaborative research, consulting/ contract research, licensing), in light of our conceptual considerations, and discuss the incentives and trade-offs involved for the participants.

TWO DIMENSIONS OF EXTERNAL KNOWLEDGE SOURCING

From the viewpoint of the firm, there are two key dimensions of external knowledge sourcing: the nature of external knowledge that is being sourced, and the way the relationship is organized.

Regarding the first dimension, Cohen and Levinthal (1989) have distinguished between two 'faces' of R&D conducted by firms: innovation and learning. While innovation refers to a firm's efforts to generate information and develop products and process along an established trajectory, learning refers to the activity of assimilating and exploiting new knowledge drawn from its environment, notably from university research or other economic sectors. In the literature, this process is often referred to as accessing 'technological opportunities' (Cohen and Levinthal 1990; Klevorick et al. 1995).

Yet firms turn to their environment not only to access technological opportunities but also to more directly support their innovating efforts. This is for instance indicated by activities such as outsourcing of R&D activities (Chiesa et al. 2004; Howells 1999; Pisano 1990). Mirroring the distinction between innovating and learning, we can therefore identify two types of external knowledge sought by firms: knowledge related to ongoing development activities, and knowledge related to technological opportunities. Sourcing knowledge required for ongoing development is essentially a 'make or buy' question (Veugelers and Cassiman 1999), balancing the risk and costs of deploying internal resources against those of external resources. The sourced knowledge is likely to be generic and non-firm-specific, including R&D tasks with specialization advantages, as for instance materials testing and process improvements (Mowery and Rosenberg 1989; Padmore et al. 1998; Veugelers and Cassiman 1999).

By contrast, access to knowledge about technological opportunities allows a firm to engage in the development of novel components, products, processes or architectures. As such knowledge is rare and unique, firms are usually not in the position to choose the 'make' option, i.e. generate such knowledge internally. Firms will usually seek exclusivity in the use of this knowledge, via intellectual property rights, or otherwise attempt to maximize appropriability via first-mover advantages, secrecy and other mechanisms (Levin et al. 1987).

Having considered the type of knowledge than can be externally sourced, we now discuss the second dimension of external knowledge sourcing: the mode of governance. Activities can be governed by a spectrum of organizational arrangements ranging from hierarchy to market (Powell 1990; Williamson 1999). From this spectrum, we exclude hierarchy, i.e. vertical integration, from our analysis as firms and universities are unlikely to consolidate into single organisations. We also exclude the other extreme, spot-market transactions, as what is being transacted between firms and universities is unlikely to take the shape of standardized and transparent goods and services.

In between these two extremes, modes of governance can range between tightly integrated partnerships involving close cooperation over longer periods of time to arm's-length relationships akin to outsourcing arrangements. For current purposes, the spectrum can be condensed into two main ways in which firms access external knowledge and expertise: partnering and contracting (Fey and Birkinshaw 2005). While partnering involves close cooperation between the parties and subsequent co-creation of assets, contracting refers to a more asymmetric relationship in which one partner acquires knowledge assets or embodied expertise held by the other partner.

Within the specific context of developing innovations, *partnering* is used by firms in situations that require close and ongoing cooperation between the involved parties. A partnering strategy is opportune when the creation of a knowledge asset involves the iterative combination of complementary competences or other assets (Gulati and Singh 1998). Equally, when the precise nature of a knowledge asset to be created is underspecified or the objective of a project is speculative and explorative, partnering will be a preferred option.

By contrast, *contracting* means a firm buys or commissions knowledge assets or expertise from other parties without having contributed to the production of those assets. In other words, contracting allows for the acquisition of knowledge assets that are already present, such as in the case of patents or other 'packaged' technology. Alternatively, knowledge assets can be commissioned according to contractually agreed specifications (Fey and Birkinshaw 2005). For instance, contracting is more likely to be used for generic, non-firm-specific R&D that allows for specialization advantages, such as routine research tasks like materials testing, and process rather than product innovations (Mowery and Rosenberg 1989).

These two dimensions, type of sourced knowledge and relationship governance, define the space of possibilities for external knowledge sourcing: one the one hand, the sourced knowledge can be about novel technological

Mode of governance		
	Partnering	Contracting
Technological opportunity	IV Exploration alliance	I Technology acquisition
Development activities	III Exploitation alliance	II Outsourced R&D

Knowledge relating to

Figure 12.1   Strategic options for external knowledge sourcing

opportunities, or it can consist in specialized knowledge and expertise required as input for ongoing innovation projects; on the other, relationships can be tightly coupled and partnership-based, or they can be contractual and akin to sourcing relationships.

This results in four different logics (Figure 12.1): technology acquisition, referring to the transfer of well-defined technology assets within a technology market (Arora et al. 2001); the outsourcing of R&D and technical development to external providers (Howells 1999); exploitation alliances, i.e. cooperation in the development and commercialization of specific product or service; exploration alliances, i.e. collaboration with the objective to explore and discover new technology (Koza and Lewin 1998; Nooteboom 2004).

In the following section, we explore various modes of university–industry interaction by mapping them onto this framework of inter-organizational relationships. We are particularly interested in what a firm can obtain by working with universities, and what this involves for how relationships are structured.

ANALYSING DIFFERENT MODES OF UNIVERSITY–INDUSTRY COLLABORATION

Firms use a variety of mechanisms for interacting with universities. For instance, Cohen et al. (2002) distinguished between the following 'channels' relevant to industrial innovation: patents, informal information exchange, publications and reports, public meetings and conferences, recently hired graduates, licences, joint or cooperative research ventures, contract research, consulting, and temporary personnel exchanges. How does this mixed list of channels map onto the above framework of external knowledge sourcing? The list contains a number of items that do not refer to inter-organizational arrangements strictly speaking and therefore need to be removed from our considerations. Notably, channels such as patents or publications effectively represent *media* rather than organizational arrangements although they can be used in conjunction with the latter. The category of 'recently hired graduates' is economically important yet it refers to a general infrastructural function of universities, as opposed to directly contributing to specific innovation processes within firms (Salter and Martin 2001). Finally, meetings and conferences represent both sources of information and opportunities to build social capital, yet will rarely in themselves be effective and sufficient in terms of knowledge sourcing. Rather, they might initiate contacts leading to closer involvement or may occur in conjunction with specific inter-organizational arrangements such as collaborative research.

The remaining categories can be synthesized into three main types of inter-organizational arrangements: licensing, research partnerships and research services.

Licensing refers to contractual agreements according to which firms acquire the right to use university-generated intellectual property against a fee.

Research partnerships are formal collaborative arrangements with the objective to cooperate on R&D activities. The terms 'collaborative research' or 'joint research' are often used to refer to these relationships. They can range from small-scale, temporary projects to larger, long-term university–industry research centres to permanent, large-scale consortia with hundreds of industrial members.

Research services are paid-for services performed by university researchers for external clients, notably consulting and contract research. Although in practice the boundaries between these two activities are blurred, consulting exploits an academic researcher's existing expertise while in contract research the industrial client requires some research to explore specific aspects of a problem. Moreover, proceeds from consulting tend to be drawn personally by academics, while contract research income is usually taken by universities and/or research groups.

There are systematic differences between industrial sectors and academic fields in terms of the predominant linking mechanisms. This goes beyond the fact that some sectors depend on science to a larger degree, with the pharmaceutical, biotechnology and chemical sectors ranging among the most 'science-intensive' sectors according to several measures (Cohen et al. 2002; Faulkner and Senker 1994; Klevorick et al. 1995). In fact, the level and modalities of university–industry relationships cannot simply be mapped onto the distinction between science-intensive sectors and those that are not. For instance, US results indicate that while partnerships (collaborative research or research joint ventures) are considered important in diverse sectors such as pharmaceuticals, steel, TV/radio and aerospace, academic consulting is highly relevant in various sectors such as food, medical equipment, petroleum, metals, search/navigational equipment and pharmaceuticals (Cohen et al. 2002). Schartinger et al. (2002) confirm this complex picture in an Austrian study by finding that among the sectors with intense industry–university interaction are, on the one hand, those with high R&D ratios (chemicals, instruments) and, on the other, sectors with low R&D ratios such as energy, basic metals, construction and agriculture. Similar evidence is provided by Meyer-Krahmer and Schmoch (1998) for Germany; they show that 'open science' channels,[3] mostly in the form of research partnerships, are predominant in the chemical industry while the mechanical engineering sector mostly relies on contract research and consulting. Given the disparities between different surveys, clear-cut general conclusions cannot be established. It appears, however, that in science-based sectors such as pharmaceuticals, biotechnology or chemicals with strong complementarities between academic research and firm-level R&D, firms tend to use both collaborative research and research services (contract research and consulting) relatively intensively while sectors relying less on scientific breakthroughs, such as mechanical engineering or software development, prefer to work with universities via contract research and consulting. Finally, licensing is predominantly used in biotechnology and pharmaceuticals, due to the nature of science-push innovation pursued in those sectors yet not necessarily at the expense of collaborative relationships between firms and universities (Cohen et al. 2002; Faulkner and Senker 1994).

In view of this diversity of university–industry relationships, in the following we discuss how they map onto the above framework of inter-organizational relationships. We begin by discussing ways in which firms can access knowledge about technological opportunities, and continue by analysing how they can solicit assistance with ongoing development projects. For each aspect, we discuss the interests of the parties involved and the constraints inherent in specific inter-organizational arrangements.

ACCESSING TECHNOLOGICAL OPPORTUNITIES

Publicly funded research constitutes a significant source of technological opportunities, particularly in some sectors, such as the pharmaceutical and chemical industries (Cohen et al. 2002; Mansfield 1995; Pavitt 1990). For firms, to access such opportunities will most probably involve engaging in inter-organizational relationships with universities. Although theoretically firms might choose to consult the openly available scientific literature – without establishing relationships with universities – this will be an unlikely avenue for accessing commercially valuable novel technologies, for several reasons. First, in the institutional environment currently encouraged by policy makers, universities are likely to protect research results that they perceive as potentially commercially exploitable. The literature even suggests that universities tend to err on the upside, i.e. patent too much, implying that a considerable proportion of university-generated intellectual property rights are never exploited (Mowery et al. 2001). Second, open science results tend to be used in industrial technology with a significant time-lag, estimated at around 20 years (Allen 1977; de Solla Price 1984). Third, novel discoveries tend to be insufficiently codified and therefore require the collaboration of the inventor (Agrawal and Henderson 2002; Zucker et al. 2002).

There are two ways in which firms can access such technological opportunities via inter-organizational relationships. A first way is contractually to acquire a body of technology developed by university researchers, usually via licensing agreements. A second way is to engage in a research partnership with a university, aiming to generate technological opportunities through close collaboration. We discuss each of these options in turn.

Licensing an invention patented or otherwise protected by a university often enables a firm to initiate new pathways of technology development, for instance when biotechnology companies license a novel molecular entity discovered in a university laboratory. Developments such as the Bayh–Dole Act in the USA and the discourse surrounding the 'entrepreneurial university' suggest that licensing is attributed a dominant role in mediating firm access to university knowledge (Etzkowitz 2003; Garnsey 2007; Slaughter and Leslie 1997). A number of industrialized countries have designed policies based on the premise that universities support innovation in industry primarily via 'deliverables', such as patented inventions (Mowery and Sampat 2005: 225). At many universities, an emerging group of professionals, technology transfer officials, operate with a remit to identify commercializable inventions, protect the university's intellectual property rights and initiate licensing deals with commercial buyers (Debackere and Veugelers 2005; Owen-Smith and Powell 2001; Siegel et al. 2003). Licensing is relevant particularly in industries where patented or otherwise protected intellectual property plays an important role, i.e. in biotechnology and information technology (Brusoni et al. 2005; Niosi 2006).

In light of our distinction made above, licensing qualifies as contracting involving the purchase of the right to use intellectual property rights from a technology owner. In terms of the type of knowledge being sourced, most licensing deals with universities refer to early-stage technologies (Colyvas et al. 2002; Thursby and Thursby 2004), i.e. access to technological opportunities, although on some occasions they might provide complements and improvements to ongoing downstream technology development projects. In terms of our framework, licensing can therefore be classified as technology acquisition.

However, in terms of governance, the picture is complicated by the fact that many licences require inventor collaboration for commercial success. Many patents written by universities represent 'embryonic' technology that requires considerable further development. For instance, a US survey of firms licensing inventions from universities showed that approximately 40 per cent of all licences require faculty involvement (Thursby and Thursby 2004). According to a survey among technology transfer offices at the top 165 US universities, even 71 per cent of licensed inventions were seen as requiring inventor cooperation for commercial success (Thursby et al. 2001). Similarly, Agrawal (2006) finds that in two-thirds of a sample of patented inventions licensed by the Massachusetts Institute of Technology (MIT), the academic inventors were actively involved in the further development of the technology within a commercial context. This evidence suggests that the need for inventor involvement is prevalent across a range of industry sectors, including biotechnology, chemistry and engineering (cf. Zucker et al. 2002).

Therefore, for the majority of university licences the contracting arrangement is supplemented by a partnering arrangement with ongoing involvement of the inventor. Licensing often includes collaborative research as part of the royalty agreement, particularly when the technology is at an early stage (Jensen and Thursby 2001). For firms, sponsored research can even be a substitute for a licensing deal when technologies are too embryonic to license or when the technology involves a platform development or process improvement (Thursby and Thursby 2004). Licensing

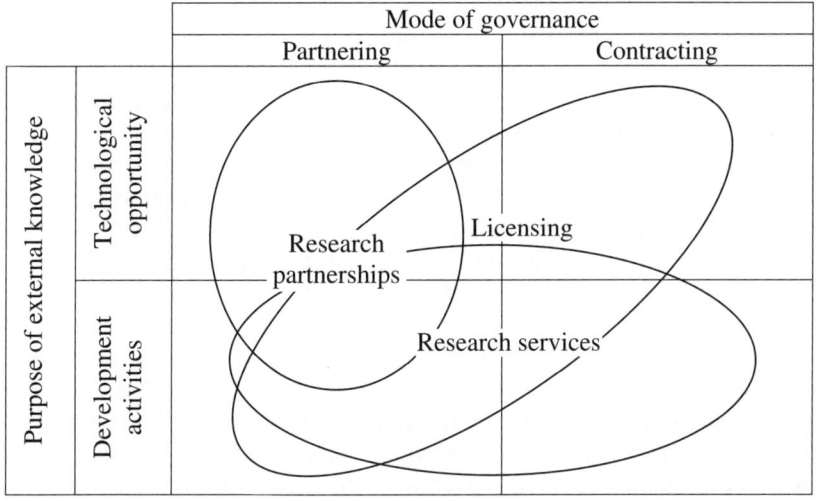

Figure 12.2   Strategic options for knowledge sourcing from universities

therefore often includes both a contracting and a partnering element (Figure 12.2).

Alternatively, academic inventors – who are usually granted royalty rights by their home institutions – can collaborate with the licensee as consultants (Agrawal 2006); (Mansfield 1995). The underlying reason for such inventor engagement is that particularly in the case of novel technologies a considerable part of related expertise is usually not available in codified form, and hence the inventor can claim exclusivity (Zucker et al. 2002). Engaging the inventor through partnership therefore represents a mechanism for firms to 'capture' rare knowledge that is too expensive to codify when its perceived value is low but gets quickly eroded by competition when its value is high (Zucker et al. 2002). This enables a firm to accelerate technology development and hence enjoy first-mover advantages before the expertise diffuses via codification. Such partnerships are likely to be structured in a way that links commercial success with monetary incentives to the inventor in order to avoid moral hazard problems (Jensen and Thursby 2001). As a result, many licensing agreements are not pure contracting relationships but will contain a partnership element in which both partners work together on the successful commercialization of the technology in question.

Research partnerships with a university provide firms with an alternative to licensing to access technological opportunities. This is particularly relevant for the large majority of sectors and disciplines where university-generated intellectual property rights play a minor economic role. Financial figures help to put the importance of research partnerships in context. For instance, in the UK, income of higher education institutions from collaborative research outstrips their income from intellectual property (mostly licensing) by a factor of 13 (DfEL 2005). Although most of these funds are contributed by public research funding organizations, these figures illustrate the extent to which external private and public organizations are involved in research projects pursued at universities.

Research partnerships are by definition partnering arrangements to which both parties bring to bear their assets and competences. They differ from licensing arrangements in that there is no pre-existing body of technology the partners view as commercially valuable enough to justify intellectual property provisions. Research partnerships are typically chosen for generating technological opportunities and therefore typically qualify as exploration alliances.

In most cases, research partnerships are subsidized by public policy programmes.[4] From the viewpoint of the firm, the availability of public funds amounts to a *de facto* subsidy that allows them to leverage its resource budgets. For this reason, work undertaken within such partnerships tends to be 'pre-competitive', i.e. without immediate commercial appropriation potential. The mobilization of public funding means that universities claim full or at least part ownership on intellectual property arising from collaborative projects (Caloghirou et al. 2001; Fontana et al. 2006; Ham and Mowery 1998; Owen-Smith 2005; Poyago-Theotoky et al. 2002).

Panagopoulos (2003) proposes an economic model according to which firms are more likely to choose research joint ventures (i.e. research partnerships) when the technology in question is novel and not well understood. Under these conditions, the opportunity cost of disclosing intellectual property and forgoing (some) intellectual property arising from collaborative research is lower and gains from knowledge 'spillovers' arising from cooperation are higher than in the case of developing and improving mature technologies. Firms will accept such agreements for upstream R&D when outcomes are highly uncertain, time frames are long term and the commercial use value of the technology in question is not proven.

Partly because of this degree of uncertainty, and partly because the interests of the participating researchers have to be served, firms engaged in research partnerships do not usually expect concrete outcomes with commercial implications. Firms' motives for engaging in such links are informed by generic benefits such as accessing students, gaining 'windows' on emerging technologies and enhancing their knowledge base rather than by the desire to develop specific commercializable innovations (Caloghirou et al. 2001; Feller 2005). As a result, firms often choose not to assess the value of these relationships via hard performance measures (Ham and Mowery 1998) and are not concerned about making a quantitative case for participation (Feller et al. 2002).

SOURCING KNOWLEDGE FOR DEVELOPMENT ACTIVITIES

Universities are used by firms not just for accessing technological opportunities but also for ongoing development requirements. Cohen et al. (2002) have shown that firms not only rely on university research as a source for new project ideas but also use it for completing existing R&D projects. Similarly, Breschi and Lissoni (2001) distinguished between two rationales informing firms' access to knowledge from universities: on one hand, by accessing innovation opportunities arising from novel ideas and technology, and, on the other, by appropriation activities aimed at developing products or processes.

For the purpose of the latter type of requirement, firms are likely to source research services from universities, notably via contract research or consulting. For ongoing development activities, a firm will be more reluctant to forfeit rights to IP than it might do in more risky and exploratory research conducted within research partnerships. For development activities already initiated, more specific, proprietary outcomes are sought and degrees of uncertainty are lower. In an analysis of US research joint ventures, Hall et al. (2001) find that firms are more likely to view IP restrictions applying to partnerships as 'insurmountable' when research output is more appropriable by firms, and when the IP characteristics of the research are more certain. In addition, public funding for downstream research is not as readily available although there is a tension in public funding between limiting support to upstream research and providing 'innovation assistance' to companies. Firms have to contribute a higher proportion or even the full project cost, hence making it likely that contract research and consulting arrangements are chosen for such projects. This means firms will exert a higher degree of control over what output is to be generated, and manage projects to tighter deadlines compared to research partnerships. In turn, this is only possible when objectives can be meaningfully specified and the uncertainty relating to the technology in question is low.

From the viewpoint of the firm, academics are expert specialists capable of resolving specific problems arising in development processes or providing scientifically grounded technological advice. Hiring them as consultants can therefore be an attractive alternative to building internal expertise that would involve major investments in human capital. In some cases, such relationships may conform to a contracting logic and hence in our terminology constitute outsourced R&D, particularly when deliverables can be concisely defined *a priori* and no actual collaboration between firm staff and academic researchers is required to produce these deliverables.

However, in other cases research services will be performed within the context of partnering arrangements between firms and universities, and hence constitute what we have defined as exploitation alliances. This is because there is a limitation on what degree academic expertise can be 'bought' from universities via a contracting transaction. Academic researchers work in an environment that rewards scientific novelty and ensuing publication activity (Owen-Smith 2005). Compared to research partnerships focusing on exploratory research, research services are less likely to generate academically valuable outputs. Research services differ from other forms of university–industry relationships in that they mobilize expertise that is commonly held within academic communities (Agrawal and Henderson 2002). Research services can therefore be seen as leveraging 'old science' (Allen 1977; Gibbons and Johnston 1974; Rosenberg 1994). They resolve problems and provide improvements rather than suggesting new project ideas or pioneering new design configurations (Gibbons 2000; Utterback 1994). Indeed, they are usually seen to be of lesser academic value (Boyer and Lewis 1984) as they do not directly contribute to research output or teaching. A UK survey on motivating factors for industry–academic collaboration suggests that barriers to establishing consultancy links are somewhat different from those for research partnerships, suggesting different incentive structures (Howells et al. 1998). This has to be judged in conjunction with the fact that few academics get involved with industry for purely financial reasons (Lee 1996). Survey evidence suggests that monetary incentives are rarely the primary driver for academics to offer consulting to outside organizations (Boyer and Lewis 1984; Jones 2000; Patton and Marver 1979).

Under what circumstances will academic researchers have an interest in engaging in exploitation alliances with firms? Academics will be interested in such alliances if they generate complementarities with their academic research. Two considerations suggest why this might occur. First, university research is not uniquely about basic science (McKelvey 1996). A considerable amount of research undertaken at universities and other public research organizations is 'applied' in the sense that it addresses technical problems and seeks technical solutions (Niiniluoto 1993). Particularly areas that grew out of professional practice, such as engineering, law and medicine, have retained a strong connection to issues of application and implementation. This is reflected in the practices deployed, for instance, by engineers compared to scientists (Allen 1977).

Second, even 'basic science' is not always removed from industrial application and product development. As Stokes (1997) has pointed out in his work on 'Pasteur's quadrant', there is a specific type of research activity that aims at resolving practical problems yet seeks fundamental scientific understanding (i.e. 'basic science'). For instance, a Pasteur logic applies to many research areas in biotechnology, computer science or aeronautical engineering. This means that research outputs are utilized, on one hand, as inputs to future research and, on the other, as inputs to ongoing technology development (Nelson 2004). As much of technological development occurs in industry, this circular relationship between science and application enables university involvement in downstream technology development activities in industry. The emphasis on the sharp contrast between open, basic science pursued at universities, and commercial, applied research pursued in industry appears therefore overstated (Mowery and Sampat 2005).

These considerations suggest that academics interested in applied science or working on Pasteur-type fundamental research have incentives to get involved in providing research services as this might inform their publicly funded research work (Mansfield 1995). This will be the case particularly if results from these activities can, for instance in generalized or anonymized form, be used in research publications. Such complementarities are likely to arise when academics are placed within social networks linking them with industrial firms on a long-term basis (Murray 2002).The embeddedness in such networks allows for the development of trust, ensuring that confidential industrial information is not 'leaked' into scientific publications while maintaining opportunities for academics to use the information for academic purposes.

In addition, for academics, the marginal cost of providing research services is often small as they already possess the required expertise and will usually not have to acquire new knowledge to carry out the assignment. In economic terms, it allows them to appropriate rents on their expertise. Many academics are therefore happy to engage in research services work, particularly if it offers some academic benefits in addition to the pecuniary reward.

CONCLUSIONS

In this chapter, we explored the variety of university–industry interactions through the lens of inter-organizational relationships. This observed variety can in part be explained by the different types of knowledge that firms seek to gain by engaging with universities: novel technological opportunities, and knowledge required for supporting ongoing technological development projects. A further source of variation lies in the characteristics of different types of relationships required to successfully mobilize this knowledge. In this respect, we distinguished between partnering, involving high levels of interdependence and collaboration, and contracting, i.e. the acquisition of knowledge or expertise in a more market-like transaction. In light of our discussion, we can draw the following conclusions.

Firms can access technological opportunities in two ways: licensing and research partnerships. While licensing allows for often exclusive access to a defined and contained body of technological knowledge, research partnerships are more uncertain in that they are pursued in the expectation of future, potentially commercially relevant research results. In light of the existing research, the licensing of university technology is arguably less important than is commonly assumed. This is illustrated both by surveys among firm representatives, as well as by figures on the proportion of university income from both licensing and research partnerships. Partly, this is because IPR play a decisive role for knowledge transfer, and innovation broadly speaking, only in selected sectors of the economy (Patel and Pavitt 1995). More importantly, university-generated technology, even if patented, is often embryonic and requires considerable further development, hence imposing considerable uncertainties on potential licensees. Even if these uncertainties are seen as acceptable, inventor involvement is often required to improve the likelihood of success (Agrawal 2006). Rather than pure contracting, partnership arrangements with the inventor and their research group are therefore often required to bring a technology to commercial fruition.

Research partnerships, by contrast, face a different set of constraints. As partnerships, they have to accommodate the interests of both the academy and industry. This means that the objectives of scientific novelty, and hence open publishing, have to be balanced against both the industrial 'direction' of the research pursued and the appropriability of the results. For this reason, the overarching majority of research partnerships receive some degree of public funding support, allowing participating firms to trade the lack of control against the lowered cost of participation. One of the implications is that IPR will usually reside with the university. In terms of appropriation, firms will therefore have to resort to accessing the generated knowledge in ways that are not mediated by IPR, or if appropriate negotiate follow-on licensing agreements.

If firms wish to have more control over both the direction of the collaborative work and appropriability, they can commission research services, i.e. contract research or consulting arrangements. This allows them to obtain assistance from academic collaborators for specific ongoing technology development. Although this may be akin to contracting R&D inputs, firms face constraints as to what they might be able to 'buy' from universities in this way. Given that they operate in an institutional realm that rewards contributions towards open science, many academics will seek complementarities with their own academic research activities when engaging in research services.

These considerations suggest that all modes of university–industry engagement are likely to contain elements of partnerships. At the same time, we argued that given institutional constraints and the nature of knowledge involved, pure contracting relationships of university technology are relatively rare. The main institutional constraint relates to the fact that the partners belong to different organizational fields characterized by specific rules, norms and objectives. This holds despite the fact that some observers identified a convergence of both fields towards a 'hybrid regime' in which university science and industrial development are increasingly integrated (Owen-Smith 2003). While the interests of firms are defined by commercial motives, their academic collaborators are to a large degree oriented towards their own success criteria, i.e. contributing to 'open science' via scientific publications (Dasgupta and David 1994).

In relating these results back to the different models of inter-organizational innovation discussed above, the cumulative learning model, as proposed by Powell et al. (1996), appears to characterize university–industry relationships better than open innovation models based on contracting relationships, as proposed by Chesbrough (2003). This view is supported by the fact that for both firms and universities there are often complementarities among various types of relationships they are engaged in. For instance, research services are often provided in conjunction with research partnerships. This allows academics to deploy their expertise for downstream problem solving within the overall context of an upstream-focused partnership that is academically fruitful. Mansfield's (1995) research suggests that academics who collaborate with industry often combine their research activities with industrial consulting. Consulting activities appear to be part of these ongoing network relationships in the same way as collaborative research (Faulkner and Senker 1994). We also argued that research services are often carried out in conjunction with licensing agreements, and that research partnerships might result in subsequent licensing. This suggests that the real unit of analysis in studying university–industry relationships might be stable networks linking academic researchers with industry staff involved in polyvalent, mutually beneficial activities (Murray 2002).

The main practical implication for firms is that if they wish to access the benefits of university research, they have to engage in some form of partnership with university researchers. Partnerships require higher investments in terms of relationship management, and are also subject to conditions imposed by the university partners. First, if collaboration with universities takes the form of partnerships rather than contracting, universities will naturally wish to own rights to exploit IPRs arising from the research. This is particularly relevant for cases where public subsidies contribute to collaboration budgets. Hence firms have to identify alternative ways of appropriating the results of their work within universities. Second, in partnerships academics will insist that at least some of the results are made available for the realm of open science, implying that considerations of commercial secrecy are at times subordinated to the publication of scientific findings. This also implies that some if not all the resources made available for a collaborative project will be deployed by scientists for activities directly related to their publishing efforts, although this might not be required for the commercial exploitation of results. Third, outputs from partnerships are likely to be of a hybrid nature in the sense that they will partly be addressed to the world of open science and partly to industrial audiences. This means firms working with universities need 'translation' competence to turn these outputs into commercially exploitable knowledge assets. Developing such absorptive capacity (Cohen and Levinthal 1990) is hence a major precondition for being able to profit from partnerships with public science organizations.

ACKNOWLEDGEMENTS

This chapter benefited from comments by Erik Stam. A previous version was presented at the EURAM conference, 16–19 May 2007, Paris.

NOTES

1. We use the term 'university' to include all types of 'public research organizations' (PROs). PROs are research organizations that are funded mainly by government, i.e. universities, public research laboratories, research institutes, etc.

2. Instrumentation includes devices and methods used for measurement, experimentation and analysis used in scientific research and R&D (Rosenberg 1992).
3. 'Open science' refers to knowledge-sharing mechanisms based on the traditional conventions in science, i.e. the free sharing of knowledge unhindered by commercial considerations (Merton 1973).
4. In Europe, the 'framework programmes' of the European Commission provide resources for collaborative projects involving universities and firms (Caloghirou et al. 2001; Larédo and Mustar 2004; Peterson and Sharp 1998). They are mirrored in the USA by federal-funded schemes such as the Advanced Technology Programme (ATP) (Hall et al. 2000), various funding instruments provided by research councils, government departments and the National Health Service in the UK (Howells et al. 1998) and joint university–industry projects within federal programmes in Germany (Schmoch 1999).

REFERENCES

Agrawal, A. (2006), 'Engaging the inventor: exploring licensing strategies for university inventions and the role of latent knowledge', *Strategic Management Journal*, **27**(1): 63–79.

Agrawal, A. and Henderson, R.M. (2002), 'Putting patents in context: exploring knowledge transfer from MIT', *Management Science*, **48**(1): 44.

Allen, T.J. (1977), *Managing the Flow of Technology: Technology Transfer and the Dissemination of Technological Information within the R&D Organization*, Cambridge, MA: MIT Press.

Arora, A., Fosfuri, A. and Gambardella, A. (2001), 'Markets for technology and their implications for corporate strategy', *Industrial and Corporate Change*, **10**(2): 419–51.

Arundel, A. and Geuna, A. (2004), 'Proximity and the use of public science by innovative European firms', *Economics of Innovation and New Technology*, **13**(6): 559–80.

Baldwin, C.Y. and Clark, K.B. (1997), 'Managing in an age of modularity', *Harvard Business Review*, **75**(5): 84–93.

Boyer, C.M. and Lewis, D.R. (1984), 'Faculty consulting: responsibility or promiscuity?', *The Journal of Higher Education*, **55**(5): 637–59.

Breschi, S. and Lissoni, F. (2001), 'Knowledge spillovers and local innovation systems: a critical survey', *Industrial and Corporate Change*, **10**(4): 975.

Brusoni, S., Marsili, O. and Salter, A. (2005), 'The role of codified sources of knowledge in innovation: empirical evidence from Dutch manufacturing', *Journal of Evolutionary Economics*, **15**(2): 211–31.

Caloghirou, Y., Tsakanikas, A. and Vonortas, N.S. (2001), 'University–industry cooperation in the context of the European framework programmes', *The Journal of Technology Transfer*, **26**(1–2): 153–61.

Chesbrough, H.W. (2003), *Open Innovation: The New Imperative for Creating and Profiting from Technology*, Boston, MA: Harvard Business School.

Chesbrough, H.W. (2006), 'Open innovation: a new paradigm for understanding industrial innovation', in H.W. Chesbrough, W. Vanhaverbeke and J. West (eds), *Open Innovation: Researching a New Paradigm*, Oxford: Oxford University Press.

Chiesa, V., Manzini, R. and Pizzurno, E. (2004), 'The externalisation of R&D activities and the growing market of product development services', *R&D Management*, **34**(1): 65–75.

Cohen, W.M. and Levinthal, D.A. (1989), 'Innovation and learning: the two faces of R&D', *The Economic Journal*, **99**(397): 569–96.

Cohen, W.M. and Levinthal, D.A. (1990), 'Absorptive capacity: a new perspective on learning and innovation', *Administrative Science Quarterly*, **35**(1): 128–52.

Cohen, W.M., Nelson, R.R. and Walsh, J.P. (2002), 'Links and impacts: the influence of public research on industrial R&D', *Management Science*, **48**(1): 1–23.

Colyvas, J., Crow, M., Gelijns, A., Mazzoleni, R., Nelson, R.R., Rosenberg, N. and Sampat, B.N. (2002), 'How do university inventions get into practice?', *Management Science*, **48**(1): 61–72.

Coombs, R., Harvey, M. and Tether, B.S. (2003), 'Analysing distributed processes of provision and innovation', *Industrial and Corporate Change*, **12**(6): 1125–55.

Dasgupta, P. and David, P.A. (1994), 'Toward a new economics of science', *Research Policy*, **23**(5): 487–521.

de Solla Price, D. (1984), 'The science/technology relationship, the craft of experimental science, and policy for the improvement of high technology innovation', *Research Policy*, **13**(1): 3–20.

Debackere, K. and Veugelers, R. (2005), 'The role of academic technology transfer organizations in improving industry science links', *Research Policy*, **34**(3): 321–42.

DfEL (2005), *Higher Education–Business and Community Interaction Survey 2002–03*, London: Department for Employment and Learning.

Etzkowitz, H. (2003), 'Research groups as "quasi-firms": the invention of the entrepreneurial university', *Research Policy*, **32**(1): 109–21.

Faulkner, W. and Senker, J. (1994), 'Making sense of diversity: public–private sector research linkage in three technologies', *Research Policy*, **23**(6): 673–95.

Feller, I. (2005), 'A historical perspective on government–university partnerships to enhance entrepreneurship and economic development', in S. Shane (ed.), *Economic Development through Entrepreneurship: Government, University and Business Linkages*, Cheltenham, UK and Northampton, MA, USA: Edward Elgar, pp. 6–28.

Feller, I., Ailes, C.P. and Roessner, J.D. (2002), 'Impacts of research universities on technological innovation in industry: evidence from engineering research centers', *Research Policy*, **31**(3): 457–74.

Fey, C.F. and Birkinshaw, J. (2005), 'External sources of knowledge, governance mode, and R&D performance', *Journal of Management*, **31**(4): 597–621.

Florida, R. and Cohen, W.M. (1999), 'Engine or infrastructure? The university role in economic development', in L.M. Branscomb, F. Kodama and R. Florida (eds), *Industrializing Knowledge: University–Industry Linkages in Japan and the United States*, MA: MIT Press, pp. 589–610.

Fontana, R., Geuna, A. and Matt, M. (2006), 'Factors affecting university–industry R&D collaboration: the importance of screening and signalling', *Research Policy*, **35**(1): 309–23.

Freeman, C. (1991), 'Networks of innovators: A synthesis of research issues', *Research Policy*, **20**(5): 499–514.

Garnsey, E. (2007), 'The entrepreneurial university: the idea and its critics', in S. Yusuf and K. Nabeshima (eds), *How Universities Promote Economic Growth*, Washington, DC: The World Bank, pp. 227–38.

Gibbons, M. (2000), 'Changing patterns of university–industry relations', *Minerva*, **38**(3): 352–61.

Gibbons, M. and Johnston, R. (1974), 'The roles of science in technological innovation', *Research Policy*, **3**(3): 220–42.

Gulati, R. and Singh, H. (1998), 'The architecture of cooperation: managing coordination costs and appropriation concerns in strategic alliances', *Administrative Science Quarterly*, **43**: 781.

Hagedoorn, J., Link, A.N. and Vonortas, N.S. (2000), 'Research partnerships', *Research Policy*, **29**(4–5): 567–86.

Hall, B.H. (2004), 'University–industry partnerships in the United States', in J.-P. Contzen, D. Gibson and M.V. Heitor (eds), *Rethinking Science Systems and Innovation Policies. Proceedings of the 6th International Conference on Technology Policy and Innovation*, Ashland, OH: Purdue University Press.

Hall, B.H., Link, A.N. and Scott, J.T. (2000), 'Universities as research partners', NBER working paper No. 7643.

Hall, B.H., Link, A.N. and Scott, J.T. (2001), 'Barriers inhibiting industry from partnering with universities: evidence from the advanced technology program', *The Journal of Technology Transfer*, **26**(1): 87–98.

Ham, R.M. and Mowery, D.C. (1998), 'Improving the effectiveness of public–private R&D collaboration: case studies at a US weapons laboratory', *Research Policy*, **26**(6): 661–75.

Harabi, N. (1998), 'Innovation through vertical relations between firms, suppliers and customers: a study of German firms', *Industry and Innovation*, **5**(2): 157–81.

Hoppe, H.C. and Ozdenoren, E. (2005), 'Intermediation in innovation', *International Journal of Industrial Organization*, **23**(5–6): 483–503.

Howells, J. (1999), 'Research and technology outsourcing', *Technology Analysis & Strategic Management*, **11**(1): 17–29.

Howells, J., Nedeva, M. and Georghiou, L. (1998), 'Industry–academic links in the UK', Manchester PREST, University of Manchester.

Iansiti, M. (1997), 'From technological potential to product performance: an empirical analysis', *Research Policy*, **26**(3): 345–65.

Jensen, R. and Thursby, M. (2001), 'Proofs and prototypes for sale: the licensing of university inventions', *American Economic Review*, **91**(1): 240–59.

Jones, L. (2000), *The Commercialization of Academic Science: Conflict of Interest for the Faculty Consultant*, doctoral dissertation, University of Minnesota.

Klevorick, A.K., Levin, R.C., Nelson, R.R. and Winter, S.G. (1995), 'On the sources and significance of interindustry differences in technological opportunities', *Research Policy*, **24**(2): 185–205.

Koza, M.P. and Lewin, A.Y. (1998), 'The co-evolution of strategic alliances', *Organization Science*, **9**(3): 255–64.

Larédo, P. and Mustar, P. (2004), 'Public sector research: a growing role in innovation systems', *Minerva*, **42**(1): 11–27.

Lee, Y.S. (1996), ' "Technology transfer" and the research university: a search for the boundaries of university–industry collaboration', *Research Policy*, **25**(6): 843–63.

Levin, R.C., Klevorick, A.K., Nelson, R.R. and Winter, S.G. (1987), 'Appropriating the returns from industrial research and development: comments and discussion', *Brookings Papers on Economic Activity*, **3**: 783–820.

Lockett, A. and Wright, M. (2005), 'Resources, capabilities, risk capital and the creation of university spin-out companies', *Research Policy*, **34**(7): 1043–57.

Mansfield, E. (1995), 'Academic research underlying industrial innovations: sources, characteristics, and financing', *The Review of Economics and Statistics*, **77**(1): 55–65.

McKelvey, M.D. (1996), *Evolutionary Innovations: The Business of Biotechnology*, Oxford and New York: Oxford University Press.

Merton, R.K. (1973), *The Sociology of Science. Theoretical and Empirical Investigations*, Chicago, IL and London: University of Chicago Press.

Meyer-Krahmer, F. and Schmoch, U. (1998), 'Science-based technologies: university–industry interactions in four fields', *Research Policy*, **27**(8): 835–51.

Mowery, D.C. and Rosenberg, N. (1989), *Technology and the Pursuit of Economic Growth*, Cambridge: Cambridge University Press.

Mowery, D.C., Nelson, R.R., Sampat, B.N. and Ziedonis, A.A. (2001), 'The growth of patenting and licensing by US universities: an assessment of the effects of the Bayh–Dole Act of 1980', *Research Policy*, **30**: 99–119.

Mowery, D.C. and Sampat, B.N. (2005), 'Universities in national innovation systems', in J. Fagerberg, D. Mowery and R. Nelson (eds), *The Oxford Handbook of Innovation*, Oxford: Oxford University Press, pp. 209–39.

Murray, F. (2002), 'Innovation as co-evolution of scientific and technological networks: exploring tissue engineering', *Research Policy*, **31**(8,9): 1389–1403.

Nelson, R.R. (2001), 'Observations on the post-Bayh–Dole rise of patenting at American universities', *The Journal of Technology Transfer*, **26**(1–2): 13–19.

Nelson, R.R. (2004), 'The market economy, and the scientific commons', *Research Policy*, **33**(3): 455–71.

Niiniluoto, I. (1993), 'The aim and structure of applied research', *Erkenntnis*, **38**(1): 1–21.

Niosi, J. (2006), 'Introduction to the Symposium: Universities as a Source of Commercial Technology', *The Journal of Technology Transfer*, **31**(4): 399–402.

Nooteboom, B. (2004), *Inter-firm Collaboration, Learning and Networks: An Integrated Approach*, London: Routledge.

Owen-Smith, J. (2003), 'From separate systems to a hybrid order: accumulative advantage across public and private science at Research One universities', *Research Policy*, **32**(6): 1081–104.

Owen-Smith, J. (2005), 'Trends and transitions in the institutional environment for public and private science', *Higher Education*, **49**(1): 91–117.

Owen-Smith, J. and Powell, W.W. (2001), 'To patent or not: faculty decisions and institutional success at technology transfer', *The Journal of Technology Transfer*, **26**(1): 99–114.

Padmore, T., Schuetze, H. and Gibson, H. (1998), 'Modeling systems of innovation: an enterprise-centered view', *Research Policy*, **26**(6): 605–24.

Panagopoulos, A. (2003), 'Understanding when universities and firms form RJVs: the importance of intellectual property protection', *International Journal of Industrial Organization*, **21**(9): 1411–33.

Patel, P. and Pavitt, K. (1995), 'Patterns of technological activity: their measurement and interpretation', in P. Stoneman (ed.), *Handbook of the Economics of Innovation and Technological Change*, Oxford: Blackwell, pp. 14–51.

Patton, C.V. and Marver, J.D. (1979), 'Paid consulting by American academics', *Educational Record*, **60**(2): 175–84.

Pavitt, K. (1990), 'What we know about the strategic management of technology', *California Management Review*, **32**: 17.

Pavitt, K. (1991), 'What makes basic research economically useful?', *Research Policy*, **20**(2): 109–19.

Perkmann, M. and Walsh, K. (2007), 'Relationship-based university–industry links and open innovation: towards a research agenda', *International Journal of Management Reviews*, **9**(4): 259–80.

Peterson, J. and Sharp, M. (1998), *Technology Policy in the European Union*, Basingstoke: Macmillan.

Phan, P.H. and Siegel, D.S. (2006), 'The effectiveness of university technology transfer: lessons learned from quantitative and qualitative research in the U.S. and the U.K.', Rensselaer Working Papers in Economics, Troy, NY.

Pisano, G.P. (1990), 'The R&D boundaries of the firm: an empirical analysis', *Administrative Science Quarterly*, **35**(1): 153–76.

Powell, W.W. (1990), 'Neither market nor hierarchy: network forms of organization', *Research in Organizational Behaviour*, **12**: 295–336.

Powell, W.W., Koput, K.W. and Smith-Doerr, L. (1996), 'Interorganizational collaboration and the locus of innovation: networks of learning in biotechnology', *Administrative Science Quarterly*, **41**(1): 116.

Poyago-Theotoky, J., Beath, J. and Siegel, D.S. (2002), 'Universities and fundamental research: reflections on the growth of university–industry partnerships', *Oxford Review of Economic Policy*, **18**(1): 10–21.

Rosenberg, N. (1992), 'Scientific instrumentation and university research', *Research Policy*, **21**(4): 381–90.

Rosenberg, N. (1994), *Exploring the Black Box: Technology, Economics, and History*, Cambridge and New York: Cambridge University Press.

Salter, A.J. and Martin, B.R. (2001), 'The economic benefits of publicly funded basic research: a critical review', *Research Policy*, **30**(3): 509–32.

Schartinger, D., Rammer, C., Fischer, M.M. and Fröhlich, J. (2002), 'Knowledge interactions between universities and industry in Austria: sectoral patterns and determinants', *Research Policy*, **31**(3): 303–28.

Schmoch, U. (1999), 'Interaction of universities and industrial enterprises in Germany and the United States – a comparison', *Industry and Innovation*, **6**(1): 51.

Shane, S.A. (2005), *Economic Development through Entrepreneurship: Government, University and Business Linkages*, Cheltenham, UK and Northampton, MA, USA: Edward Elgar.

Siegel, D.S., Waldman, D. and Link, A. (2003), 'Assessing the impact of organizational practices on the relative productivity of university technology transfer offices: an exploratory study', *Research Policy*, **32**(1): 27–48.

Slaughter, S. and Leslie, L.L. (1997), *Academic Capitalism: Politics, Policies and the Entrepreneurial University*, Baltimore, MD: Johns Hopkins University Press.

Stokes, D.E. (1997), *Pasteur's Quadrant: Basic science and technological innovation*, Washington, DC: Brookings Institution Press.

Thursby, J.G. and Thursby, M.C. (2004), 'Are faculty critical? Their role in university–industry licensing', *Contemporary Economic Policy*, **22**(2): 162–78.

Thursby, J.G.A., Jensen, R.A. and Thursby, M.C.A. (2001), 'Objectives, characteristics and outcomes of university licensing: a survey of major US universities', *The Journal of Technology Transfer*, **26**(1): 59–72.

Utterback, J.M. (1994), *Mastering the Dynamics of Innovation: How Companies can Seize Opportunities in the Face of Technological Change*, Boston, MA: Harvard Business School Press.

Veugelers, R. and Cassiman, B. (1999), 'Make and buy in innovation strategies: evidence from Belgian manufacturing firms', *Research Policy*, **28**(1): 63–80.

Von Hippel, E. (1987), *The Sources of Innovation*, New York: Oxford University Press.

Williamson, O.E. (1999), *The Mechanisms of Governance*, Oxford: Oxford University Press.

Zucker, L.G., Darby, M.R. and Armstrong, J.S. (2002), 'Commercializing knowledge: university science, knowledge capture, and firm performance in biotechnology', *Management Science*, **48**(1): 138–53.

13. What are the factors that drive the engagement of academic researchers in knowledge transfer activities? Some reflections for future research

Pablo D'Este and Andy Neely

1. INTRODUCTION

This chapter provides a review and commentary on the literature concerning knowledge transfer, with the objective of highlighting the major streams of research and constructing a general framework of the research in this area. Additionally, the chapter provides original data to complement some of the points raised by the literature review.

The review presented in this chapter shows that the literature on knowledge transfer activities between university and business remains fragmented and inconclusive in two crucial respects.[1] First, there is a lack of understanding both about *who* in academia interacts with industry, and *why* they interact. Second, there is the contested question of the impact of knowledge transfer activities on both the quality and quantity of academic research and innovation.

These two issues are particularly important for those responsible for formulating policy in the area of business and university collaboration. On the one hand, a better understanding of the mechanisms that shape the inclination of researchers to interact with industry, and the most pervasive channels through which such interactions occur, should contribute to the formulation of more nuanced and effective policies. On the other hand, by understanding the implications that university–business interactions have not only for business innovation, but also for the quality and quantity of academic research, we will be better positioned to infer some normative implications regarding which policies are most beneficial to encourage university–business interactions, and which policies could have unintended, counterproductive effects.

The next sections are devoted to a discussion of these issues in more detail. Section 2 provides a background to the overall topic of university–business interactions. Section 3 discusses in more detail the factors that might shape the academic researchers' involvement with industry, paying particular attention to the high degree of heterogeneity among researchers in their degree of engagement in knowledge transfer activities. Section 4 discusses the impact that university–business interactions is expected to have on both industrial innovation and academic research. Finally, Section 5 concludes by setting up some areas for future research.

2.   SETTING THE SCENE

Much of what universities do can be conceived as knowledge transfer activities: from teaching students to publication of articles as a result of research work. However, since the late 1970s, knowledge transfer has come to take a special meaning: the flow of knowledge, know-how and technology from university to the wider society – outside academic environments. While this form of knowledge transfer has a well-documented history that can be traced at least to the mid-1800s (Rosenberg and Nelson 1994), it is in the last two decades that this notion of knowledge transfer has been increasingly perceived as a university core mission in its own right, often seen as additional to teaching and research (Lee 1996; Molas-Gallart et al. 2002).

One of the reasons for this shift in focus was a slowdown in productivity growth in the 1970s and early 1980s, which led to a re-examination of science and technology policies in most industrialized countries. The necessity to reinvigorate technological innovation brought to the forefront of the policy debate the perception that the huge scientific and technological resources commanded by universities were insufficiently exploited in the service of national industrial competitiveness (Lee 1996; Florida 1999). In addition to this, there has been a sea change in support for science throughout the 1980s and 1990s. Governments have introduced more competitive funding allocation channels and universities have been compelled to establish tighter relationships with industry in their search for research funding (Geuna 2001). As a consequence, most industrialized countries have experienced a surge of policies designed to increase the flow of knowledge, know-how and people to industry and society at large.

The UK has been particularly active in embracing policies to encourage knowledge transfer activities. For instance, the Department of Trade and Industry (DTI) has set up a number of programmes to promote knowledge transfer and innovation, such as the Higher Education Innovation Fund – a permanent third stream of funding for universities to build capacity in the university sector for knowledge transfer; the LINK programme to support research collaborations between firms and universities; and Knowledge Transfer Partnerships to support graduates to work on innovative projects in firms, with staff from the company and the academy jointly supervising the graduate, among other initiatives. As a consequence the number of UK university science parks has increased from two in 1972 to 46 in 1999 (Siegel et al. 2003a). Also, the UK University Commercialisation Survey concludes that technology transfer is thriving in the UK, as reflected by the increasing number of universities that have commercialization office activities (23 universities before 1990, 116 universities by 2002) and by the number of spin-out companies created per $1 billion sponsored research expenditures compared to the USA – 50 in the UK for 15 in the USA, in 2002 (UNICO et al. 2003). Nevertheless, the *Lambert Review* (2003), a review commissioned by HM Treasury, DTI and DfES, concluded, among other things, that although UK universities have made real progress in their efforts to work with business, there still remain significant gains to be made by improving those efforts further. The *Lambert Review* points, for instance, to the need for further clarification about ownership of intellectual property, especially when there is funding from business; and for universities to place less emphasis on spin-outs and more on licensing, among other recommendations.

Given the prominence of debates about university–business interaction, it is not surprising that the empirical literature relating to knowledge transfer has flourished in recent times. Evaluations of the impact of academic research show that a substantial proportion of industrial innovation in high-technology industries has been based on academic research (Mansfield 1991, 1995; Cohen et al. 2002). Scholars have examined the main factors that lead firms to actively engage in university–industry R&D partnerships or draw upon universities as important sources of information for their innovative activities (Veugelers and Cassiman 2003; Mohnen and Hoareau 2002; Arundel and Geuna 2004; Laursen and Salter 2004).

From the perspective of the university, empirical studies have provided evidence about the large variety of modes of university–industry interaction (e.g. informal contacts, contract research, R&D partnerships, personnel exchanges, etc.), although these studies largely concentrate on patenting, licensing and spin-off activities (Di Gregorio and Shane 2003; Friedman and Silberman 2003). Scholars have also examined the most important factors of university institutional features that are associated with different forms of interaction, both from the perspective of the university and/or the department (Di Gregorio and Shane 2003; Friedman and Silberman 2003; Schartinger et al. 2001; Tornquist and Kallsen 1994), and from the perspective of individual academic researchers as the unit of analysis (Landry et al. 2005; Bercovitz and Feldman 2003; Agrawal and Henderson 2002; Louis et al. 2001).

However, while the empirical literature has been growing, several areas deserve further attention. Most crucial among these, as mentioned in the introduction, is the question of the factors that shape the emergence of a highly heterogeneous population of academic researchers in terms of their engagement in interaction with industry: who in the academy interacts with industry, and why some researchers interact so heavily while others interact very little or not at all. The other set of questions regards the impact of university–industry interactions on the business innovation and the quality and quantity of academic output. This latter question is particularly contested, with some scholars arguing that increased commercialization may restrict the diffusion of scientific research (Louis et al. 2001; Tijssen 2004; Murray and Stern 2007), while others suggest that increased engagement enhances the quantity or quality of scholarly output (Van Looy et al. 2004; Breschi et al. 2005). We discuss these issues in more detail in the sections that follow.

3.   HETEROGENEITY OF RESEARCHERS: WHO INTERACTS AND WHY

3.1   Preliminary Evidence on Heterogeneity among Academic Researchers

Previous research, based largely on data provided by patenting records, has shown that the proportion of academic researchers engaged in science–technology interactions is small, and that the distributions of researchers in terms of the number of linkages established with non-academic partners is highly skewed, with a few researchers accumulating a very large number of linkages. For instance, Balconi et al. (2004), examining Italian academic inventors – i.e. those university researchers whose name appears on one or more patents – show that the proportion of academic researchers involved in patenting (as inventors) ranges from 1 per cent of staff in physics, to 5 per cent in industrial engineering and 9 per cent in chemistry (including biotechnology and pharmaceuticals).[2] Also, looking at the patents generated by academic researchers at two MIT departments, Agrawal and Henderson (2002) show that the distribution of academic researchers in terms of patenting is highly skewed: first, about 44 per cent of the academic researchers have never patented over the period examined (1983–97); and second, about 30 per cent of academic researchers have been involved in one to three patents, and less than 15 per cent have been granted more than five patents.

However, there is little evidence on the heterogeneous behaviour of academic researchers in connection with 'softer' forms of interaction with industry, such as joint research collaborations, consultancy work, contract research, or other forms of interactions centred on the facilitation of the process of knowledge flow rather than on the allocation or exploitation of intellectual property rights. Agrawal and Henderson (2002) and D'Este et al. (2005) (among others) have shown that, compared to other channels of interaction, patenting is relatively infrequent. Moreover, there is increasing evidence that many academic researchers circumvent the formal channels of technology transfer established by the university (Link et al. 2006). It is then reasonable to believe that the proportion of academic researchers involved in, for instance, collaborative research is much wider than that of academic researchers involved in patenting-related activities (among other things because collaborative research may embrace the achievement of a wide variety of outputs, many of them not necessarily related to the generation of patents). Consequently, looking only at researchers who engage in patenting activities may lead us to focus on too small a range of academic researchers.

To start filling this gap, the work presented here focuses on one of such 'softer' forms of knowledge transfer: research collaborations. This chapter presents evidence based on collaborative research between university researchers and third parties, as reflected in the records of collaborative grants awarded by the Engineering and Physical Sciences Research Council (EPSRC).[3]

EPSRC encourages the partnership between researchers and the potential users and beneficiaries of research, where collaboration can help the progress and take-up of research results. Partners may include people working in industry, commerce, government agencies, local authorities, public bodies, National Health Service (NHS) trusts, non-profit organizations, research and technology organizations or the service sector. As a result, almost 45 per cent of EPSRC-funded research grants involve partnerships with industry or other stakeholders.

This section considers the two types of grants awarded by the EPSRC: those that involve collaborative partners and those that do not. Nevertheless, particular attention is paid to collaborative projects, where university researchers collaborate with third parties in the course of the research project, and in which collaborators provide either funds or in-kind support (or a combination of both) to the joint project.

The EPSRC remit covers the engineering and physical sciences, but it also funds related research outside its remit. Over the period 1991–2003, it has granted about 20 000 grants (including both collaborative and non-collaborative grants). Of these, about 38 per cent corresponded to projects in the areas of physics, chemistry and mathematics, while about 48 per cent of the total amount of grants corresponded to projects in the areas of engineering, including chemical, civil, electrical and electronic, mechanical, aero and manufacturing, metallurgy and materials, computer science, and general engineering. The remaining 14 per cent of grants extend over research related to social science, biomedical and architecture.

We examine the researchers who appear as principal investigators in EPSRC-awarded grants. The list of awarded researchers can be seen as a fair representation of active researchers in the fields of chemistry, physics, mathematics and engineering, where researchers have the EPSRC as their main source of funding for research. As Table 13.1 shows, about 45 per cent of EPSRC-awarded researchers have been engaged in at least one collaborative grant. This is a comparatively high percentage compared to the proportion of researchers who report having been involved in patenting activities, as reported by studies such as Agrawal and Henderson (2002), or more closely related to the case of EPSRC grant recipients examined here, the figures shown by D'Este and Patel (2007).

Moreover, Table 13.1 shows that the proportion of researchers who engage in collaborative research with third parties largely varies across disciplines, ranging from a lowest 15 per cent in mathematics, to a highest 63 per cent in metallurgy and materials. As expected, engineering-based disciplines show larger proportions of researchers engaging in collaborative

Table 13.1   Proportion of researchers who engage in at least one collaborative research project (period 1991–2003)

Discipline	% of researchers engaged in collaborative research	Total number of researchers
Metallurgy & materials	63.1	331
Civil engineering	62.7	399
Mechanical & manufacturing eng.	62.0	841
General engineering	59.8	497
Chemical engineering	56.0	266
Electrical & electronic eng.	54.0	833
Physics	40.8	914
Computer science	37.5	833
Chemistry	36.0	1110
Mathematics	15.5	824
Total	45.1	6848

Note: 'Researchers' refers to EPSRC grant recipients of either collaborative or non-collaborative grants.

research with third parties than disciplines such as chemistry, physics and mathematics – since in these three latter cases a larger proportion of research is likely to prioritize advance in fundamental understanding as compared to research driven by considerations of use and application. Researchers in computer science show a profile that is closer to researchers in physics and chemistry rather than to the pattern followed by engineering disciplines.

Finally, it is important to note that, regarding collaborative grants, partners from industry (i.e. companies) are the most frequent type. Indeed, over 90 per cent of collaborative grants in engineering fields have at least one industrial partner, while the percentage is above 80 per cent for chemistry and physics. The second most frequent collaborative partners are 'Public and private research organizations' – on average, about 25 per cent of collaborative projects have at least one public or private research organization in the case of chemistry, physics and mathematics, while the percentage drops to 12 per cent in the case of engineering fields. Government organizations, such as UK ministries, government agencies or councils, are the third most frequent partner, with about 8 per cent of collaborative grants having at least one organization of this type.

Figure 13.1 shows the distribution of researchers according to the number of collaborative partnerships over the period 1991–2003.[4] The first thing to highlight is the high heterogeneity among university researchers in terms of the degree of their engagement in collaborative research with third parties. As the figure shows, the distribution of academic researchers

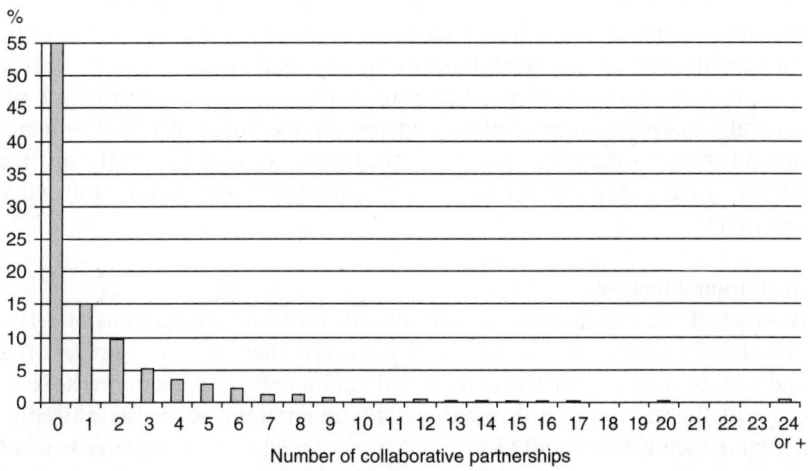

Figure 13.1   Number of collaborative partnerships by researchers

by number of links with partners in collaborative research activities is extremely skewed. Most academic researchers interact with very few partners, and a few academic researchers interact with a very large number of partners. For instance, while 55 per cent of researchers have had no interaction with third parties in their research projects, 25 per cent of researchers have interacted with one or two partners, and only 11 per cent interact with five or more.[5] Moreover, this skewed distribution of academic researchers, regarding collaborative partnerships, is common in all the disciplines examined here, even though chemistry, physics and mathematics show a much more skewed distribution of academic researchers than engineering fields.

3.2 Factors Explaining Heterogeneity

The previous section has highlighted that there is huge degree of heterogeneity among academic researchers in respect of the extent to which they engage with third parties (e.g. potential users of research), as measured by the number of research collaborative partners. The question that naturally arises is: what are the factors that drive such heterogeneity? Why do some academic researchers interact so heavily while others hardly interact at all?

The literature on knowledge transfer has proposed a number of complementary explanatory arguments to shed some light on these questions, first, regarding the importance of institutional features in shaping the likelihood that a researcher engages in knowledge transfer; second, regarding the motivations that drive academic researchers to engage with industry and other types of research partners; and third, regarding the type of individual skills required to sustain networks of collaborative partners. While there is a growing empirical literature on these issues, there is yet no general consensus In this subsection, we discuss some of the arguments raised in the literature with respect to the factors accounting for the diverse patterns of engagement in knowledge transfer activities among university researchers.

Institutional features
It is reasonable to contend that attitudes towards knowledge transfer activities may be shaped by the organizational specificities of the universities and departments to which researchers are affiliated. Drawing upon the sociological literature on embeddedness to understand academic entrepreneurship (see Kenney and Goe 2004), it is argued that the involvement of academic researchers in technology transfer is influenced by the institutions in which the individuals conduct their research and teaching activities.

Following Kenney and Goe (2004), we consider in particular three institutional layers that may influence the researcher's participation in entrepreneurial activities: the department, the university, and the scientific discipline to which the researcher belongs.

As Becher and Kogan (1992) argue, the department is the basic unit in traditional university structures, in the sense that university departments can be considered as units that have a corporate life of their own (i.e. a specific administrative unit) and a capacity to determine professional values (i.e. maintenance and development of a particular area of academic expertise). In this sense, the practices established by university departments may strongly influence the disposition of researchers to set up networks with potential users of their research. For instance, a high proportion of departmental research income from industry may reflect an environment favourable to interaction with industry. Moreover, previous research has highlighted the influence exerted by the cohort of colleagues on a particular researcher's behaviour in regard to commercialization activities, and particularly by the observed behaviour of those in leadership positions. For instance, Bercovitz and Feldman (2003) show that department chairs who are actively involved in knowledge transfer activities may contribute to legitimate these activities among department faculty by signalling that such activities are considered as highly valuable.

The policies implemented by the university administration and the culture of the university should also influence the disposition of faculty towards entrepreneurial activities. For instance, Link and Siegel (2005) and Friedman and Silberman (2003) report that universities allocating a higher percentage of royalty payments to faculty members are more productive in technology transfer activities. Also, Jensen et al. (2003) have shown that the incentive structures implemented by technology transfer offices (TTOs) strongly influence the researchers' disclosure of inventions; while Feldman et al. (2002) show that the age of TTOs – i.e. how long TTOs have been in existence – is positively related to equity positions in companies as payments for the use of university intellectual property.

Finally, Owen-Smith and Powell (2001) have shown, by comparing faculty in life sciences and physical sciences, that cultural norms across scientific fields may also be critical in shaping the faculty involvement in entrepreneurial activities. As Kenney and Goe (2004) argue, academic researchers who belong to the same scientific discipline have a common set of perceptions and practices that are likely to influence their degree of engagement in knowledge transfer activities, conforming a community of practice across university boundaries. As we have shown in Table 13.1, strong differences exist across scientific disciplines in regard to the propensity to have at least one collaborative partnership, further supporting the claim that different scientific communities have different norms and practices.

However, in addition to institutional features, the individual characteristics of researchers are also likely to exert an important influence on the disposition of researchers to participate in knowledge transfer activities. Indeed, in the light of the evidence of the preceding section, regarding individual heterogeneity among academic researchers in their involvement in collaborative research with third parties, it is reasonable to argue that individual characteristics are important factors in accounting for such variation in behaviour. To this issue we now turn.

Motivations and the type of research

Howells et al. (1998), Meyer-Krahmer and Schmoch (1998) and D'Este et al. (2005) (among others) have shown that university researchers choose to interact with industry for a diverse set of reasons. These include access to additional research income, applicability of research, access to industry skills and facilities, and keeping abreast of industry problems. It is unlikely that any single form of interaction satisfies such a wide range of motivations. For instance, consultancies may raise additional income, but have little effect on the need to access industry skills and facilities. On the other hand, joint research agreements will enable researchers to access these skills and facilities, but they may have little influence on satisfying other types of motivations. This implies that researchers motivated to interact with industry are likely to do so through a variety of forms rather than via a single mechanism. Engaging in a variety of channels would enable researchers to reap both larger pecuniary (e.g. research income) and non-pecuniary (e.g. satisfaction from seeing research brought into application) returns (see D'Este and Patel 2007).

Several individual features may shape the incentives of researchers to interact in knowledge transfer activities. First, the motivation to engage with industry partners is likely to be related to the type of research in which researches are involved, and more specifically, the type of questions addressed by researchers. In other words, since interaction among academic researchers with non-academic partners is not exclusive of applied fields of science, but rather takes place across all scientific disciplines, engagement is likely to be related to the type of research conducted by academic researchers rather than exclusively to the discipline to which researchers belong. As discussed by Stokes (1997), many researchers across disciplines, in both basic and applied fields, draw their inspiration from applied needs. Stokes shows that a high proportion of basic research is driven both by considerations of fundamental understanding and consideration of use, and he argues that researchers who are motivated by these two missions in their research activities – i.e. the Pasteur's quadrant type of researchers – are likely to play a role in the technological return from the knowledge they generate, either as consultants, industry employees, entrepreneurs or mentors of researchers who enter industry.

The above argument on the research missions that motivate research finds further resonance in the argument on 'boundary-spanning' researchers. Researchers who cross disciplinary boundaries in their research activities are likely to be exposed to multiple research perspectives and methods. This cross-disciplinary fertilization may make these researchers particularly susceptible to the evaluation and integration of quests that expand beyond the boundaries of improving understanding alone, leading them to embrace considerations of use and application. Consequently, one would expect that researchers who are more interdisciplinary in their research activities are likely to engage more actively in knowledge transfer activities. For instance, Bercovitz and Feldman (2003) find that there is a positive relationship between boundary-spanning individuals (as measured by whether researchers were appointed as faculty members to multiple departments) and faculty participation in technology transfer (as measured by whether the researcher files an invention disclosure).

Second, drawing upon the human capital argument that researchers allocate time between different types of activities – i.e. research, teaching and knowledge transfer activities – to maximize utllity (e.g. financial rewards) over the academic career, the incentives to engage in interactions with industry may vary over the career life cycle of university researchers (Levin and Stephan 1991). For instance, researchers may be particularly driven by excelling in research productivity at the early stages of the academic career, as an investment-motivated strategy to secure future financial rewards, while researchers at later stages of the academic career – e.g. having been rewarded with professorial status – may be more inclined to devote larger portions of time to knowledge transfer activities compared to younger peers.

However, some studies contend that the relationship between seniority and inclination to engagement in knowledge transfer activities runs in the opposite direction. For instance, Bercovitz and Feldman (2003) argue that the growing acceptance of the role of scientists as entrepreneurs in academic institutions may have caused a vintage effect in which the closer in time a researcher completes her training (e.g. PhD), the more likely she is to have adopted an attitude towards interaction with industry that conceives such interaction as an inherent part of the research mission. In contrast, this argument runs, the earlier an individual completes her training, the more likely she is to have been exposed to, and consequently adopted, the traditional norms of science that do not favour knowledge transfer activities.

Whether the relationship runs in one direction or the other might be influenced by the specific environment of the researcher and the channel of interaction being examined. More precisely, the options confronted by the researcher in terms of whether knowledge transfer activities are perceived as complementary or substitutive with the mission of teaching and research may largely depend on the culture dominant in her relevant environment – e.g. the field of science or the academic organization in which the researcher is active – and on the specific type of mechanism used for the interaction with non-academic partners.

Finally, the probability of interaction is also likely to be influenced by the extent to which the researcher has had some experience of interaction in the past. For instance, previous experience of collaborative research with industry may have an impact on the probability of interacting further in the future, since a large number of interactions in the past may point to the formation of a personal network of relationships with industry (or other third parties), which is generally built upon mutual trust and thus is likely to endure over time. This argument leads us to the final topic of this section: the acquisition of the personal (and organizational) capabilities conducive to the formation of collaborative networks.

The development of integration skills
In addressing the question of heterogeneity in the extent to which university researchers engage in knowledge transfer activities, it is important to analyse why certain researchers become prominent players in setting up and sustaining a network of interactions with industry. There are a number of aspects that we consider significant but largely neglected in the existing literature.

While most of the research on the nature of public–private strategic partnerships (Peters et al. 1998; Breschi and Cusmano 2004) has focused on the analysis of the network structure and on the network-related features of actors (e.g. degree centrality of the actors), comparatively less is known about the extent to which there is a strong (or weak) connection between the actor's role in the network and his/her background characteristics. To examine the extent to which certain characteristics of the individual researchers do influence the prominence achieved by researchers within science and technology interactions generally, and within the R&D collaborative research network between university and industry in particular, is a fundamental question for research.

There are several reasons for focusing on the characteristics of the actors playing a central role in networks. First, as stated by Noyons (2004), to consider the 'actors' as the unit of analysis (as opposed to the network structure itself) is relevant since the actors constitute one of the crucial elements that may be directly affected by policy. This does not mean that the characteristics of the networks are unimportant. Rather the opposite: what is argued here is that the emphasis is placed on the interplay between the characteristics of actors (individuals and/or institutions) and those of the network formed.

Second, by examining the characteristics of the main network actors we may get a better understanding about the capabilities necessary to make science and technology networks persistent over time. The persistency of networks is likely to be related to the development of capabilities, at both the individual and institutional level, to effectively integrate the worlds of research and application – what we call 'integration' capabilities.

In this sense, a focus on network actors should allow us to investigate what are the individual skills (and organizational capabilities) required to most effectively integrate scientific research with a proper understanding of the context of application of research. For instance, it becomes relevant to examine the combination of attitudes and experience that academic researchers need in order to develop their integration skills. Early training that encourages researchers to engage with users is likely to create attitudes favourable to industrial engagement and help establish a network of partners that is likely to be reinforced over time (Bercovitz and Feldman 2003; Casper and Murray 2005). However, while attitudes favourable to industrial engagement are an important requirement, they may not be enough: a certain level of exposure to knowledge transfer activities is likely to be indispensable to develop this type of skill. Such exposure to knowledge transfer activities should contribute, for instance, to increasing the researcher's capacity to balance and align conflicting interests arising from the distinct system of incentives between academia (governed by 'open science' norms) and industry (governed by 'proprietary technology' norms) (D'Este and Patel 2007).

4. IMPACT OF SCIENCE–TECHNOLOGY INTERACTIONS ON ACADEMIC RESEARCH AND INNOVATION

A growing body of policy and academic literature has welcomed the arrival of the 'entrepreneurial university', by arguing that the development of 'third stream activities' or its 'third mission' will become a 'new engine of growth' (Etzkowitz and Leydesdorff 2000; OECD 2002). However, there are still many unanswered questions regarding what the implications of encouraging knowledge transfer activities are for the nature of both: (a) innovation in business, and (b) academic research. This section briefly examines these two issues.

4.1 Impact of Knowledge Transfer Activities on Innovation[6]

The theoretical rationale for government support of knowledge transfer activities is based on the argument that the relationships between university and business are a mechanism to attenuate innovation-related market failures, in particular those related to underinvestment in basic research efforts and coordination problems in the presence of major uncertainties. In this sense, university–industry links can be extremely valuable mechanisms to ensure that: (a) benefits from potential applications from fundamental research are perceived by collaborative partners (applications that otherwise would have gone unrealized); (b) tacit knowledge is satisfactorily transmitted through frequent and close interaction between university faculty and industrial scientists; and (c) complementary skills can be exploited to mutual advantage (Poyago-Theotoky et al. 2002). In brief, university–industry links have the potential to accelerate technological diffusion, and establish new research agendas.

As research conducted by Mansfield (1991, 1995) has shown, universities play a major role in originating and promoting the diffusion of knowledge and techniques that contribute to industrial innovation, as indicated by the finding that, between 1975 and 1985 about 10 per cent of the new products and processes in US high-technology industries were based directly on recent academic research. Moreover, as shown by Cohen et al. (2002), industry uses a variety of mechanisms to access the university system (from patents and publications to informal interaction, consulting and recent hires), and industry benefits from interaction with university not only to generate new ideas but also to complete existing R&D projects.

However, there is a considerable inter-industry variation in the propensity of firms to draw from universities in their innovative activities (Klevorick et al. 1995; Laursen and Salter 2004). Results suggest that firms in sectors characterized by high levels of investment in R&D have a higher propensity to draw from sources of information from universities. For instance, while there is a large proportion of firms using universities as a source of information for innovative activities in the chemical industry, the proportion is very low in industries such as paper and printing or textiles.

Additionally, results from innovation surveys, such as the Community and Innovation Survey (CIS), report that only a small fraction of firms consider information from universities as being *highly important* for their innovative activities. For instance, based on information from the UK CIS-4, only about 20 per cent of firms report attaching some value to information obtained directly from universities, compared to 64 per cent of enterprises reporting that they attach some value to information coming from suppliers or 65 per cent that report attaching value to the information coming from clients and customers. Moreover, only 2 per cent assess information sources from universities as being highly important to the enterprise's innovation activities. Nevertheless, as the DTI (2006) Report on the CIS-4 argues, those enterprises that collaborate with universities show better performance compared to those that do not collaborate, suggesting a benefit that may flow from business–university collaboration.[7]

Also, the DTI (2006) Report shows that for those companies that collaborate with local universities, the university is a less important source of information that it is for those who collaborate with universities outside the region and/or outside the UK. However, this evidence may highly depend on the industrial sector examined: for instance, Mansfield and Lee (1996) show that firms conducting applied research tended to support R&D at local universities (i.e. less than 100 miles away). Moreover, as Mansfield and Lee (1996) show, it is not only leading-edge research departments that attract the attention of businesses; low-ranked research departments can also be strong recipients of research income from industry.

In brief, more research is required to disentangle what are the most frequently channels used by university and business to interact with each other, how the portfolio of channels used may differ across industries, and with which universities/departments companies choose to interact (e.g. close versus distant/high rated versus low rated in terms of quality of research).

Impact of Knowledge Transfer Activities on Academic Research
In recent years many scholars have raised concerns about whether policy initiatives to support technology transfer at universities have swung too far towards commercialization, and whether knowledge transfer activities at universities are consistent with the traditional university missions of teaching and research. Some authors have argued that universities are better producers of talent (via education and graduates) than of technology (Florida 1999; Salter et al. 2000; Pavitt 2001). Still others have pointed out that the changes in university behaviour associated with the new incentive structures encouraging knowledge transfer activities could have long-term negative unintended consequences on the culture of open science (Dasgupta and David 1994; Geuna 1999, 2001).

Dasgupta and David (1994) and Geuna (1999) (among others) argue that keeping the results of scientific research open for all to use renders large economic benefits both for the advancement of science itself (since reliability of scientific knowledge depends on the results of scientific research being laid in the public domain for testing and further development; Merton 1957 and Ziman 1991), and for the advance of technology (since, as firms report, publications and conferences are one of the most important channels through which companies gain access to the outcomes of university research: Cohen et al. 2002 and Agrawal and Henderson 2002). Therefore an increasing shift towards knowledge transfer activities, and in particularly towards patenting, may jeopardize these benefits. This is mainly for the reasons outlined below.

First, scientists may have an incentive to postpone or disregard publication in open science, since disclosing information may severely impair the claim for novelty in patent applications. Second, a focus on knowledge transfer (and patenting in particular) may divert researchers from research trajectories characterized by basic research, with long-term research commitments and loosely defined goals, towards research trajectories characterized by applied research, often mediated by the short-term commitments imposed by industrial partners. Finally, the zeal to encourage IP management at universities may lead to protracted negotiations with industry, likely to deter business from collaborating with university and, even more damaging for the goal of knowledge transfer, causing industry to perceive university researchers as direct competitors with their own research efforts.

However, the empirical evidence examining the impact of commercialization on the quality, quantity and direction (i.e. shift from basic to applied) of academic research remains inconclusive. While some studies have shown that commercialization of academic research is associated with both delays in publication and refusal to share research results upon request (Blumenthal et al. 1996, and 1997; Louis et al. 2001), other studies have found little or no support for the argument that commercialization has come at the expense of placing knowledge in the public domain. For instance, Agrawal and Henderson (2002) have found that while patenting activity does not appear to be significantly related to publishing activity, publication counts are a reasonable predictor of the 'importance' of a researcher's publications, as measured by citations. Van Looy et al. (2004) found that researchers' involvement in contract research seems to stimulate their scientific productivity, while Stephan et al. (2007) found patents to be positively and significantly related to the number of publications. Also, Calderini and Franzoni (2004) and Breschi et al. (2005) conclude that Italian academic inventors are more productive than researchers who are not involved in patenting activities.

There are several reasons why studies on the relationship between engagement in knowledge transfer activities and research productivity may generate conflicting results. First, in some types of research, patenting and publishing may be complementary rather than substitutive, as a consequence of the fact that research oriented to advance technologies and user-oriented goals often contributes to address gaps in fundamental understanding (i.e. the 'Pasteur's quadrant' types of research; Breschi et al.

2005; Stephan et al. 2007; Van Looy et al. 2004). In these circumstances, research is likely to generate results that are both patentable and publishable, and thus patenting is unlikely to come at the expense of publication productivity or quality of research. Second, faculty interaction with industry is often reported to be a source of inspiration for expanding lines of enquiry in fundamental research (Mansfield 1995; Siegel et al. 2003b), and a source of research funding or sophisticated equipment necessary for exploratory research (Breschi et al. 2005). Third, engagement in knowledge transfer activities can increase the academic status and visibility of researchers, attracting both public research funding and ties with industry (Owen-Smith and Powell 2001). Finally, the effectiveness of technology transfer offices in raising awareness for commercial exploitation and facilitating the patent application process can play a crucial role in influencing the faculty perception about the personal and professional benefits of patenting, and contributing to the creation of an environment that favours entrepreneurial science, and the engagement in both high-quality research and patenting (Owen-Smith and Powell 2001).

As a consequence, the relationship between knowledge transfer activities and the quality of academic research is likely to be contingent on a number of environmental factors. Under certain circumstances the two may reinforce each other, while under others they may come at the expense of each other. More research is required to better understand what conditions are more conducive to generating complementarities and what are more conducive to creating substitution effects.

5.   SUMMARY AND COMMENTS ON POTENTIAL AVENUES FOR FUTURE RESEARCH

As this review has highlighted, while the empirical literature on university–industry knowledge transfer has been growing at a fast pace recently, many issues remain unsolved. These issues can be grouped as follows.

First, while a large proportion of the existing literature is based on data provided by patent records, there is comparatively little evidence on the behaviour of academic researchers in connection with 'softer' forms of interaction with industry, such as joint research collaborations or consultancy work. This is particularly important in the light of some preliminary evidence highlighting that: (a) the proportion of academic researchers involved in patenting is small; and (b) compared to other channels of interaction, patenting is a relatively infrequent channel.

Second, an important area of research that remains widely open concerns the factors that shape both the inclination of academic researchers to interact with business, and the huge heterogeneity among academics in terms of their degree of engagement in knowledge transfer activities. That is: why some academic researchers engage so heavily while others, the large majority, interact very little or not at all. The main issues that this review has highlighted as deserving further attention in addressing this question can be summarized as follows.

On the one hand, we highlighted the influence exerted by the institutions where academics conduct their research activities: departments, universities and the scientific communities to which researchers belong. While a large volume of literature has addressed the role played by technology transfer offices in facilitating technology transfer, we still know very little about what are the institutional settings most conducive to knowledge transfer activities. This is particularly important given that a large volume of knowledge transfer appears to occur outside the remit of the technology transfer offices.

On the other hand, this review has highlighted the importance of individual characteristics of researchers as important factors behind their propensity to, and degree to which they will, engage in knowledge transfer activities. We still need to better understand the motivations that drive researchers to interact with industrial partners (as well as other types of research partners, such as charities, local government agencies, hospitals, etc.), and to engage in types of research highly inspired by potential applications. Additionally, this review has highlighted that we still know very little about the factors that lead to the development of the individual skills necessary to effectively integrate the worlds of scientific research and application.

The third set of unsolved issues regards the impact of knowledge transfer activities on both the innovative activities conducted by business and the nature of academic research. While in this review we have mainly discussed the latter, we have also highlighted that on the former there is much more we need to know about (a) whether channels of interaction significantly differ by industry; and (b) what type of universities or departments are more likely to attract the attention of business.

In regard to the impact of knowledge transfer activities on the nature of academic research, this review has pointed out that there is little agreement among scholars on whether an increase in the engagement in knowledge transfer activities is causing a shift in the direction of research towards more applied (rather than fundamental) research, or a lower quality and/or quantity of research. It is reasonable to conjecture that such relationships are likely to be contingent on a number of environmental factors. Further research should disentangle the circumstances that are more conducive to complementarities between knowledge transfer activities and academic research, and those that are more conducive to substitution effects.

Overall, by looking mainly at the issue of knowledge transfer from the perspective of the academy, and academic researchers in particular, this chapter has aimed to highlight the complex nature of this topic, stressing the range of crucial questions that remain open. We hope that this chapter thereby contributes to signalling future research avenues that will shed new light on informing policy making in regard to the most appropriate mechanisms to facilitate and encourage knowledge transfer.

NOTES

1. The words 'industry' and 'business', even though they have different meanings, will be used interchangeably throughout this chapter.
2. Also, Balconi et al. (2004) show that the distribution of academic inventors according to the number of linkages with co-inventors they have worked with (mostly, non-academic inventors) is highly skewed, with about 50 per cent of academic inventors having three or fewer acquaintances (i.e. linkages with other inventors). However, they also show that the distribution of researchers by the number of acquaintances is much less skewed for academic inventors than is the case for non-academic inventors, indicating both that academic inventors tend to work in larger teams compared to non-academic inventors, and also that they tend to work for a larger number of patent applicants (i.e. different organizations).
3. The EPSRC is one of the UK research councils responsible for administering funding for research and innovation activities in the UK. It distributes some 23 per cent of the total UK science budget (about £500 million a year). This and the Medical Research Council (MRC) are the two largest councils in terms of the volume of research funded (accounting for about 44 per cent of the total science budget). The EPSRC is responsible for funding research in the areas of engineering and physical sciences, including chemistry, mathematics and computer science, and it welcomes research proposals that span the remits of other research councils (such as research projects in biology, social science, or medical-related research).
4. Collaborative partnerships refer to the count of partners with which researchers collaborate within the context of EPSRC collaborative grants. In those cases in which a researcher interacts with the same partner in several projects, partnerships refer to the number of occasions the researcher participates with the same organizational partner throughout different collaborative grants.
5. A small fraction of researchers did engage very heavily, since 2 per cent of researchers (about 130 researchers) established partnerships with 15 or more partners over the period 1991–2003.
6. The issue of 'impact of knowledge transfer on innovation' is not discussed in detail here, since this is done in other chapters in this volume.
7. It is important to highlight that the impact of universities on innovation activities is more complex and subtle than what is expressed via direct collaboration between university and businesses. As Florida (1999) argues, universities are crucial pieces of the infrastructure of the knowledge economy as suppliers of talent to society. If that is one, if not the main, contribution of universities to the innovative capacity of businesses, then formal linkages between university and business are likely to capture only a part, arguably a small one, of the overall contribution of universities to economic development and innovation.

REFERENCES

Agrawal, A. and Henderson, R. (2002), 'Putting patents in context: exploring knowledge transfer from MIT', *Management Science*, **48**(1): 44–60.

Arundel, A. and Geuna, A. (2004), 'Proximity and the use of public science by innovative European firms', *Economics of Innovation and New Technology*, **13**(6): 559–80.

Balconi, M., Breschi, S. and Lissoni, F. (2004), 'Networks of inventors and the role of academia: an exploration of Italian patent data', *Research Policy*, **33**: 127–45.

Becher, T. and Kogan, M. (1992), *Process and Structure in Higher Education*, 2nd edn, London: Routledge.

Bercovitz, J. and Feldman, M. (2003), 'Technology transfer and the academic department: who participates and why?', paper presented at the DRUID Summer Conference 2003, Copenhagen, 12–14 June.

Blumenthal, D., Campbell, E.G., Anderson, M.S., Causino, N. and Louis, K.S. (1996), 'Withholding research results in academic lifescience: evidence from a national survey of faculty', *Journal of the American Medical Association*, **277**(15): 1224–8.

Breschi, S. and Cusmano, L. (2004), 'Unveiling the texture of a European research area: emergence of oligarchic networks under EU Framework Programmes', *International Journal of Technology Management*, **27**(8): 747–72.

Breschi, S., Lissoni, F. and Montobbio, F. (2005), 'The scientific productivity of academic inventors: new evidence from Italian data', CESPRI Working Paper 168.

Calderini, M. and Franzoni, C. (2004), 'Is academic patenting detrimental to high quality research? An empirical analysis of the relationship between scientific careers and patent applications', CESPRI Working Paper 162.

Casper, S. and Murray, F. (2005), 'Careers and clusters: analyzing the career network dynamic of biotechnology clusters', *Journal of Engineering and Technology Management*, **22**: 51–74.

Cohen, W.M., Nelson, R.R. and Walsh, J.P. (2002), 'Links and impacts: the influence of public research on industrial R&D', *Management Science*, **48**(1): 1–23.

DTI (2006), 'Innovation in the UK: indicators and insights', DTI Occasional Papers No. 6, July.

Dasgupta, P. and David, P.A. (1994), 'Towards a new economics of science', *Research Policy*, **23**(5): 487–521.

D'Este, P. and Patel, P. (2007), 'University–industry linkages in the UK: what are the factors underlying the variety of interactions with industry?', *Research Policy*, **36**(9): 1295–1313.

D'Este, P., Nesta, L. and Patel, P. (2005), 'Analysis of university–industry research collaborations in the UK: preliminary results of a survey of university researchers', SPRU Report, May, http://www.sussex.ac.uk/spru/documents/deste_report.pdf.

Di Gregorio, D. and Shane, S. (2003), 'Why do some universities generate more start-ups than others?', *Research Policy*, **32**(2): 209–27.

Etzkowitz, H. and Leydesdorff, L. (2000), 'The dynamics of innovation: from National Systems and "Mode 2" to a triple helix of university–industry–government relations', *Research Policy*, **29**: 109–23.

Feldman, M., Feller, I., Bercovitz, J. and Burton, R. (2002), 'Equity and the technology transfer strategies of American research universities', *Management Science*, **48**(1): 105–21.

Florida, R. (1999), 'The role of university: leveraging talent, not technology', *Issues on Science and Technology*, **15**(4): 67–73.

Friedman, J. and Silberman, J. (2003), 'University technology transfer: do incentives, management, and location matter?', *Journal of Technology Transfer*, **28**: 17–30.

Geuna, A. (1999), 'Patterns of university research in Europe', in A. Gambardella and F. Malerba (eds), *The Organisation of Economic Innovation in Europe*, Cambridge: Cambridge University Press, pp. 367–89.

Geuna, A. (2001), 'The changing rationale for European university research funding: are there negative unintended consequences?', *Journal of Economic Issues*, **35**(5): 607–32.

HM Treasury (2003), *Lambert Review of Business–University Collaboration*, Final Report, December.

Howells, J., Nedeva, M. and Georghiou, L. (1998), 'Industry–academic links in the UK', Report to HEFCE, PREST, University of Manchester, www.hefce.ac.uk.

Jensen, R., Thursby, J.G. and Thursby, M.C. (2003), 'The disclosure and licensing of university inventions: the best we can do with the s**t we get to work with', *International Journal of Industrial Organization*, **21**: 1271–300.

Kenney, M. and Goe, W.R. (2004), 'The role of social embeddedness in professorial entrepreneurship: a comparison of electrical engineering and computer science at UC Berkeley and Stanford', *Research Policy*, **33**: 691–707.

Klevorick, A.K., Levin, R.C., Nelson, R.R. and Winter, S.G. (1995), 'On the sources and significance of interindustry differences in technological opportunities', *Research Policy*, **24**: 185–205.

Landry, R., Nabil, A. and Ouimet, M. (2005), 'A resource-based approach to knowledge-transfer: evidence from Canadian university researchers in natural sciences and engineering', paper presented at the DRUID Tenth Anniversary Summer Conference, Copenhagen, Denmark, 27–29 June.

Laursen, K. and Salter, A. (2004), 'Searching high and low: what types of firms use universities as source of innovation?', *Research Policy*, **33**: 1201–15.

Lee, Y.S. (1996), ' "Technology transfer" and the research university: a search for the boundaries of university–industry collaboration', *Research Policy*, **25**: 843–63.

Levin, S.G. and Stephan, P. (1991), 'Research productivity over the life cycle: evidence for academic scientists', *The American Economic Review*, **81**(1): 114–32.

Link, A.N. and Siegel, D.S. (2005), 'Generating science-based growth: an econometric analysis of the impact of organizational incentives on university–industry technology transfer', *European Journal of Finance*, **11**: 169–82.

Link, A.N., Siegel, D.S. and Bozeman, B. (2006), 'An empirical analysis of the propensity of academics to engage in informal university technology transfer', NBER Working Paper.

Louis, K.S., Jones, L.M., Anderson, M.S., Blumenthal, D. and Campbell, E.G. (2001), 'Entrepreneurship, secrecy, and productivity: a comparison of clinical and non-clinical faculty', *Journal of Technology Transfer*, **26**(3): 233–45.

Mansfield, E. (1991), 'Academic research and industrial innovation', *Research Policy*, **20**: 1–12.

Mansfield, E. (1995), 'Academic research underlying industrial innovations: sources, characteristics, and financing', *The Review of Economics and Statistics*, **77**(1): 55–65.

Mansfield, E. and Lee, J.Y. (1996), 'The modern university: contributor to industrial innovation and recipient of industrial R&D support', *Research Policy*, **25**: 1047–58.

Merton, R.K. (1957), 'Priorities in scientific discovery', *American Sociological Review*, **22**: 635–59.

Meyer-Krahmer, F. and Schmoch, U. (1998), 'Science-based technologies: university–industry interactions in four fields', *Research Policy*, **27**(8): 835–51.

Mohnen, P. and Hoareau, C. (2002), 'What type of enterprise forges close links with universities and government labs? Evidence from the CIS 2', MERIT–Infonomics Research Memorandum Series 2002-008. Maastricht: MERIT.

Molas-Gallart, J., Salter, A., Patel, P., Scott, A. and Duran, X. (2002), 'Measuring third stream activities', Final Report to the Russell Group of Universities, SPRU, University of Sussex.

Murray, F. and Stern, S. (2007), 'Do formal intellectual property rights hinder the free flow of scientific knowledge? An empirical test of the anti-commons hypothesis', *Journal of Economic Behavior & Organization*, **63**(4): 648–87.

Noyons, E.C.M. (2004), 'Science maps within a science policy context', in H.F. Moed et al. (eds), *Handbook of Quantitative Science and Technology Research. The Use of Publication and Patent Statistics in Studies of S&T Systems*, London: Kluwer Academic Publishers, pp. 237–56.

OECD (2002), *Benchmarking Industry–Science Relationships*, Paris: OECD.

Owen-Smith, J. and Powell, W.W. (2001), 'To patent or not: faculty decision and institutional success at technology transfer', *Journal of Technology Transfer*, **26**: 99–114.

Pavitt, K. (2001), 'Public policies to support basic research: what can the rest of the world learn from US theory and practice? (And what they should not learn)', *Industrial and Corporate Change*, **10**(3): 761–79.

Peters, L., Groenewegen, P. and Fiebelkorn, N. (1998), 'A comparison of networks between industry and public sector research in materials technology and biotechnology', *Research Policy*, **27**: 255–71.

Poyago-Theotoky, J., Beath, J. and Siegel, D.S. (2002), 'Universities and fundamental research: reflections on the growth of university–industry partnerships', *Oxford Review of Economic Policy*, **18**(1): 10–21.

Rosenberg, N. and Nelson, R.R. (1994), 'American universities and technical advance in industry', *Research Policy*, **23**: 323–48.

Salter, A., D'Este, P., Martin, B., Geuna, A., Scott, A., Pavitt, K., Patel, P. and Nightingale, P. (2000), *Talent, not Technology: Publicly Funded Research and Innovation in the UK*, London: Committee of Vice-Chancellors and Principals and the Higher Education Funding Council of England.

Schartinger, D., Schibany, A. and Gassler, H. (2001), 'Interactive relations between university and firms: empirical evidence for Austria', *Journal of Technology Transfer*, **26**: 255–68.

Siegel, D.S. and Zervos, V. (2002), 'Strategic research partnership and economic performance: empirical issues', *Science and Public Policy*, **29**(5): 331–43.

Siegel, D.S., Westhead, P. and Wright, M. (2003a), 'Assessing the impact of university science parks on research productivity: exploratory firm-level evidence from the United Kingdom', *International Journal of Industrial Organization*, **21**: 1357–69.

Siegel, D.S., Waldman, D.A., Atwater, L.E. and Link, A.N. (2003b), 'Commercial knowledge transfers from universities to firms: improving the effectiveness of university–industry collaboration', *Journal of Higher Technology Management Research*, **14**: 111–33.

Stephan, P.E., Gurmu, S., Sumell, A.J. and Black, G. (2007), 'Who's patenting in the university? Evidence from the Survey of Doctorate recipients', *Economics of Innovation and New Technology*, **16**(2): 71–99.

Stokes, D.E. (1997), *Pasteur's Quadrant. Basic Science and Technological Innovation*, Washington, DC: Brookings Institution Press.

Tijssen, R.J.W. (2004), 'Is the commercialisation of scientific research affecting the production of public knowledge? Global trends in the output of corporate research articles', *Research Policy*, **33**: 709–33.

Tornquist, K.M. and Kallsen, L.A. (1994), 'Out of the ivory tower: characteristics of institutions meeting the research needs of industry', *Journal of Higher Education*, **65**(5): 523–39.

UNICO, NUBS and AURIL (2003), *UK University Commercialisation Survey: Financial Year 2002*, AURIL–Nottingham University Business School (NUBS)–UNICO.

Van Looy, B., Ranga, M., Callaert, J., Debackere, K. and Zimmermann, E. (2004), 'Combining entrepreneurial and scientific performance in academia: towards a compounded and reciprocal Mathew-effect?', *Research Policy*, **33**: 425–41.

Veugelers, R. and Cassiman, B. (2003), 'R&D cooperation between firms and universities. Some empirical evidence from Belgian manufacturing', CEPR Discussion Papers, 3951.

Ziman, J. (1991), 'Academic science as a system of markets', *Higher Education Quarterly*, **45**(1): 57–68.

14. Exploring the role of geographic proximity in shaping university–industry interaction

Kate Bishop, Toke Reichstein and Ammon Salter

INTRODUCTION

This chapter explores the literature on the effects of geographic proximity on university–industry interaction. Despite rapid advances in communication technology, numerous studies have demonstrated that geographic proximity continues to play an important role in shaping economic behaviour, especially the formation of university–industry linkages (e.g. Feldman 1994). Geographic proximity is central to university–industry interaction because it facilitates the exchange of personal knowledge through geographically bounded social networks (see Maskell and Malmberg 1999; Storper 2004; Asheim and Gertler 2005) and there are powerful reasons to suggest that it will continue to be influential in shaping relationships between universities and industrial firms in the future. However, the impact of geographic proximity on university–industry links is not always positive: university–industry linkages can be enhanced and constrained by physical proximity.

The goal of this chapter is to review the existing empirical studies on university–industry links, focusing on the geographical dimension in these relationships. This will allow us to identify a number of gaps in the literature, leading to the development of a research agenda.

MAPPING AND CHARACTERIZING UNIVERSITY–INDUSTRY INTERACTIONS

There is a broad and active tradition of research on the antecedents to and consequences of university–industry interaction (Salter and Martin 2001; Pavitt 1991; Shane 2004; Cohen et al. 2002). This research tradition highlights the subtle, complex and multifaceted role of universities in the economic system. It draws on a large body of knowledge on the historical development of universities and technological developments in different areas of industry practice (Rosenberg and Nelson 1994; Mowery and Rosenberg 1989). Given the size and breadth of this literature, we focus on the role of geographic distance in shaping patterns of university interaction, while outlining some of the key elements in the wider research tradition.

Henderson et al. (1998) produced a seminal study on the geography of university–industry interaction which examined the economic benefits of university research in the USA using patent application data for the period 1965–88. They found that the number of university patents had increased dramatically and, based on a random sample of patents, was receiving almost 25 per cent more citations on average. However, this advantage began to be eroded by the 1980s. Part of this decline was due to the increasing number of low-quality university patents, resulting from the incentives for universities to patent provided by the Bayh–Dole Act. Overall, Henderson and colleagues (ibid.: 126) found that most of the economic benefits from university research were due to private sector inventions that built upon the scientific and engineering base created by universities, rather than commercial inventions generated directly by universities.

A later empirical study by Mowery et al. (2001) explores the effects of the Bayh–Dole Act at three leading universities. Their evidence reveals that the Act is only one of many important factors behind the rise of university patenting and licensing. Moreover, they found that the Act had no real impact on the content of academic research.

In an attempt to measure more directly the contribution of academic research to industrial innovation Mansfield (1991, 1998) studied the rates of return to publicly funded research. These studies explore the role of academic research in industrial innovation, based on US data collected for 1986–94 and 1975–85 on firm sales based on recent academic research. The results of the first study reveal that the estimated rate of return from academic research is 28 per cent. The key results from the 1998 study confirm the earlier 1991 study and show that over 10 per cent of the new products and processes introduced could not have been developed (without substantial delay) without the benefit of recent academic research. The one clear difference between the two studies is that over time there has been a decrease in the average time lag between academic research results and the first commercial introduction of a new product or process, suggesting that the ability of firms to commercialize academic finding is improving. Using a large sample of manufacturing firms, Beise and Stahl (1999) found that approximately 5 per cent of new product sales could not have been developed without academic research, and that large firms are more likely than small firms to take advantage of universities.

As Nelson (1986) suggested, universities rarely generate new technologies by themselves. However, university research can have a positive impact on the productivity of private research and development (R&D) activities and enhances technological opportunities. Building on Nelson (1959), Salter and Martin (2001) present a classification of the economic benefits from publicly funded research that goes beyond the usual assumption that research is simply a source of useful information. These benefits include: an increased stock of useful knowledge; the training of skilled graduates; the creation of new scientific instrumentation and methodologies; the formation of networks; increased capacity for scientific and technological problem solving; and the creation of new firms. An empirical example of the diversity of links between universities and industry is contained in Meyer-Krahmer and Schmoch's (1998) study. This study combines European Patent Office (EPO) data with information from a survey of universities to illustrate the 'two-way' interaction between universities and industry. For example, academic researchers gain funding, knowledge and flexibility through collaboration with industry. However, for industry to benefit from this two-directional flow of knowledge and informal discussions, well-developed 'absorptive capacity' is necessary (Cohen and Levinthal 1990). Absorptive capacity can be defined as the ability of firms to successfully acquire the research results of others, and of universities to respond to new problems and communicate within and beyond the borders of scientific disciplines.

Building on these findings, Bercovitz and Feldman (2006: 176) report that university–industry relationships are heterogeneous and can involve sponsored research, licensing, spin-offs and the hiring of research students. Due to the evolutionary nature of these relations, transactions may occur sequentially to reinforce the commercialization process. They propose a conceptual framework to model the role of universities in the innovation system, which considers the objective functions of academic scientists, the university organizational structures and processes, firms' characteristics and the legal and policy environment as determinants of collaboration. Schartinger et al. (2002) attempted to account for the variation in the patterns of knowledge interaction between a field of science and a sector of economic activity, using a large dataset of Austrian university research projects. Their results reveal some interesting findings: the size of a scientific field and a sector of economic activity (measured by research personnel), and the knowledge proximity between the field of science and the economic sector, explain a significant part of the variance in interaction shares between industry sectors and fields of science. They examined the influence of geographic distance but found it to be insignificant in most models, with the exception of contract research where spatial distance negatively impacts on the interaction. On the industry side, they found that a high R&D intensity and high employment dynamics positively influence a firm's propensity to engage in knowledge interactions with universities. In terms of university characteristics, they found that the length of experience in contract research and the scientific quality of the research positively affected the propensity of a university to interact.

Large-scale surveys of industrial R&D laboratories are a major component of the literature on university–industry linkages. A key study is the Yale Survey, which was conducted in 1982. This study shows there are strong industry differences in the relationship between academic research and industrial practice. In some industries the links are tight, whereas in others they are weak or non-existent (Klevorick et al. 1995). In a follow-up study, Cohen et al. (2002) report on the Carnegie Mellon Survey, which asked R&D managers what types of public research they used for technical advance. This study found that the proportion of public research outputs exploited in industrial R&D projects was roughly equal for new R&D projects and the completion of existing projects. Public research also provided instruments and techniques developed by university and government labs; these were declared useful in 22 per cent of industrial projects. In terms of public research prototypes, only 8.3 per cent of industrial research projects made use of this public research output, although they are frequently used in the glass and motor/generator industries.

In a UK study Laursen and Salter (2004) extended the existing work on university and industry interaction drawing on a large-scale cross-industry dataset, the UK Innovation Survey. They considered key structural firm variables such as firm size, firm age, and R&D expenditures, and also firms' search strategies, reflecting the *openness* of a firm to external sources of knowledge (see also Chesbrough 2003). Their analysis reveals that openness in the firm's external search strategy is positively linked to exploiting university research. Also firm size positively impacts on the propensity of a firm to seek knowledge from universities. Laursen and Salter conclude by noting that a firm's search strategy influences the propensity to use university knowledge and information and, thus, future research should focus on managerial choice and search strategy. Building on this research by Laursen and Salter, Fontana et al. (2006) considered the role of firms in gathering information from and revealing knowledge to external sources. They examined the impact of *searching*, *screening* and *signalling* on the incidence and extent of formal collaborative R&D projects between innovative firms and public research organizations (PROs). Their econometric analysis is based on the results of the KNOW survey carried out in seven

EU countries. Their results suggest that larger firms that are heavily
involved in R&D activities are involved in a higher number of R&D projects with PROs, compared to small firms. They also consider the legal
status of the firm and confirm that firms belonging to large units tend to
collaborate more than independent firms. With regard to the openness variables, searching, as measured by the mean percentage of new products and
processes in collaboration with external partners, is insignificant, revealing
that searching does not affect the number of collaborations between firms
and PROs. However, screening, as measured by firms looking at publications and participating in subsidized projects, positively influences the
number of collaborations with PROs. The results were similar for patenting, the proxy for signalling.

A number of studies have tried to determine whether interacting with
universities and academic researchers has a positive or negative effect on
firm performance. Link and Rees (1990) explore both the determinants of
university–industry collaboration and its subsequent impact on firm performance. They examine the propensity of firms to engage in research relationships with universities and the impact of university-based relationships
on the returns to R&D. First, they hypothesize that firm size will have an
impact upon the decision to collaborate, and second, that firm size impacts
on R&D efficiency: large firms will suffer from a diseconomies of scale issue
owing to the degree of bureaucratization in the innovation decision-
making process, which inhibits inventiveness. The study shows that the
probability of participating in a university research programme increases
with firm size. They also found that the estimated returns to R&D in firms
that collaborate with universities are more than twice that of firms that do
not. Their analysis shows that although large firms are more likely to
exploit university relationships, small firms are better able to utilize these
linkages in terms of R&D returns.

In a path-breaking study, Zucker et al. (2002) examine one particular
aspect of university–industry interaction – that of joint publication involving an industry scientist and an academic or 'star scientist' from a top
university. To test this they use data from the Institute of Scientific
Information and Genbank, an online reference file. Their analysis reveals
that for a sample of US biotechnology firms, publications co-authored by
firm scientists and top university scientists increase the number and citation
rate of firm patents. This confirms some of their earlier work on actual
work ties at the laboratory bench level between firm scientists and academics and firm performance in the biotechnology sector.

A follow-up study by George et al. (2002) examines the impact of university–firm alliances on innovative outputs and on financial performance
in the biotechnology sector. They find that firms with university linkages

have lower R&D expenditure and a high number of patents granted, although the relationships between alliances and financial performance was insignificant. Their study raises an interesting point about the quality of linkages. They measure the total number of a firm's university alliances, the total federal R&D funding received, the type of linkage and the content of linkages (technology versus marketing).

Along the same lines, Motohashi (2005) examines the impact of university–firm collaboration on R&D productivity for 800 new technology-based firms in Japan. Their results show that the incidence of university collaboration has a positive effect on R&D productivity, measured by the number of patents developed. This relationship is especially strong for smaller firms. One of the methodological issues they mention in their analysis is the possible lag between collaboration and firms' R&D productivity.

UNIVERSITY–INDUSTRY INTERACTION AND GEOGRAPHIC PROXIMITY

Feldman (1994) observed that it is widely accepted that innovation involves external sources of knowledge and, as a result, the spatial boundaries of innovation have widened. Thus geography provides organizations with a diverse knowledge source, which plays a central role in new product commercialization. In similar vein, Dosi (1988) and Lundvall (1988) reported that innovation may have a strong geographic dimension due to the specific and cumulative nature of knowledge-based innovative inputs. Gertler et al. (2000) examine the extent to which the institutional context and local setting determine the innovative activities of manufacturing firms in Ontario, Canada. Despite recent globalization trends they posit that the characteristics of the home market will have an impact upon sectoral specialization and technological activity. Thus they expect that the regional and local economy will have an influence on a firm's innovative practices and, as a result, firms will seek to embed themselves in learning-rich regions. For example, firms that are clustered within a region often share a *common regional culture*, which can facilitate the process of social learning. This knowledge transmission can be assisted by the creation of regional institutions, which can help to develop rules and conventions governing firms' behaviour and interaction (ibid.: 694). The results from their study of indigenous and foreign-owned manufacturing firms reveal that the home region is the main site at which firms engage in local learning through interaction. This view is consistent with Maskell and Malmberg (1999), who discuss the role of geographic proximity in promoting knowledge creation and diffusion. They argue that some types of knowledge creation require close, tight relationships, which are easier and cheaper to achieve locally. Also, at the local level firms are able to engage in *collective learning*, involving the exchange of both partly codified and tacit knowledge.

Breschi (2000) identifies some 'empirical regularities' in the geographical distribution of innovative activities. This work hypothesizes that the relationship between innovation and spatial proximity is mediated by specific industry and technology conditions, embodied in technological regimes (Malerba and Orsenigo 1990: 215). The dimensions of technological regimes include appropriability, cumulativeness, opportunity and the knowledge base, all of which help to shape modes of learning, competition and selection processes that determine the number and types of innovative actors across sectors. The hypotheses are tested using patent data on the UK, France, Germany and Italy from the EPO. On the basis of this analysis, Breschi concludes that spatial patterns of innovation differ systematically across technological classes and, therefore, that technological regimes play a fundamental role in shaping spatial patterns of innovation across countries. Furthermore, spatial cumulativeness is a key explanatory variable of spatial concentration of innovative activities and regional technological performance.

Other work has examined the relationship between firms' innovative activities and local research efforts. For instance, Jaffe's (1989) seminal study claims that geographic proximity to a source of knowledge, such as a university or another firm, can be useful in capturing spillover benefits. Jaffe's study examines the extent to which spatially mediated R&D spillovers influence the generation of increased innovative output (measured by corporate patents) at the US state level. One of the key results from this study was that corporate patenting responds positively to knowledge spillovers from universities. The study also provides some evidence of the importance of geographically mediated commercial spillovers from university research, with the impact being statistically strongest in the pharmaceuticals industry.

Feldman (1994) builds on Jaffe's (1989) study, adopting an alternative measure of innovative output – new product citations, collected from the US Small Business Administration innovation citation database: and two new innovative inputs – related industry and business service presence, which reflect regional innovation capacity. The analysis reveals that innovation is a function of an area's technical infrastructure; innovation is positively related to geographic concentration of industrial and university R&D expenditures and to the presence of related industry and business services. This suggests that the co-location of complementary resources can provide economies of scope, which encourage innovation and product commercialization.

The literature on knowledge spillovers raises an important issue: could innovative activities in some industries cluster geographically more than in other industries due to the spatial concentration of the location of production? Audretsch and Feldman (1996) try to address this question by controlling for geographic concentration of production location. In order to measure the extent to which innovative and production activity are spatially concentrated, they calculate Gini coefficients, as recommended by Krugman (1991). They consider three sources of economic knowledge – industry R&D, skilled labour; and size of the pool of basic science for a specific industry – as key influences on the geographic concentration of innovation. In order to account for the use of basic science by industry, academic departments are assigned to industries based on a survey of industry R&D managers. Audretsch and Feldman consider other explanatory variables: dependence of an industry on natural resource inputs and the capital intensity of an industry. Their key results show that, even when controlling for geographic concentration of the location of production, the three knowledge-generating variables described above still impact significantly on the propensity for innovative activities to cluster spatially. In terms of dependence on natural resources, the results show that the proportion of natural resources in industry inputs influences the geographic concentration of production. Lastly, they found that industries tend to be less geographically concentrated when scale economies play a more important role (Audretsch and Feldman 1996: 636).

Extending this focus on the geography of university–industry interaction, Mansfield and Lee (1996) examined the impact of geographic distance and university quality in their US study of the university contribution to industrial innovation. They hypothesize that the probability that a firm funds academic R&D at a particular university is inversely related to the distance between the firm and university. A shorter distance implies easier and cheaper interaction between academic and firm personnel, which may encourage a firm to deal with local universities. They also consider the effect of the quality of university faculty (as measured by the National Academy of Sciences) on the propensity of a firm to support academic R&D. In some cases, increases in faculty quality may not be worth the additional costs, or high-quality universities may enforce tight regulations on industrial support. In their industry-wide sample of firms, Mansfield and Lee found that distance does influence a firm's decision to support academic R&D. Distance is particularly important for universities with only 'adequate to good or marginal' facilities: for these universities the chances of industry support are quite low if they are more than 100 miles from a firm.

Another example of work that attempts more formally to model distance in collaborations is Katz (1994). Katz maintains that scientific collaboration is a social process and therefore informal communication is crucial, and is more likely to occur when the collaborators are geographically close to one another. Katz claims that other studies fail to isolate geographic effects from other factors and therefore he develops a non-distorting method for analysing the effect of geographical proximity on university collaborations in the UK, Canada and Australia. By using data from the Science Citation Index, Katz creates a collaboration and publication matrix for a given set of universities. The distance between collaborators is calculated via radial distance matrices, which contain the distance, measured in miles between pairs of universities within a country. These matrices are then applied to exponential regression analysis, which reveals that collaboration between universities decreases exponentially with distance, and is therefore more likely to occur when partners are close. This result provides evidence to support the hypothesis that informal communication is vital for research collaborations, and that distance can be a hindrance.

The importance of geography in influencing university–industry interactions can be examined using survey data on industrial firms' attitudes to public research. A typical study in this tradition is Arundel and Geuna (2004), which examines how the role of distance in determining firm–PRO collaboration is mediated by the type of knowledge sought. For instance, when useful knowledge is in codified form (patents and publications) the importance of being physically close to public science should decline, but increase when the useful knowledge is available only in tacit form, making personal contact necessary. In order to test these hypotheses they use the Policies, Appropriability and Competitiveness for European Enterprises (PACE) survey, which covers some of Europe's largest R&D-intensive firms. They use the responses from this survey to develop an index of proximity defined by the importance firms give to knowledge obtained from domestic versus foreign sources. Their analysis finds that the proximity effect declines with the increase in R&D expenditure (which they use as a proxy for firm size), but increases with the quality and availability of domestic outputs from public science. Their results also confirm that those firms seeking codified knowledge are less likely to find geographic proximity of importance.

Fabrizio (2006) considers the relationship between the number of non-patent citations in a firm's patents and the firm's basic research focus and collaborative efforts using patent application data from the US Patent and Trademarks Office. The results illustrate that pharmaceutical and biotechnology firms investing more in university scientists (measured by percentage of publications co-authored with a university) cite more public science in their patented innovations. Interestingly, the results reveal some potential diminishing returns to additional collaborations with universities, suggesting that to exploit the results of public scientific research, firms need to develop internal scientific ability and expertise. Fabrizio's study includes a measure of the minimum distance to a university and, as expected, finds that greater distance from a research university is associated with lower exploitation of public science in the firm's innovations.

Agrawal (2006) explores the hypothesis that licensing strategies that directly engage the inventor will impact on the likelihood and degree of commercialization success, using a dataset of 124 licensing agreements from MIT inventors. Surprisingly, no relationship was found between agreements made with firms in the local area (within 50 miles of MIT) and commercial success; however, the author reports that this is probably due to the large incidence of local licences.

In a recent UK study, Abramovsky et al. (2006) explore the relationship between the location of private sector R&D labs and university research departments. They exploit ONS (Office for National Statistics) establishment-level UK Business Enterprise R&D data to develop measures of business sector R&D activity, along with data from the UK Research Assessment Exercise (RAE), to gather information on university research quality. They construct measures of the presence of business sector R&D activity at post-code level for 111 postcode areas. Their empirical results provide strong evidence for co-location in pharmaceuticals R&D, which is disproportionately located near to relevant university research, especially RAE 5 or 5* rated chemistry departments. There is also some evidence for co-location of lower-level departments and machinery and communications equipment firms.

In summary, the literature suggests that geographic proximity can play an important role in shaping university–industry interactions. Geographic proximity can lower the costs of exchange and facilitate more effective knowledge sharing between individuals. In part, these advantages are related to similarities and common interests between individuals and organizations located in the same area, but they also arise from regional cultural factors, shared histories and institutional arrangements. Despite this overall view, the mechanisms that moderate the effects of geographic proximity remain undertheorized and underdeveloped. More research is required to substantiate the findings of the studies cited above, and especially to provide an insight into what triggers interaction and how geographic proximity may enhance or discourage interaction. Despite the fact that universities are often seen as a 'strategic asset in the knowledge economy', the number of UK firms that use universities as a source of information remains low (Tether and Swann 2003). This pattern also holds across other European countries. Therefore the question of what determines or shapes university–firm interaction is a pertinent one for researchers and policy makers attempting to encourage this activity.

There are several shortcomings in the literature on geographic proximity and university–industry interactions discussed above. First, many of the studies referred to do not account for sectoral differences and, so far, most research effort has focused on a small number of highly specialized sectors, such as biotechnology. It would be unwise to generalize based on sectors where the links between research and industry practice are by nature and tradition relatively strong. We lack evidence on the importance of geographic proximity for a variety of sectors, although there is an increasing number of studies that are adopting this approach. Second, as Bercovitz and Feldman (2006) note, few studies take the firm as their focus. The attitudes and managerial practices within firms in terms of universities and how these are shaped by geographic factors remain an open question. As Laursen and Salter (2004) point out, managerial decisions about how to organize their firms' innovative activities have traditionally received little attention in the literature on university–industry interaction. There is potential to import concepts from managerial studies about the nature of technological search and to use these concepts to achieve a better appreciation of antecedents to and consequences of managerial decision making in this area. Third, most measures of geographic proximity are relatively simple and incomplete. It is difficult to get access to detailed geographic data on university–industry interaction. Proximity measures are often confined to administrative boundaries, frequently arbitrarily created by governments and with little meaning for the firms that inhabit these spaces. For example, in the UK, the South East region is an administrative convenience that extends around London, stretching from Oxford to Brighton, with little or no physical or administrative cohesion. The use of state-level data in the USA is a common approach, but is subject to biases based on the size of states and their physical location. Indeed, how geographic distance should be measured remains an open question: by kilometre, by city, by administrative region or by country. As yet, there are few standardized measures of distance in the field and, therefore, studies of the importance of geographic distance tend to present a mixed and unreliable picture.

FUTURE RESEARCH QUESTIONS

Given this situation, there is a range of new and interesting research questions that future research may seek to tackle. These include refining and extending the measures and treatments of geographic distance, allowing nuanced and refined portraits of the effect of geography on university–industry interaction. Using detailed data on firm location and exploring their relationships with local universities would make it possible to explore how small distances may shape the likelihood of collaboration between firms and universities. It would be useful to know how the research activities of these universities may shape local firms – through the movement of skilled labour, access to recent research, or through face-to-face contacts.

Also, we still know little about how the different activities of universities may shape firm-level innovation. To date, most attention in the literature has focused on universities' research outputs. Does the increasing use by universities of formal intellectual property protection mechanisms, such as patents, have a positive or negative effect on the likelihood that local firms will interact with the university sector? The costs and benefits of university patents themselves remain the subject of considerable debate (see Geuna and Nesta 2006). Yet, little is known about what effect the movement towards greater use of intellectual property protection by universities, especially in the UK, is having on firms' attitudes to working with universities. There is evidence from the USA to suggest that high levels of formal intellectual property protection can act as a significant barrier to formal and informal interaction. To date, we know little about what are the real effects of these changes in universities on the incidence and depth of university–industry interaction.

Finally, there is the question of how the firms' use of open and distributed models of the innovation process might reshape the role of geographic proximity on university–industry collaboration (von Hippel 2005). External sources of knowledge for innovation may come from other firms, clients, industry associations or universities, reflecting the multi-actor character of the innovation process. Chesbrough et al. (2006) suggest several methods for utilizing these external knowledge sources, for example, by imitation of a competitor; use of intermediate markets as an incentive for innovation; and, perhaps most pertinent to university–industry interaction, use of alliances as a method of identifying and incorporating external knowledge into the innovation process. For some, this movement towards open and distributed innovation processes represents a fundamental departure from the traditional vertically integrated model, whereby innovation was seen to stem from internal R&D activities (see Chesborough et al. 2006). Others (Helfat 2006) remain sceptical. As yet, however, the role of universities in open and distributed models of innovation remains a relevant and underexplored research area. In theory, if firms are increasingly looking outwards for ideas and technology, there is significant opportunity to enhance the direct links between universities and industry practice. At the same time, universities are more frequently seeking to demonstrate their usefulness to society by engaging with problems relevant to industry practice. In some cases, universities are becoming commercial actors in their own right, seeking to capture rents from their intellectual property through patents and spin-offs. We know little about the influence that these changes in the corporate and university sectors will have, or how geographic proximity may moderate or shape these patterns of exchange.

Changes in the university system do not take place in isolation, and new managerial approaches, focusing on open and distributed innovation, may profoundly alter the nature and types of university–industry interactions that will take place in the future. How these changes in firm behaviour will reshape the university system as it seeks to respond to the opportunities, and tries itself to become an active exploiter of its intellectual property, is unknown. However, they provide fertile ground for new research to peel back and explore both the changes in the nature of university research and in patterns of industrial innovation.

OUR KEY MESSAGES

1. Despite technological advances, proximity still matters for some types of interaction, especially university–industry collaboration.
2. The role of the university in the economic system is complex, subtle and often varied.
3. We should explore how the research activities of universities may shape the behaviour of local firms through various channels: the movement of skilled labour, access to recent research, or face-to-face contacts.

REFERENCES

Abramovsky, L., Harrison, R. and Simpson, H. (2006), 'University research and the location of business R&D', IFS Working Paper 07/02.

Agrawal, A. (2006), 'Engaging the inventor: exploring licensing strategies for university inventors and the role of late knowledge', *Strategic Management Journal*, **27**: 63–79.

Arundel, A. and Geuna, A. (2004), 'Proximity and the use of public science by innovative European firms', *Economics of Innovation and New Technology*, **13**(6): 559–80.

Asheim, B. and Gertler, M. (2005), 'The geography of innovation: regional innovation systems', in J. Fagerberg, D. Mowery and R. Nelson (eds), *The Oxford Handbook of Innovation*, Oxford: Oxford University Press, pp. 291–317.

Audretsch, D. and Feldman, M. (1996), 'R&D spillovers and the geography of innovation and production', *American Economic Review*, **86**(3): 630–40.

Beise, M. and Stahl, H. (1999), 'Public research and industrial innovations in Germany', *Research Policy*, **28**: 397–422.

Bercovitz, J. and Feldman, M. (2006), 'Entrepreneurial universities and technology transfer: a conceptual framework for understanding knowledge based economic development', *Journal of Technology Transfer*, **31**: 175–88.

Breschi, S. (2000), 'A geography of innovation: a cross sector analysis', *Regional Studies*, **34**(3): 213–29.

Chesbrough, H. (2003), *Open Innovation: The New Perspective for Creating and Profiting from Technology*, Boston, MA: Harvard Business School Press.

Chesbrough, H., Vanhaverbeke, W. and West, J. (eds) (2006), *Open Innovation: Researching a New Paradigm*, Oxford: Oxford University Press.

Cohen, W. and Levinthal, D. (1990), 'Absorptive capacity: a new perspective on learning and innovation', *Administrative Science Quarterly*, **35**(1): 128–52, *Special Issue: Technology, Organisations, and Innovation*.

Cohen, W., Nelson, R. and Walsh, J. (2002), 'Links and impacts: the influence of public research on industrial R&D', *Management Science*, **48**(1): 1–23.

Dosi, G. (1988), 'The nature of the innovation process', in G. Dosi, R. Nelson, G. Silverberg, C. Freeman and L. Soete (eds), *Technical Change and Economic Theory*, London: Pinter, pp. 221–38.

Fabrizio, K. (2006), 'The use of university research in firm innovation', in H. Chesbrough, W. Vanhaverbeke and J. West (eds), *Open Innovation; Researching a New Paradigm*, Oxford: Oxford University Press, pp. 134–60.

Feldman, M. (1994), *The Geography of Innovation*, Dordrecht, The Netherlands: Kluwer Academic Publishers.

Fontana, R., Geuna, A. and Matt, M. (2006), 'Factors affecting university–industry R&D projects: the importance of searching, screening and signalling', *Research Policy*, **35**: 309–23.

George, G., Zahra, S. and Wood, R. (2002), 'The effects of business–university alliances on innovative performance: a study of publicly traded biotechnology companies', *Journal of Business Venturing*, **17**: 577–609.

Gertler, M., Wolfe, D. and Garkut, D. (2000), 'No place like home? The embeddedness of innovation in a regional economy', *Review of International Political Economy*, **7**(4): 688–718.

Geuna, A. and Nesta, L. (2006), 'University patenting and its effects on academic research: the emerging European evidence', *Research Policy*, **35**(6): 790–807.

Helfat, C. (2006), 'Open innovation: the new imperative for creating and profiting from technology', *Academy of Management Perspectives*, **20**(2): 86.

Henderson, R., Jaffe, A. and Trajtenberg, M. (1998), 'Universities as a source of commercial technology: a detailed analysis of patenting, 1965–1988', *The Review of Economics and Statistics*, **80**(1): 119–27.

Jaffe, A. (1989), 'Real effects of academic research', *American Economic Review*, **79**: 957–70.

Katz, J. (1994), 'Geographical proximity and scientific collaboration', *Scientometrics*, **31**(1): 31–43.

Klevorick, A.K., Levin, R.C., Nelson, R. and Winter, S. (1995), 'On the sources and significance of inter-industry differences in technological opportunities', *Research Policy*, **24**: 185–205.

Krugman, P. (1991), *Geography and Trade*, Cambridge, MA: MIT Press.

Laursen, K. and Salter, A. (2004), 'Searching high and low: what types of firms use universities as a source of innovation?', *Research Policy*, **33**: 1201–15.

Link, A.N. and Rees, J. (1990), 'Firm size, university based research, and the returns to R&D', *Small Business Economics*, **2**: 25–31.

Lundvall, B. (1988), 'Innovation as an interactive process: user producer relations', in G. Dosi, R. Nelson, G. Silverberg, C. Freeman and L. Soete (eds), *Technical Change and Economic Theory*, London: Pinter, pp. 349–69.

Malerba, F. and Orsenigo, L. (1990), 'Technological regimes and patterns of innovation: a theoretical and empirical investigation of the Italian case', in A. Heertje and M. Perlman (eds), *Evolving Technologies and Market Structure*, Ann Arbor: Michigan University Press, pp. 283–306.

Mansfield, E. (1991), 'Academic research and industrial innovation', *Research Policy*, **20**: 1–12.

Mansfield, E. (1998), 'Academic research and industrial innovation: an update of empirical findings', *Research Policy*, **26**: 773–6.

Mansfield, E. and Lee, J. (1996), 'The modern university: contributor to industrial innovation and recipient of industrial R&D support', *Research Policy*, **25**: 1047–58.

Maskell, P. and Malmberg, A. (1999), 'Localised learning and industrial competitiveness', *Cambridge Journal of Economics*, **23**: 167–85.

Meyer-Krahmer, F. and Schmoch, U. (1998), 'Science based technologies: university–industry interactions in four fields', *Research Policy*, **27**: 835–51.

Motohashi, K. (2005), 'University–industry collaborations in Japan: the role of new technology based firms in transforming the National Innovation System', *Research Policy*, **34**: 583–94.

Mowery, D. and Rosenberg, D. (1989), *Technology and the Pursuit of Economic Growth*, Cambridge: Cambridge University Press.

Mowery, D., Nelson, R., Sampat, B. and Ziedonis, A. (2001), 'The growth of patenting and licensing by UD universities: an assessment of the effects of the Bayh Dole Act of 1980', *Research Policy*, **30**: 99–119.

Pavitt, K. (1991), 'What makes basic research economically useful?', *Research Policy*, **20**(2): 109–19.

Nelson, R. (1959), 'The simple economics of basic scientific research', *Journal of Political Economy*, **67**: 297–306.

Nelson, R. (1986), 'Institutions supporting technical advance in industry', *American Economic Review*, Proceedings, **76**: 186–9.

Rosenberg, N. and Nelson, R. (1994), 'American universities and technical advance in industry', *Research Policy*, **23**(3): 323–48.

Salter, A. and Martin, B. (2001), 'The economic benefits of publicly funded basic research: a critical review', *Research Policy*, **30**: 509–32.

Schartinger, D., Rammer, C., Fischer, M. and Fröhlich, J. (2002), 'Knowledge interactions between universities and industry in Austria: sectoral patterns and determinants', *Research Policy*, **31**: 303–28.

Shane, S. (2004), *Academic Entrepreneurship: University Spinoffs and Wealth Creation*, Cheltenham, UK and Northampton, MA, USA: Edward Elgar.

Storper, M. (2004), 'Buzz: face-to-face contact and the urban economy', *Journal of Economic Geography*, **4**(3): 351–70.

Tether, B. and Swann, P. (2003), 'Sourcing science: the use by industry of the UK science base for innovation; evidence from the UK's Innovation Survey', CRIC Discussion Paper, No. 64.

Von Hippel, E. (2005), *Democratizing Innovation*, Cambridge, MA: MIT Press.

Zucker, L., Darby, M. and Armstrong, J. (2002), 'Commercializing knowledge: university science, knowledge capture, and firm performance in biotechnology', *Management Science*, **48**(1): 138–53.

15. Enhancing the flow of knowledge to innovation: challenges for university-based knowledge transfer systems

Hossein Sharifi, Weisheng Liu,
Brian McCaul and Dennis Kehoe

1. INTRODUCTION

Innovation through the creation, diffusion and application of knowledge has increasingly become recognized as a crucial driver for economic growth, social evolution (OECD 1999, 2002; Foray and Lundvall 1966; DTI 2003), and a primary source of competitive advantage (Dutta 1997) in the global market. In this context, these changing elements have also triggered a substantial evolution in the process of innovation and knowledge diffusion (Robertson 1967), characterized by networking, integration, flexibility and just-in-time information processing (Freeman 1994; Wonglimpiyarat and Yuberk 2005). Innovation systems have therefore been evolving in theory and practice at great speed, indicated by the number of new models that have emerged in the past few years. Current debate on innovation theory is dominated by the fifth-generation innovation model proposed by Rothwell (1994) and the 'open innovation' paradigm propounded by Henry Chesbrough (Chesbrough 2003a, 2003b, 2006).

On the other hand, 'innovation diffusion' is defined as 'a process by which innovation is communicated through certain channels over time among members of a social system' (Rogers 1995). This social system includes several key players, such as the knowledge adopter, originator and intermediary agents. To accommodate the brisk pace of innovation, innovation diffusion calls for the input of components (technology, management etc.) of innovation from a broader array of players. In respect of technological innovations and perhaps spin-outs with regard to business model innovations, universities, which have traditionally been viewed as support structures and originators of innovation (EC 2003), are increasingly seen by industry and government as one of the critical sources of knowledge and ideas (Fabrizio 2006; DTI 2002; Kelly et al. 2006).

Theoretical and empirical work in innovation economics suggests that the use of scientific knowledge by setting up and maintaining good industry–university relations positively affects innovation performance (Mansfield 1991; Mansfield and Lee 1996; OECD 2002). Fuelled by the notion that streamlined interaction between science and industry is important, but not sufficient, for the success of innovation activities and ultimate economic growth, industry–university links have become a central concern of many government policies in recent years (Office of Science and Technology 1993; Polt 2001). The empirical evidence shows an intensification of the knowledge flow from universities to industry in the new century (EC 2003; Polt 2001; Lambert 2003; Lundvall 1992), and universities have increasingly been recognized as major players not only for originating but also for promoting the diffusion of knowledge and techniques that contribute to industrial innovations (Feller 1990; Henderson et al. 1998). However, the same empirical evidence has equally elucidated the significant institutional barriers to the commercialization of university researches, which could lead to low efficiency of university–industry transfers.

From an institutional perspective, establishing a specialized technology transfer office within the university which serves as the primary intermediary (Howells 2006; Hoppe and Ozdenoren 2005), or plays the boundary role (Tushman 1977) in university–industry technology/knowledge transfer, has been viewed as an instrumental means for developing relations with industry (Bercovitz et al. 2001; Schaettgen and Werp 1996). Over the past two decades, many universities and research institutes in the UK have developed institutional structures that are specifically charged with handling every aspect of technology transfer activities (Lambert 2003; Polt 2001).

Since the termination of the British Technology Group's (BTG) monopoly in the 1980s, UK-based knowledge transfer offices (UKTOs) have evolved from the simple business liaison functions to a more comprehensive range of activities to cover most of the university knowledge transfer operations. With around 20 years of development, the UK-based UKTO has gained significant recognition, as well as equal criticism, from university, government and industry grounded mainly on the figures of fact extracted from these surveys. For example, most UK-based UKTOs are still not able to cover their costs in comparison to some UKTOs in the USA that have gained great financial benefits (Boone 2006). The *Lambert Review* (2003: 5) points out that 'A barrier to commercialising university IP lies in the variable quality of technology transfer offices. Most universities run their own technology transfer operations, but only a few have a strong enough research base to be able to build high-quality offices on their own.'

Concerns from various entities and in particular policy makers have been expressed over the significant institutional barriers to the commercialization of university research. As a result, the research for improving the university–industry link through university knowledge supply and knowledge transfer (KT) capabilities has received considerable attention. Issues include whether such agencies' intermediation would lead to the monopolization of knowledge flows and become a barrier to the creation of a positive knowledge culture, and what institutional role and definition should be considered for these offices. Governments, while supportive of the establishment of UKTOs, have expressed their concerns over the results achieved (HEFCE 2004; NESTA 2006).The roles and economic impact of such offices have been examined and analysed in various studies and surveys (Conesa et al. 2005; Graff et al. 2002; Carlsson and Fridh 2002; Siegel et al.2003, 2004; Thursby et al. 2001) to reveal critical enablers and barriers to managing university–industry knowledge flow. Studies have led to proposals for improving the position, practice and management of these offices, but the major question of whether systems devised in universities to facilitate and promote the science and knowledge base into socio-economic value and competitive edge for the nation remains to be satisfied.

Specifically very little attention seems to have been given to the root causes of the relative failure of the industry–science–relationship (ISR) system, knowledge-to-innovation system, or, in short, university-based knowledge transfer systems (normally via UKTOs). These causes may be the same that are now being addressed in the innovation systems within business and industry, and have driven them to accept and embrace open innovation (Chesbrough 2003b). Universities are part of the changing circumstances and, as well as being participants in the process of innovation, perhaps they should apply the same principles that are proving to be the basis for successful innovation in industry. This potential does not seem to have been fully understood nor addressed by universities and policy makers.

Issues such as the definition, role, impact, position, practice, management, evaluation and classification of these entities within the context of the innovation for knowledge process now need serious attention within the new context of open innovation and an integrated approach in terms of value chain management. The characteristics that would make such organizations fit for emerging global circumstances, how adopting modern innovations views and systems could be realized and what are the requirements of a successful migration of university-based knowledge base to adapt to the new criteria should be addressed. This chapter examines these issues in order to identify challenge ahead, and proposes a conceptual model for this purpose.

2. KNOWLEDGE FLOW TRANSFER SYSTEMS AND INNOVATION MODELS

To understand university-based systems of knowledge transfer as part of the national innovation systems (Lundvall 1992) it is important to consider the evolution of such models and systems, and the evolving roles of universities as well as government in the process. Etzkowitz et al. (2000) suggested a new location for the research and technology transfer in the form of a 'triple helix' of university–industry–government relations. They believe that the increasing role for the state in funding science within the general context of the economy is well established now.

Technology (knowledge) transfer is a complex process, involving diffusion of basic research and its ultimate commercialization. Several authors have presented frameworks and models of knowledge transfer. In reviewing models by Cooley (1987), Cohen and Levinthal (1990), Trott et al. (1995), and Horton (1999), two main streams can be distinguished: discrete models which describe nodes and the discrete steps each goes through; and process models which describe knowledge transfer by separate processes undertaken by each. In terms of process, Rothwell (1994) defines the generations of innovation process, which might be grouped into linear and non-linear innovation process. According to Godin (2005), the linear model is 'one of the first frameworks' (historically) for understanding science and technology interaction. The knowledge value chain is another popular model that is often utilized for the analysis of the flow of knowledge. Most knowledge value chain models (Lee and Yang 2000; Wang and Ahmed 2005) are derived from Porter's value chain model, and can reveal the target and management of a specific organization's knowledge.

Between the 1950s and 1980s the 'linear model' mode of innovation assumed a central role for R&D in the development of new ideas (products or processes), whereby research in universities is followed directly by its application in industry, which postulates that innovation begins with research and ends with diffusion and application (Bush 1945; Mowery 1983). Later, Gibbons et al. (1994) argued that some parts of the university systems are outmoded. They suggested that the traditional model of segregated knowledge is being replaced by a more fluid and dispersed model in which universities have become an important actor among many overlapping, interdependent knowledge producers, thus questioning the distinction between academic and non-academic work. In addition, Dasgupta and David (1994) advocated the 'republic of science' to contrast norms of disclosure and dissemination of research in industry and academia. Stokes (1997) and Roger and Pielke (1996) argued that much academic research involves problem-focused basic research, linked to specific societal or industrial challenges; that is, university research is a complex mix of basic and applied work. Further, Etzkowitz and Leydesdorff (Etzkowitz et al. 2000; Leydesdorff and Etzkowitz 1998; Leydesdorff 2001) suggested that universities are freeing themselves from public control to become actors in their own right in the knowledge marketplace. They describe the emergence of the 'entrepreneurial university', which raises significant funds from the private sector and acts as a spur for economic development.

Correspondingly, government is the superpower in the world of university technology transfer (Grady and Pratt 2005). On one hand, governments tend to guarantee the stable interactions and exchange of knowledge from university to industry (Lambert 2003). On the other hand, governments have intervened in the technology arena to address market failures (Polt 2001). For instance, from the 'demand' side, when industrial companies underinvest in R&D due to the existence of 'spillover', which limits their ability to fully appropriate returns, or due to the uncertainty associated with innovation, governments have used measures aimed at increasing the volume of R&D without giving enough consideration to improving the effectiveness and efficiency of existing R&D (Lambert 2003). From the 'supply' side, governments also make policy amendments (Ploszek 2006) or introduce financial support, such as HEIF 1 to 3 in the UK, to propel the university to engage in the third-leg activities – knowledge transfer (HEFCE 2005). Governments need and will play an integrating role in managing knowledge on an economy-wide basis by making technology and innovation policy an integral part of overall economic policy (DES 2003).

The new knowledge transfer systems are based on the modern innovation theories mainly rooted in work by economist J. Schumpeter (1939, 1989), which was set around promoting entrepreneurship as the key element in introducing innovations into the economy, and dependence of research on resource availability. His works have been extended (Carter and Williams 1957, 1958, 1959) to highlight the importance of contacts with the world of science, and to the increasing interdependence of science and technology (Freeman 1974; Mansfield 1980; Rosenberg 1990, 1992). The post-neo-Schumpeterian research also demonstrates that the ability to make use of external sources of scientific expertise and advice was one of the main determinants of success (van de Vrande et al. 2006; Kirschbaum 2005).

Rothwell's 'fifth-generation of innovation' model was aimed at accommodating and explaining the intensification of the innovation process brought about by the application of the new management toolkit (Rothwell 1994). The other recent paradigm of 'open innovation', coined by Chesbrough (2003a), has become dominant in business areas as well as in academic and theoretical research. 'Open innovation' is established on some legacy of innovation theories and reveals recent evolutions in innovation. The idea is based on the low strategic impact and hence importance of internal R&D in industrial innovation systems and changes in the way companies generate ideas and bring them to market: a move from 'close innovation' to 'open innovation'. In a recent article Chesborough (2006) defines the main points of differentiation for open innovation, relative to prior theories:

1. Equal importance given to external knowledge, in comparison with internal knowledge
2. The centrality of the business model in converting R&D into commercial value
3. Type I and type II measurement errors (in relation to the business model) in evaluating R&D projects
4. The purposive outbound flows of knowledge and technology
5. The abundant underlying knowledge landscape
6. The proactive and nuanced role of IP management
7. The rise of innovation intermediaries
8. New metrics for assessing innovation capability and performance.

In the meantime Chesborough reflects on a growing trend in knowledge transfers from science to industry which highlights the importance of the organizations that can transfer technology to market. He also points to the implications of the idea for policy, including the need to rethink the sources of fundamental breakthroughs in the future, the institutions needed (education, university research, start-up formation, disciplines and IP), and finally considers to what extent policy should promote diffusion as opposed to protecting invention activity in this new environment.

3. UNIVERSITY KNOWLEDGE TRANSFER OFFICES

UKTOs originated from the evolution of the entrepreneurial university and the third stream of university activities. In the USA, the Bayh–Dole Act (1980), the Stevenson–Wydler Act (1980), and the Federal Technology Transfer Act (1985) led to a fundamental change in the way scientific discoveries at universities and federal laboratories were commercially exploited. Since then, the number of US universities that engage in technology transfer and licensing has increased eightfold, to more than 200, and the volume of university patents has increased fourfold (Mowery and Shane 2002). Over the past two decades, many universities and research institutes in the UK have developed institutional structures that are specifically in charge of handling every aspect of technology transfer activities (Lambert 2003; Polt 2001). The specific institutional arrangement has varied greatly, ranging from university-controlled off-campus technology brokers and technology incubators for university spin-offs, to university-managed units integrated to the overall university administration. The UKTO in its broadest sense has emerged as an important player within universities and generally plays a central role in identifying technologies with commercial potential, assisting researchers to patent their inventions, packaging the technology appropriately so as to attract industry, developing strategies to market such technologies, and leading the licensing negotiations with potential licensees (Etzkowitz 2003; Allan 2001).

Since the termination of BTG monopoly, the major formality of UKTO changed from a business liaison office to add the functions of licensing and patenting. The linear model of university-to-industry knowledge flow, through the invention of organizational mechanisms to move early-stage inventions from academic research into use, could potentially resolve the so-called 'European paradox' (EC 2002; Dosi et al. 2005), the gap between R&D spending and lack of expected contribution to economic growth. In the transformation from research to entrepreneurial university, the UKTOs' role expanded from a narrow focus on IP protection to a broader role in the innovation system. An indicator of change is the movement of the UKTO from the periphery to the core of the academic enterprise through administrative restructuring and change in attitude among faculty and administrators of technology transfer from a merely tolerated activity to an encouraged and prestigious academic task (Wright et al. 2004).

Along with the rising concern over the university–industry knowledge flow within government and universities, the formation, operations, performance and evolution of these intermediaries have become a subject of academic and practitioner research studies, as well as government watch and scrutiny in the light of the financial inflow of public funds directed to support and enhance the process of innovation from knowledge through universities. In the UK, professional bodies for support, study and monitoring of UKTOs such as AURIL (Association for University Research and Industry Links), UNICO (The University Companies Association), HE–BCI (The Higher Education–Business and Community Interaction Survey), UKTO study at NUBS (Nottingham University Business School) as well as units in the UK DTI (Department of Trade and Industry) have been involved in this. Information on the activities of the UKTOs, usually captured using quantitative measures, is provided by these organizations. Such data have been used by a wider community of researchers in studying UKTOs to explore and analyse activities of such offices in the knowledge transfer process and university–industry relations, mainly providing a narrow view that leaves out the broader emerging reality of UKTO activities and their roles, which are either not fully quantifiable or not yet attended to. Examples of this stream of work are Siegel et al. (2003, 2004), Carlsson and Fridh (2002), Bercovitz et al. (2001), Thursby et al. (2001); Thursby and Thursby (2002) and Rogers et al. (2000), who have mainly focused on obtaining and analysing qualitative data, in many cases complemented by quantitative data based on government-led surveys. The typical studies mentioned aim to determine the effectiveness of UKTOs in achieving additional income for the universities (i.e. royalty fees, equity shares) and contributing to regional development through creation of high-tech start-ups and improvement of the technological capacity of existing firms.

Furthermore, commissioned studies by government have also been carried out to investigate concerns over the achievements of UKTOs from the investments and impacts expected. The *Lambert Review* (Lambert 2003), as an example, was published in 2003 with some substantial conclusions. In the report, although attention was paid to the global changes and new trends in innovation systems and approaches, the conclusion was that 'The main challenge for the UK is not about how to increase the supply of commercial ideas from the universities into business. Instead, the question is about how to raise the overall level of demand by business for research from all sources', identifying the problem as the demand side. Praising the culture change in the UK's universities towards taking a more active and broader role in the regional and national economy, the report suggests that much more attention is being paid to governance and management issues in the knowledge transfer from university to industry. The report goes on to identify the problem with the process of IP commercialization and knowledge transfer to lie in the variable quality of technology transfer offices, and the ability of universities to build high-quality offices.

Other aspects of UKTOs have also been the subject of research. Studies have been conducted on economic modelling for the roles of UKTOs in the university–industry technology transfer. Lach and Schankerman (2003) provide support for the importance of inventors' royalty shares for university performance in terms of inventions and licence income. Hoppe and Ozdenoren (2002) present a theoretical model to explore the conditions under which innovation intermediaries (i.e. UKTOs) emerge to reduce the uncertainty problem. They suggest that UKTOs have the motivation to invest in expertise to locate new inventions, and sort profitable from unprofitable ones. While the UKTOs will reduce the uncertainty problem, there is still a high probability of inefficient outcomes due to coordination failure. This type of models is based on broader literature on intermediation to solve the problem of asymmetric information on the quality of technology between licensor and licensee. Biglaiser (1993), Howells (2006) and Lizzeri (1999) have investigated the role of intermediaries for information manipulation between buyer and seller. Their results indicate that a UKTO is often able to benefit from its capacity to pool the inventions across research units and build a reputation within universities. It will have a motivation to shelve some of the projects, thus raising the buyer's (firm's) beliefs about expected quality. It may result in less but more valuable innovations being sold at higher prices. This UKTO reputation can explain the importance of critical size for UKTOs in order to be successful as well as the stylized fact that UKTOs may lead to fewer licensing agreements, but higher income from innovation transfers. However, this model offers only a partial view on the rationale for such intermediaries.

Barriers to the flow of knowledge and technology transfer has also been a widely studied subject. European Commission and Federal Ministry of Economy and Labour (2001) point out that such barriers include information asymmetry, incompatible objectives, high financial cost and uncertainty of income. Siegel et al. (2004) summarized the barriers in several categories, including: lack of understanding regarding university corporate or scientific norms and environments; insufficient rewards for university researchers; bureaucracy and inflexibility of university administrators; insufficient resources devoted to technology transfer by universities; poor marketing/technical/negotiation skills of UKTOs; universities' over-aggressive approach to exercising IP rights; unrealistic expectations of faculty members/administrators regarding the value of their IP; and the 'public domain' mentality of universities. Jones-Evans et al. (1999) analysed the barriers of university–industry knowledge flow and found that the barriers include lack of financial support, cultural conflict, lack of academic recognition of technology commercialization, lack of defining entrepreneurship in society and lack of industry demand.

Sceptical views have also been expressed by workers in this area. Mowery et al. (2003, 2004) argue that the UKTOs are not only superfluous to effective technology transfer but are impeding the free flow of knowledge from university to industry. Nelson has argued that companies were well aware of what was coming out of university research without any advertising or pushing from university offices. In these cases, the holding of intellectual property rights by the university is unlikely to have facilitated technology transfer, but rather probably made it more difficult by imposing transaction costs on firms that wanted to further develop that technology.

4. CHALLENGES FOR UNIVERSITIES' KNOWLEDGE TRANSFER CENTRED AT UKTOS

From the discussion above, it is evident that universities (or higher education generally) are now facing a great challenge resulting from a fast and extensive range of changes in their business environment, and in particular issues related to policy and government. The background to such challenges can be summarized as follows:

- Increasing expectations from society and the government for universities to play a more active and institutional role in transforming the knowledge/science base into economic value and competitive advantage.
- A paradigm shift in strategic management, organization, and innovation theories and approaches within the business environment.
- A legacy of inefficient and slow-response organization and process management as well as a non-supportive culture within universities.
- Inappropriate governance mechanisms including performance measurement for knowledge transfer and innovation activities.

The real challenges, however, could only be understood well and formulated properly if discussed within a suitable theoretical context to bring the problems into a reference framework. A number of theoretical grounds have been identified for this purpose which we believe can be used to address and formulate the challenges in universities' knowledge transfer process. The theories, which include stakeholder, value chain, and innovation theories, will be discussed briefly in the following, leading to a conceptual framework in the next section.

The Stakeholder-oriented UKTO

The stakeholder theory and approach entered the academic area of strategic management about two decades ago. Suggesting that addressing stakeholders' interest would lead to better performance (Freeman 1984), the approach provides a framework for identifying various stakeholders, their position and orientation in order to define and direct the value to all interest-holding entities. The classic definition of a stakeholder is 'any group or individual who can affect or is affected by the achievement of the organisation's objectives' (ibid.: 25). That is, the term 'stakeholders' refers to those people who have or perceive they have some stake in the future success of an organization or organizational unit. It is therefore imperative to have a clear idea of who these people are, and their needs and expectations. Their points of view and expectations should also be considered in developing strategic goals and objectives. If they have a stake in the output of the process, they should have a stake in the input to the process. If a stakeholder group is important to achieving organizational objectives, the organization should actively manage the relationship with that stakeholder group.

It is believed that the UKTO-focused knowledge transfer process is initially multi-stakeholder-oriented, which mainly involves stakeholders such as the government, industry and university (EC 2003), defined as the 'primary' stakeholders (Kaler 2003), who have a sound relationship with and great influence on the organization. Other stakeholders, deemed 'secondary' stakeholders, are situated at the boundary of UKTOs, and have relatively less influence on the UKTOs' activities. In this context, the complexity of coordinating those stakeholders becomes the main mission for UKTOs.

Value Chain-based Management of University Knowledge Transfer

The value chain concept has been widely used in many areas to overcome the weaknesses of the traditional static view of management and has been bounded to deal with dynamic linkages between value-adding activities that in a collective perspective go beyond any specific party or sector. The value chain approach, made popular mainly by Porter (1985) has entered the literature on innovation and knowledge transfer to take it beyond firm-specific analysis. Concentrating on interlinkages, it allows uncovering of the dynamic flow of economic, organizational and coercive activities between different players on a global scale. Lundquist (2003) suggests that technology transfer happens in technology and product-development value chains, where the value chain, as the series of steps taken for this purpose, is driven by value that is consciously structured, led and managed to constantly increase the win–win value of relationships with customers and others in the marketplace. Attention to knowledge as a source of competitive advantage has led to the proposition of models for knowledge value in terms of economic–value (Porter 1985; Lee and Yang 2000; Wang and Ahmed 2005) or less market-based and more social-context-based (Bozeman and Rogers 2002).

The value chain is conceptualized and modelled in various ways. However, most of the existing models share some principles, including the existence of various stakeholders interlinked through relationships, oriented towards a direction (buyer or producer), and usually biased or controlled by a dominant party. The role of governance, mainly determined and applied by the dominant player or players, has found an important place in work on value chains. The governance mechanism's role is multiple, including protecting the chain from competition by identifying and implementing barriers to entry, supporting the upgrading of the chain's members and the whole chain's position and appropriation of value or rents generated in the process of value chain activities.

UKTOs as the intermediaries within university for the knowledge transfer process can play a more effective role if they direct their activities towards value creation through building a better fit between relationships and knowledge. This in fact shapes one of the major challenges for UKTO management. Universities as a main source of knowledge in which many parties hold stakes can be viewed and managed as a value chain, with centrality of the UKTO as the intermediary for the transfer of knowledge to economic and social capital. Centring the strategic orientation of the UKTO on bridging between stakeholders based on a business-model-driven approach will define a unique role for the UKTO in driving the university science/knowledge base to value creation.

Innovation Challenge: From Random to Open Innovation

Universities have come a long way in finding and enhancing their roles in the development of the national and regional economy. With support from the government through new funding schemes such as third-stream and regional development frameworks, universities have made considerable efforts in changing the culture of knowledge transfer and commercialization of IP by paying attention to management issues through erecting UKTOs, and in this they seem to have had some success, as supported by research (Lambert 2003). Recent reports from HEFCE also support the distances that universities have come on the road to becoming major players in the national and global innovation game (HEFCE 2005). The research by Wright et al. (2004) suggests a change of culture in universities from being quite varied and hostile to a more positive attitude towards entrepreneurship. EIRMA's report (2005) suggests that European universities are proactively resolving to play a more active role in the innovation process and implementing policies and resources to do this effectively. The Innovation Survey 2005 by CBI concludes that collaboration between business and universities on innovation-related work is increasing, yet many companies say such collaborations are not their most effective.

The EIRMA's report (2005) suggests, however, that for the universities to benefit from open innovation they need to introduce changes in the way that knowledge generated is handled and protected. And so there seems to be a long way to go for universities to play the role they are expected to. Considering that nations are increasingly concerned with their economic competitive advantage, and that knowledge may be the only lasting source of competitive advantage (Hitt et al. 1998), technology transfer could play a strategic role in national competitiveness. The UK's system of innovation in principle suffers from the same problem of the 'European paradox' (EC 2002). Although the Lambert Report associated the problem, to a great extent, with the low efficiency of UKTOs as a barrier to the smooth knowledge flow from university to industry, the problem seems to be rooted in the more fundamental issue of understanding and embracing new forms of

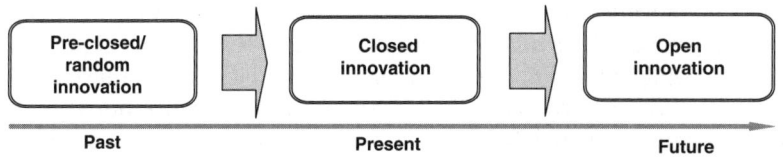

Figure 15.1   Moving from random to open innovation

innovation, emerging partly due to the increasing complexity of industrial innovation processes, in which innovative business models and new forms of organization are core elements. Review of the literature and reports from public bodies governing university knowledge transfer suggests that universities have now moved from a 'random or accidental' system of innovation prior to structured approaches such as close innovation to one which, at best, can equate to the outmoded closed innovation paradigm. With the well-grounded emerging theory of 'open innovation' it seems that the next destination for the universities is to move to an environment of such capacity and characteristics (Figure 15.1).

Universities, UKTOs and Open Innovation

The concept of open innovation has already received attention from the European knowledge transfer community, and there is evidence of some influence of the concept in the related communities' views, approaches and even policy and practical recommendations. Traces of reference to the concept can be found in the Lambert Review (2003) and in the Responsible Partnerships Initiative (EIRMA 2005). This partly stems from the fact that open innovation advocates have explicitly referred to universities as key sources of IP and technology for adoption by innovation firms (Chesbrough 2006), which can be appreciated by the knowledge transfer community who are tasked with commercializing university IP.

However, as mentioned before, the move within the university knowledge transfer community and process has just started to understand the situation and is barely fitting itself to the 'closed innovation' model. Therefore there is potentially much more to be gained from the concept of open innovation by universities and the knowledge transfer community than simply viewing open innovation as a convenient environment in which universities might increase industrial research funding and licensing opportunities, or as a means of outsourcing R&D activity to the university sector. An organizational and strategic response to open innovation might improve the role of the university as a player in the open innovation chain itself, for broader social benefit.

Open innovation may offer important insights for both the business and social/regenerative agendas within universities. Universities, while having certain significant additional social functions and differing value chains to those of a typical corporation, do share certain key organizational characteristics with industrial R&D organizations. Even where the social and value propositions of the university sector differ from those of business, it is also possible that proper understanding and appreciation of open innovation would contribute to the efficacy of the universities as actors in the move of innovation paradigm to an open innovation process.

Theoretically it is plausible to use open innovation as a model for redefining the university knowledge transfer process. With reference to eight characteristics of open versus closed innovation (Chesbrough 2006), we intend to analyse the relevance of the concept to university knowledge transfer and the extent of its influence on practices to provide a new vision for extending the role of the university knowledge base in the promotion and improvement of the national innovation system contributing to the economic and social competitiveness of the country. The eight areas of difference between open and closed innovation are re-packaged in five areas, a brief account of which as regards universities and their knowledge transfer system is provided in the following.

Equal importance given to external knowledge, in comparison with internal knowledge

This issue is widely applicable to universities as potential repositories and origins of knowledge and innovative ideas, and will affect many aspects of universities' knowledge transfer activities. To start with academe itself, while universities differ greatly from businesses in terms of external collaboration, and academics are inherently inclined to collaborate externally, the potential tendency towards what Chesbrough calls the 'Not Invented Here' syndrome (Chesbrough 2006) can hinder the connectivity expected. This can happen in two ways: first, it is unlikely that research projects acquire IP (other than with the transition of an academic) to meet the needs of a specific R&D programme; second, it remains difficult to encourage and coordinate cross-disciplinary collaboration (or more specifically cross-departmental). Therefore the UKTO strategy and practice could be redirected to assist in facilitating connectivity with R&D activities (within other university departments and in other universities) relevant to achieving viable commercial exploitation.

The principles of open innovation could be found even more easily applicable to the practice of knowledge transfer, such as exploitation of knowledge and capability. Whether universities' knowledge transfer strategies are supportive and capable of seeking a wider array of expertise from external resources (consultants, intermediaries, venture capital angels and other UKTO practitioners), and not limit itself to structured public funds such as 'Third-mission funding' is a different approach for universities.

Developing absorptive capacity through engaging in and stimulating networks (networking capability) for exchange of IP and the development of exploitation models is another aspect that universities can and should address. Adding to this is the capability to continuously scan for and adopt technologies that might provide a competitive advantage for the successful exploitation of internally generated IP or business models. Likewise, adopting externally developed and tested business models might be as important to successful knowledge transfer. In short, the ability to create successful ventures will depend on external resources and partnerships, including obtaining venture capital funding, as well as resources and guidance from potential end users in the R&D process to access certain knowledge assets and the market.

The centrality of the business model in converting R&D into commercial value

Chesbrough suggests that the main role of a business model is to create a heuristic and simplified cognitive map from the technical domain of inputs to the social domain of outputs (Chesbrough and Rosenbloom 2002). This will involve a process of selection in which alternatives should be filtered for the best outcome, and in this process there is a danger of missing better business models because it conflicts with the firm's existing model. A business model can therefore be a 'double-edged sword' (Osterwalder et al. 2005).

It is therefore possible to assume that this concept also applies to large and complex organizations such as universities. Ability to use various business models for different sectors, fields and opportunities, as well as for various social value propositions, can be critical for university knowledge transfer. At the same time it is possible to suggest that the level of flexibility in terms of ranges of business models which such organizations can adopt is higher than is usual in industry. For instance, a range of business models relevant to what universities do could include, and of course is not limited to: spin-out, start-ups, licensing, direct services such as research and consultancy, publications, and regional engagement. Still, within each of these models is a further broad range of business models relevant for creating maximum value. To create new models, experimenting with alternative business models should be considered; this will require creation of a process to explore the social domain far more thoroughly (Chesbrough 2003a).

The purposive outbound flows of knowledge and technology and proactive management of IP

This mainly relates to the issue of significant wastage of IP in universities which is often patented but not exploited and therefore new routes to 'spin out' IP are desperately needed. For this purpose it is critical for universities to be aware of this fact and to set their IP management system/process focused on a 'business model'. As has been supported by research in this area, UK UKTOs have developed significant expertise in the management and protection of IP. However, it is debatable that policy and practice are as proactive and nuanced as the open innovation concept might suggest as necessary. Problems could range from when no consideration of the business model exists in the process of IP protection, to where a linear stage-gate process is adopted that is unlikely to be as open to the adoption of new business models. It is not difficult to find examples of how UKTO practice is opposed to the characteristics of open innovation. Certainly the notion that the generator of a business model (as opposed to, or in addition to, the inventor) should be rewarded is alien in universities' knowledge transfer, and it would also be considered unorthodox to actively seek external IP and patents to unlock exploitation of internally generated IP. To unlock this and for universities to adapt the innovation system to the open innovation model the following are required:

- To ensure the utilization of IP (versus waste) in proportion to what is exploited, not left on the shelf, and not just turned into publication.
- Reasonably rapid exploitation (Markman et al. 2005).
- To actively seek external resources to unlock exploitation of internally generated IP by bringing in other IP and patents, and a party that has a business model to create, capture and appropriate value from the patent.
- 'Conflict of interest' to be resolved by introducing incentives for both innovation and commercialization (through adopting a proactive IP management) rather than solely invention.
- The IP management process to be designed as a proactive, effective and flexible process in which legal issues and integration with business model are balanced.

Alternative innovation intermediaries

The role of intermediaries in innovation has already been well addressed by researchers, and the concept has long been practised in universities through erection of KTO/TTOs. However, the role of intermediaries has changed in the wider environment of innovation and they are now playing a more direct role (Chesbrough 2003a). Stakeholders in the emerged intermediate markets can transact in ways that were usually internal processes. Recognizing this fact, and inclusion of any possible channel that fits the universities' knowledge transfer business model is a necessary move for understanding and adopting open innovation.

New metrics for assessing innovation capability and performance
Classical and typically matrix form frameworks adopted by and sometimes imposed on universities for measuring knowledge transfer and their social and economic impacts will not support the approach to achieving new capabilities, maximum benefit for all stakeholders, and open innovation. Value of knowledge and IP should be measured against the business model and the generated benefit to stakeholders, utilization of the IP, and networking. Measures should also be extended to include all stakeholders in the value chain. To take a comprehensive view, a combination of tangible and intangible measures should be adopted, to include social, business and academic (personal performance) values (capital).

5. A CONCEPTUAL FRAMEWORK FOR UNIVERSITY-BASED KNOWLEDGE TRANSFER

UKTOs, primarily representing universities, now face some grand challenges in answering the call from policy makers to play an essential role in turning the potential knowledge and science base to competitive advantage for the nation. At the same time, with the increasing changes in the global business environment and fundamental shifts in theories and concepts of innovation, the response required from UKTOs does not seem to lie in the usual areas of improvement, such as efficiency and professionalism. Turning the innovation culture within the university-based knowledge transfer process from one of accidental connection to a business model based on open innovation, managing complex transfer arrangements, and at the same time maintaining and improving the core process of generating knowledge and science within the universities are some main issues that must enter the agendas of managing such institutes.

Based on the discussion above, this section presents a conceptual model that suggests a new approach to managing the process of knowledge transfer in universities centred in UKTOs. Figure 15.2 represents the model, bringing together the elements discussed in this chapter. 'Open innovation' is considered to be the new reference model for structuring the process and organization of knowledge transfer within universities. The objectives of open innovation are also understood to be realized if the approach to the management of the process is based on a value chain model, and at the

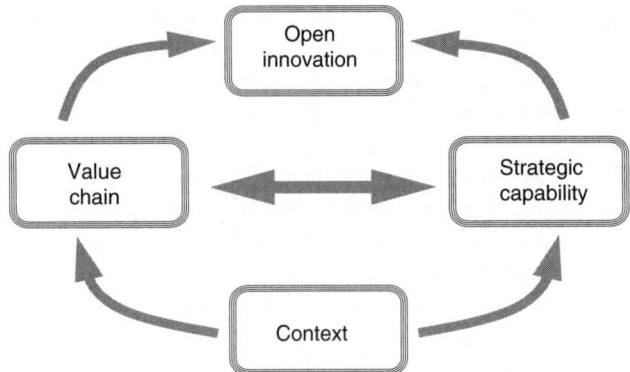

Figure 15.2   Innovation challenge for university-based knowledge transfer

same time supported by strategic and dynamic capabilities that give the UKTO the strengths and capabilities required for fulfilling its roles and responding to changes. The underlying support element on which these elements should be developed and managed is the provision of the required context for achieving the strategic objectives. The context – the interrelated external influential factors that may enable or hinder UKTO operations – also poses challenges for the organizational components and mechanisms supporting the technology transfer operations.

The model can be used in revisiting the role and activities of UKTOs and as a reference for identifying the strategic and operational gaps within the universities' knowledge transfer process. It can also be used to develop models for better practice and redefining or re-engineering the process.

6.   CONCLUSION

For universities to play a more significant role in the innovation game and to contribute more to national competitiveness in the light of complex and fast-paced global changes, the challenge is enormous. The preliminary study shows that it is not just a question of size, location and structure of the office or competence of the staff, but it may be about a fundamental misunderstanding of the process and theory of innovation that should be applied to universities. The solution is perhaps not in providing more training or doing and encouraging more R&D alone. It is the use of innovation in transforming R&D into exploitable products and services that is the heart of the problem. Studies of universities' knowledge transfer process and the responsible offices are not scarce, but they mainly lack a proper theoretical background for understanding innovation within the context of universities. Although universities are typically large and old organizations with considerable inertia obstructing fundamental and quick changes, at the same time they contain a vast potential of knowledge, economic and social capital, as well as diversity which could enable them to migrate to a new mindset in approaching innovation and knowledge transfer. Applying emerging theories in innovation and resource management such as open innovation, stakeholder theory and value chain management introduces a new ground for understanding the role, position and operations of universities in knowledge transfer.

REFERENCES

Allan, M.F. (2001), 'A review of best practices in university technology licensing offices', *AUTM Journal*, **XIII**: 57–68.

Bercovitz, Janet, Feldman, Maryann, Feller, Irwin and Burton, Richard (2001), 'Organizational structure as a determinant of academic patent and licensing behavior: an exploratory study of Duke, Johns Hopkins, and Pennsylvania State Universities', *Journal of Technology Transfer*, **26**: 21–35.

Biglaiser, G. (1993), 'Middlemen as experts', *RAND Journal of Economics*, **24**: 212–23.

Boone, J. (2006), 'Leading universities see six fold return on technology transfers', *Financial Times* (London), 20 September, p. 8.

Bozeman, B. and Rogers, J.D. (2002), 'A churn model of scientific knowledge value: Internet researchers as a knowledge value collective', *Research Policy*, **31**: 769–94.

Bush, V. (1945), *Science: The Endless Frontier*, http://www.nsf.gov/od/lpa/nsf50/vbush1945.htm, accessed 30 October 2006.

Carlsson, B. and Fridh, A.-C. (2002), 'Technology transfer in United States universities: survey and statistics analysis', *Journal of Evolutionary Economics*, **12**: 199–232.

Carter, C.F. and Williams, B.R. (1957), *Industry and Technical Progress*, Oxford: Oxford University Press.

Carter, C.F. and Williams, B.R. (1958), *Investment in Innovation*, Oxford: Oxford University Press.

Carter, C.F. and Williams, B.R. (1959), *Science and Industry*, London: Oxford University Press.

Chesbrough, H.W. (2003a), *Open Innovation: The New Imperative for Creating and Profiting from Technology*, Boston, MA: Harvard Business School Press.

Chesbrough, H.W. (2003b), 'The era of open innovation', *MIT Sloan Management Review*, **44**(3): 35–41.

Chesbrough, H.W. (2006), 'Open innovation: a new paradigm for understanding industrial innovation', in H.W. Chesbrough, W. Vanhaverbeke and J. West (eds),

Open Innovation: Researching a New Paradigm, Oxford: Oxford University Press, pp. 1–12.

Chesbrough, H. and Rosenbloom, R. (2002), 'The role of the business model in capturing value from innovation: evidence from Xerox Corporation's technology spin-off companies', *Industrial and Corporate Change*, **11**(3): 529–55.

Cohen, W.M. and Levinthal, D.A. (1990), 'Absorptive capacity: a new perspective on learning and innovation', *Administrative Science Quarterly*, **35**: 128–52.

Conesa, F., Castro, E. and Zarata, M.E. (2005), *ProTon Europe Annual Survey Financial Year 2004*, Belgium: Proton.

Cooley, M. (1987), *Architect or Bee: The Human Price of Technology*, London: The Hogarth Press.

Dasgupta, P. and David, P.A. (1994), 'Toward a new economics of science', *Research Policy*, **23**, 487–521.

Department for Education and Skills (2003), *The Future of Higher Education*, London: HMSO, CM5735.

Department of Trade and Industry (2002), *Investing in Innovation: A Strategy for Science*, engineering and technology, London: DTI.

Department of Trade and Industry (2003), *Innovation Report: Competing in the Global Economy*: the innovation challenge, London: DTI.

Dosi, G., Llerena, P. and Labini, M.S. (2005), *Science–Technology–Industry Links and the 'European Paradox': Some Notes on the Dynamics of Scientific and Technological Research in Europe*, Working paper of Laboratory of Economics and Management in Sant'Anna School of Advanced Studies.

Dutta, S. (1997), 'Strategies for implementing knowledge based systems', *IEEE Transactions on Engineering Management*, **44**(1): 9–90.

Etzkowitz, H. (2003), 'Innovation in innovation: the triple helix of university–industry–government relationship', *Social Science Information*, **42**(3): 293–337.

Etzkowitz, H., Webster, A., Gebhardt, C. and Terra, B. (2000), 'The future of the university and the university of the future: evolution of ivory tower to entrepreneurial paradigm', *Research Policy*, **29**: 313–30.

European Commission (2002), *Report on Research and Development*, Economic Policy Committee, DG ECFIN, Working Group on Research and Development.

European Commission, Directorate-General for Research (2003), *Third European Report on Science and Technology Indicators: Towards a Knowledge-Based Economy*, Brussels: European Commission.

European Commission and Federal Ministry of Economy and Labour Austria (2001), *Benchmarking Industry–Science Relations: The Role of Framework Conditions*, Vienna and Mannheim: European Commission, Enterprise DG and Federal Ministry of Economy and Labour.

European Industrial Research Management Association (2005), *Responsible Partnering: Joining Forces in a World of Open Innovation – A Guide to Better Practices for Collaborative Research between Science and Industry*, http://www.eirma.asso.fr/f3/local_links.php?action=jump&id=796, accessed 10 October 2006.

Fabrizio, K.R. (2006), 'The use of university research in firm innovation', in H. Chesbrough, W. Vanhaverbeke and J. West (eds), *Open Innovation: Researching a New Paradigm*, Oxford: Oxford University Press, pp. 134–60.

Feller, I. (1990), 'Universities as engines of R&D based economic growth: they think they can', *Research Policy*, **19**: 335–48.

Foray, D. and Lundvall, B.D. (1996), 'The knowledge-based economy: from the economics of knowledge to the learning economy', in *Employment and Growth in the Knowledge-based Economy*, Paris: OECD, pp. 11–32.

Freeman, C. (1974), *The Economics of Industrial Innovation* (2nd edn 1982), London: Frances Pinter.

Freeman, C. (1984), *Strategic Management: A stakeholder approach*, Boston, MA: Pitman.

Freeman, C. (1994), 'The economics of technology change', *Cambridge Journal of Economics*, **18**: 463–514.

Gibbons, M., Limoges, C., Nowotny, H., Schwartzman, P., Scott, P. and Trow, M. (1994), *The New Production of Knowledge*, London: Sage.

Godin, B. (2005), 'The linear model of innovation: historical construction of an analytical framework', Project on history and sociology of S&T statistics, working paper 30.

Grady, R. and Pratt, J. (2005), 'The UK technology transfer system: calls for stronger links between higher education and industry', *Journal of Technology Transfer*, **25**: 205–11.

Graff, G., Heiman, A. and Zilberman, D. (2002), 'University research and office of technology transfer', *California Management Review*, **45**(1): 88–115.

HEFCE (2004), *Higher Education–Business and Community Interaction Survey 2002–2003*, London: HEFCE.

HEFCE (2005), *Report on Higher Education–Business and Community Interaction Survey*, http://www.hefce.ac.uk/Pubs/HEFCE/2005/05_07/, accessed 10 October 2006.

Henderson, R., Jaffe, A.B. and Trajtenberg, M. (1998), 'Universities as a source of commercial technology: a detailed analysis of university patenting: 1965–1988', *Review of Economics and Statistics*, **65**: 119–27.

Hitt, M.A., Keats, B.W. and DeMarie, S.M. (1998), 'Navigating in the new competitive landscape: building strategic flexibility and competitive advantage in the 21st century', *Academy of Management Executive*, **12**(4): 22–43.

Hoppe, H.C. and Ozdenoren, E. (2002), 'Intermediation in invention: the role of technology transfer offices', *International Journal of Industrial Organization*, **23**: 483–503.

Hoppe, H.C. and Ozdenoren, E. (2005), 'Intermediation in innovation', *International Journal of Organization*, **23**: 483–503.

Horton, A.M. (1999), 'A simple guide to successful foresight', *Foresight*, **1**(1): 5–9.

Howells, J. (2006), 'Intermediation and the role of intermediaries in innovation', *Research Policy*, **35**: 715–28.

Jones-Evans, D., Klofsten, M., Andersson, E. and Pandya, D. (1999), 'Creating a bridge between university and industry in small European countries: the role of industrial liaison office', *R&D Management*, **29**(1): 47–56.

Kaler, J. (2003), 'Differentiating stakeholder theories', *Journal of Business Ethics*, **46**: 71–83.

Kelly, U., McLellan, D. and McNicoll, Iain (2006), *The Economic Impact of UK Higher Education Institutions: A Report for Universities UK*, London: Universities UK.

Kirschbaum, R. (2005), 'Open innovation in practice', *Research–Technology Management*, **48**(4): 24–8.

Lach, S. and Schankerman, M. (2003), 'Royalty sharing and technology licensing in universities', *Journal of the European Economic Association*, **2**(2–3): 252–64.

Lambert, R. (2003), *Lambert Review of Business University Collaboration: Final Report*, London: HMSO.

Lee, C.C. and Yang, J. (2000), 'Knowledge value chain', *Journal of Management Development*, **19**(9): 783–93.

Leydesdorff, L. (2001), *Knowledge-Based Innovation Systems and the Model of a Triple Helix of University–Industry–Government Relations*, Proceedings of conference, New Economic Windows: New Paradigms for the New Millennium, Salerno, Italy, September.

Leydesdorff, L. and Etzkowitz, H. (1998), 'The triple helix as a model for innovation studies', *Science and Public Policy*, **25**(3): 195–203.

Lizzeri, A. (1999), 'Information revelation and certification intermediaries', *RAND Journal of Economics*, **30**: 214–31.

Lundquist, G. (2003), 'A rich vision of technology transfer technology value management', *Journal of Technology Transfer*, **28**: 265–84.

Lundvall, B. (1992), *National Systems of Innovation: Towards a Theory of Innovation and Interactive Learning*, London: Frances Printer.

Mansfield, E. (1980), 'Basic research and productivity increase in manufacturing', *American Economic Review*, **70**: 863–73.

Mansfield, E. (1991), 'Academic research and industrial inventions', *Research Policy*, **20**: 1–12.

Mansfield, E. and Lee, J.Y. (1996), 'The modern university: contributor to industrial invention and recipient of industrial R&D support', *Research Policy*, **25**: 1047–58.

Market & Opinion Research International (2005), *Innovation Survey for CBI*, http://www.cbi.org.uk/ndbs/Press.nsf, accessed 10 October 2006.

Markman, G.D., Gianiodis, P.T., Phan, P.H. and Balkin, D.B. (2005), 'Innovation speed: transferring university technology to market', *Research Policy*, **34**: 1058–75.

Mowery, D.C. (1983), 'Economic theory and government technology policy', *Research Policy*, **16**: 27–43.

Mowery, D.C. and Shane, S. (2002), Introduction to the Special Issue on University Entrepreneurship and Technology Transfer, *Management Science*, **48**(1).

Mowery, D.C., Nelson, R.R., Sampat, B.N. and Ziedonis, A.A. (2003), 'The growth of patenting and licensing by US universities: an assessment of the effects of the Bayh–Dole Act of 1980', *Research Policy*, **30**: 99–119.

Mowery, D.C., Nelson, R.R., Sampat, B.N. and Ziedonis, A.A. (2004), *Ivory Tower and Industrial Innovation: University–Industry Technology Transfer Before and after the Bayh–Dole Act*, Stanford, CA: Stanford Business Books.

NESTA (2006), *The Innovation Gap: Why Policy Needs to Reflect the Reality of Innovation in the UK*, NESTA, http://www.nesta.org.uk/assets/pdf/innovation_gap_report.pdf, accessed 23 November 2006.

OECD (1999), *Managing National Innovation Systems*, Paris: OECD.

OECD (2002a), *Dynamising National Innovation Systems*, Paris: OECD.

OECD (2002b), Draft final report on the strategic use of intellectual property by public research organisations in OECD countries, DSTI/STP(2002) 42/REV1, Paris.

Office of Science and Technology (1993), *Realizing Our Potential: A Strategy for Science, Engineering and Technology*, CM 2250, London: HMSO.

Osterwalder, A., Pigneur, Y. and Tucci, C. (2005), 'Clarifying business models: origins, present, and future of the concept', *Communication of AIS*, **15**.

Ploszek, R. (2006), *Reform of Higher Education Research Assessment and Funding*, London: Royal Academy of Engineering.

Porter, M. (1985), *Competitive Advantage: Creating and Sustaining Superior Performance*, New York: Free Press.

Robertson, T.S. (1967), 'The process of innovation and diffusion of innovation', *Journal of Marketing*, **31**(1): 14–19.

Roger, A. and Pielke, J. (1996), *Asking the Right Questions: Atmospheric Sciences Research and Societal Needs*, http://sciencepolicy.colorado.edu/admin/publication_files/resource-145-1997.13.pdf, accessed 30 October 2006.

Rogers, E. (1995), *Diffusion of Innovations*, New York: The Free Press.

Rogers, E., Yin, M.J. and Hoffmann, J. (2000), 'Assessing the effectiveness of technology transfer offices at U.S. research universities', *The Journal of the Association of University Technology Managers*, **8**: 47–80.

Rosenberg, N. (1990), 'Why do firms do basic research with their own money?', *Research Policy*, **19**(2): 165–75.

Rosenberg, N. (1992), 'Scientific instrumentation and university research', *Research Policy*, **21**(4): 381–90.

Rothwell, R. (1994), 'Towards the fifth-generation innovation process', *International Marketing Review*, **11**(1): 7–31.

Schaettgen, M. and Werp, R. (1996), *Good Practice in Transfer University Technology to Industry*, European Commission, EIMS No. 16.

Schumpeter, A.J. (1939), *Business Cycles: A Theoretical, Historical and Statistical Analysis of the Capitalist Process*, New York: McGraw-Hill.

Schumpeter, A.J. (1989), *Essays on Entrepreneurs, Innovations, Business Cycles, and the Evolution of Capitalism*, New Brunswick, NJ: Transaction Books.

Siegel, D.S., Waldman, D. and Link, A. (2003), 'Assessing the impact of organizational practices on the relative productivity of university technology transfer offices: an exploratory study', *Research Policy*, **32**: 27–48.

Siegel, D.S., Waldman, D.A., Atwater, L.E. and Link, A.N. (2004), 'Toward a model of the effective transfer of scientific knowledge from academicians to practitioners: qualitative evidence from the commercialization of university technologies', *International Journal of Technology Management*, **21**: 115–42.

Stokes, D.E. (1997), *Pasteur's Quadrant: Basic Science and Technological Innovation*, Washington, DC: Brookings Institution Press.

Thursby, J.G. and Thursby, M.C. (2002), 'Who is selling the ivory tower? Sources of growth in university licensing', *Management Science*, **48**(1): 90–104.

Thursby, J.G., Jensen, R. and Thursby, M.C. (2001), 'Objectives, characteristics and outcomes of university licensing: a survey of major U.S. universities', *Journal of Technology Transfer*, **26**: 59–72.

Trott, P., Cordey-Hayes, M. and Seaton, R.A.F. (1995), 'Inward technology transfer as an interactive process', *Technovation*, **15**(1): 25–43.

Tushman, M.L. (1977), 'Special boundary roles in innovation process', *Administrative Science Quarterly*, **21**(4): 587–605.

van de Vrande, V., Lemmens, C. and Vanhaverbeke, W. (2006), 'Choosing governance modes for external technology sourcing', *R&D Management*, **36**(3): 347–63.

Wang, C.L. and Ahmed, P.K. (2005), 'The knowledge value chain: a pragmatic knowledge implementation network', in *Handbook of Business Strategy*, Emerald Group Publishing Ltd, pp. 321–6.

Wonglimpiyarat, J. and Yuberk, N. (2005), 'In support of innovation management and Roger's innovation diffusion theory', *Government Information Quarterly*, **12**: 411–22.
Wright, M., Birley, S. and Mosey, S. (2004), 'Entrepreneurship and university technology transfer', *Journal of Technology Transfer*, **29**: 235–46.

16. Enabling information infrastructures and technologies

Roula Michaelides and Dennis Kehoe

1. INTRODUCTION

The advent of a knowledge economy has been a driver for change in organizations, where individuals persistently utilize and draw on a wealth of knowledge to devise new ideas, solutions and products for a rapidly changing global marketplace. Innovation and product development cycles have seen a dramatic acceleration in response to customer demand.

Science and innovation are at the heart of business transformation, since technology itself is viewed as a vehicle for enabling globalization and fostering the ability to innovate. New ideas offer new direction: they boost commerce, create new products and markets while improving efficiency by delivering benefits to companies, customers and society.

The UK government's response to the global economy challenges is to create the best possible environment for science and innovation in the UK that would facilitate a seamless connection between a world-class science base and businesses. This would be achieved through the provision of supporting measures to grow new knowledge-based firms and take advantage of commercial opportunities arising from research (Science and Innovation framework 2006), According to the Science and Innovation framework (2006), in order to create an effective ecosystem for innovation the UK government has put forward three action areas:

1. Improvement of the strategic management of investment in science and innovation, to ensure that the UK's science and innovation system is more responsive to economic and public policy priorities. This also includes the more effective coordination of different funding mechanisms.
2. Enabling the appropriate brokering mechanisms to encourage greater collaboration between industry and the research base, and assisting businesses and the science base to interact in a range of ways to suit their needs.

3. To make science, technology, engineering and materials (STEM) subjects more attractive to students, to ensure a highly skilled and diverse workforce to drive future innovation and growth.

In line with the second action area, this chapter reviews the role of information and communication technologies (ICT) as collaboration enablers. The UK government has successfully launched in 2001 the e-science (2001) research framework to bring large computing power to dispersed research collaborations working with large datasets.

Two of the most significant developments in science in terms of innovation management are the development of large-scale computing grid technologies to support global science and the development of research collaborative environments/communities based on the open innovation paradigm. Open innovation is by definition focused on establishing ties between research/scientific/innovative groups and other organizations. The enabling impact of grid computing is discussed in Section 2 and collaborative communities and technologies are considered in Section 3.

The eight characteristics of open innovation as identified by Chesbrough et al. (2006) are: importance of external knowledge; significance of establishing a business model; Type I and II errors in commercializing R&D projects; creation of value though outbound flows of knowledge and technology; facilitation of an abundant underlying knowledge landscape; proactive intellectual property (IP) management; enabling innovation intermediaries; and establishing new metric for innovation capability. Grid computing exhibits most of these characteristics, particularly in providing an abundant underlying landscape for creating research/scientific communities as well as establishing infrastructure for knowledge flows and external contribution. Knowledge communities and collaboration technologies also demonstrate most of the open innovation characteristics, such as enabling external knowledge access, facilitating outbound flow of knowledge and technologies, and displaying the role of innovation intermediaries.

Developments both in the grid computing area as well as the collaborative technologies and knowledge communities have seen a major contribution by the open source community, itself an exemplar of open innovation (West and Gallagher 2006). Open source demonstrates two key elements of open innovation: shared rights to use the technology and a collaborative development of that technology (ibid.). The interesting research element here is that collaborative Web 2.0 technologies discussed in Section 2 have themselves emerged from open innovation processes as adopted by the open source communities. This chapter examines each of the above areas and addresses the current research challenges through the review of related literature.

2. GRID COMPUTING TECHNOLOGIES SUPPORTING E-SCIENCE

Over the last few years grid computing has evolved from a niche technology associated with scientific research and technical computing into a business technology of increased commercial acceptance (Kourpas 2006). Large-scale science and engineering is becoming increasingly intricate not only because of the complexity of the large datasets needed for simulation or modelling purposes, but also because of the geographically distributed nature of research collaborations. Scientists are increasingly using computing tools to enable, automate and visualize experimental processes. When scientists use these techniques, the term e-science is used to describe the work and the participating scientists are termed e-scientists (McGough et al. 2006).

Intense problem-solving computation, however, is not only needed by research and science communities, but is also demanded by worldwide manufacturing, biotechnology and finance businesses. The business needs to employ complex computational intensive solutions are driving up costs and operational overheads of the technology environments (Joseph and Fellenstein 2004). Grid technologies present a viable solution to these computing industrial needs, with the provision of many computing resources working together in advanced innovative collaborations across technical communities.

Grid computing is a standards-based application/resource-sharing architecture that makes it possible for heterogeneous systems and applications to share, compute and store information resources transparently (Clabby 2004).

In traditional non-grid environments, information infrastructures are organized in 'silos' as individual applications are run separately with their dedicated internal resources. Grid computing, however, provides an architecture flexible enough for many software applications to run efficiently in parallel, thus allowing the storage, sharing and analysis of large diverse volumes of data. In operational terms grid computing ensures that cross-organizational people have access to information at the right time, which can improve decision making, employee productivity and collaboration. The grid is fundamentally about virtualization of both information and workload.

This new innovative approach to computing can be likened to a large power 'utility' grid that provides power to homes and businesses daily. Grid technology connects regional and national computer networks (or individual grids), eventually creating a universally available source of computing power that improves resource utilization and reduces costs, while maintaining a flexible infrastructure that can cope with changing business demands, yet remain reliable, resilient and secure (Joseph and Fellenstein 2003).

Grid types are classified depending on their application focus. *Computing grids* focus on exploiting unused computing power; *data grids* focus on data analysis; *collaborative grids* focus on modelling, visualization and the sharing of large, graphics-intensive files between collaborative groups (that are often geographically dispersed); *utility grids* focus on using computing power when necessary; and *enterprise optimization grids* focus on providing increased computing resources and better storage systems utilization for enterprises that are trying to better leverage their investments in computer systems and storage.

Industrial sectors that have adopted grid technologies are identified by Clabby (2004):

- aerospace and automotive, mainly for collaborative design and modelling applications. As an example, BMW uses SGI Origin servers for stochastic crash simulation – a technique that considers the behaviour of a population of vehicles rather than a single deterministic vehicle used in testing.
- Architecture, civil engineering and construction sectors.
- Electronics industries, primarily for design and testing.
- Energy industries for oil and gas exploration where companies use grids for data-intensive seismic processing and imaging applications (these applications help energy companies analyse geological information to determine, for instance, where to drill for oil/gas). As an example, Compagnie Générale de Géophysique (CGG), supplier of services and products to the worldwide oil and gas industry, has deployed 512 grid-clustered Dell servers running RedHat Linux in the UK to process data in finding new oil fields around the world.
- Finance/insurance/real-estate services sectors mainly focusing on securities and brokerage – especially for stock/portfolio analysis and risk management. For example, Deutsche Bank Group uses Platform LSF software to streamline portfolio risk analysis.
- Life sciences, biotechnology and pharmaceutical industries, where almost every major drug company is making use of compute grids for drug discovery. An example is Novartis Pharmaceuticals, which has deployed United Devices grid products to speed in-silico research projects to discover new prescription drugs.
- Manufacturing industries requiring inter/intra-team collaborative design as well as intensive process management.
- Media/entertainment to generate digital animation.

- Utilities sector to improve efficiency while dealing with peaks of demand. For instance, Hewitt Associates, a large human resource consulting and outsourcing firm, operates one of the best-known examples of an enterprise optimization grid.

A comprehensive list of well-established, ongoing grid applications is available at the Grid Research Integration Deployment and Support Centre (2006). Some of the largest and best-known grid configurations are briefly highlighted in Table 16.1.

In the UK in 2001, the e-science (2001) programme (led by UK Research Council programmes and a core programme to develop and broker generic technology solutions) was launched. Nine e-science centres established within the UK include the national centre, Northern Ireland centre, Cambridge e-science, Wales e-science, London e-science, North West e-science, North East e-science, Oxford e-science and Southampton e-science.

e-science research is made possible through modern distributed computer power. Scientists working in collaboration may be dispersed throughout the country or the world, but can work together on large datasets, using terascale computing resources and sharing computer-based research tools. The Internet and the grid are the main infrastructures supporting this. Researchers believe that consistent and inexpensive access to high-powered computing facilities, databases, sensors (and collaborating colleagues) will transform science and human society, much as mainframe and personal computers have done.

Table 16.1   Large grid applications

Grid application	Objective
SETI@Home project	Searching for extraterrestrial life
Mersenne project	Worldwide mathematics research
NASA Information Power Grid	Linking researchers and industry with NASA scientists
EU DataGrid, the EuroGrid	Building next-generation computing infrastructures for intensive computation, analysis of shared large-scale databases, from hundreds of TeraBytes to PetaBytes, across distributed scientific communities in Europe
Oxford e-Science grid	Providing very large-scale computing power to large datasets of globally distributed researchers and scientists

Source: Grid Research Integration Deployment and Research Centre (2006).

3. COMMUNICATION TECHNOLOGIES AND INFRASTRUCTURES SUPPORTING KNOWLEDGE COMMUNITIES

Innovation usually involves day-to-day participation in a flow of knowledge that consists of organizational data and information as well as exchange of ideas with other individuals who have expertise related to the same area of work. Knowledge is nowadays considered to be a fundamental asset of the organization (Teece 1998). Knowledge and information flows have long been identified as key determinants of successful innovation and new product development processes (Brown and Eisenhardt 1995; Tidd et al. 1997; Rothwell 1994).

The rapid diffusion of Internet-based networking technologies has not only provided the necessary infrastructure to enable information flows, but has also resulted in a proliferation of knowledge communities operating across organizational borders, independently of time and location. Information networks allow a large number of users to systematically share ideas, create distributed learning systems (Sproull and Kiesler 1991), and enable virtual teams to execute the innovation process (Qureshi 2000; Kessler 2003; Rad and Levin 2003). ICT infrastructures, particularly in relation to innovation generation, are considered by Dodgson et al. (2005) as beneficial for providing improved connectivity speed, enhanced processing power and speed, thus resulting in technological advances in an organization.

Dahlbom (2000) formulates the correlation between information technologies and social networks by arguing that information technologies provide more than a foundation for improving productivity; they enable 'social structures to be formed, re-formed and dissolved in a continuous process of networking'. Internet connectivity and the development of new information standards have enabled an open and almost cost-free exchange of information between users/actors in any market (Evans and Wurster 1999; Shapiro and Varian 1998).

Kubicek and Wagner (2002) produced a historical analysis of community networks to provide an insight into how these social networks evolved over time with varying technology infrastructures and the kinds of services such applications make available. A conclusion to their analysis is that a standard for the design or development of community information systems does not exist. This leads to the assumption that the precise impact of differing forms of collaboration infrastructures is not well understood. While the pervasiveness of Internet technologies has enabled the creation of networked communities, it has also made it increasingly difficult for people to know the scope and range of their 'virtual' social networks.

Harrison and Zappen (2005) take this further by stating that since there is no standard for designing community networks, each instance of computerized community information system can be seen as a trial for easing tensions between hardware/software infrastructure, design of the particular application or system, user needs and the resources that support these efforts.

Existing research proposes a number of distinctive properties that information infrastructures supporting knowledge systems should exhibit (Abecker et al. 1999):

- Active support by means of intelligent searching facilities so that users become aware of rapidly changing information relevant to their work. Thus the information system infrastructures should actively offer interesting knowledge.
- Integrative functions by means of supporting heterogeneous data formats, informal representations (texts, memos, minutes of meetings, documentation, business letters, graphics and drawings) and knowledge embedded ('materialized') in artefacts/representations of work (e.g. a product design).
- Self-adaptiveness and self-organization by means of building user behaviour memory by automatically gathering and storing interesting facts and information in an unobtrusive manner.

The holistic approach of adapting modern Internet infrastructures and collaborative technologies with social networks and communities of practice presents both opportunities and challenges in the modern era.

3.1 Knowledge Flows and Communities of Practice

Over the last ten years in the academic literature there has been a growing interest in communities of practice (CoPs) as a method for transferring and generating knowledge within organizations. The MySpace phenomenon and the increased use of virtual communities (VCs) and CoPs by large international organizations such as IBM and Procter and Gamble (P&G) confirms the importance of VCs in today's society and the global economy. CoPs are found in many organizations and use different names, such as 'learning communities' at Hewlett-Packard, 'family groups' at Xerox Corporation, 'thematic groups' at the World Bank, 'peer groups' at British Petroleum, and 'knowledge networks' at IBM Global Services, but they remain similar in general intent (Gongla and Rizzuto 2001).

A CoP is defined as a group of people who come together to learn from each other by sharing knowledge and experiences about the activities in which they are engaged (Wenger 1998, 2001). CoPs can be physically located, locally networked (e.g. within a company via an intranet), virtual (i.e. networked across distance) or, as often happens, a hybrid of these.

The success of CoPs could be attributed to the fact that collaboration of cross-functional teams with diverse occupational and intellectual backgrounds increases the likelihood of combining knowledge in novel ways (Nonaka and Takeuchi 1995). In such teams, the amount and diversity of information available to members is amplified, thus facilitating different alternatives and various perspectives (Brown and Eisenhardt 1995). However, the real value of technology in supporting knowledge transfer in collaborative communities is not yet fully understood. Many knowledge transfer applications found in literature are inherently partial towards information processing rather than knowledge flows (Cross and Baird 2000; McDermott 1999; Luftman and Brier 1999). While explicit knowledge can be effectively codified into databases or electronic documents, the same task can be much harder when knowledge is mostly tacit (Brown and Duguid 2000a).

The characteristics of tacit and explicit knowledge are briefly outlined in Table 16.2. CoPs support both types of knowledge exchange but they have a special role in tacit knowledge exchange. Tacit knowledge interaction methods such as story-telling, anecdotes, impromptu comments and opinions occur naturally in many CoPs (Prusak 2001). The rigidity of hierarchical organizational structures that tend to constrain informal communication in many work environments are less prominent in CoPs. As a collaborative community evolves and matures, its members share expertise and tacit knowledge, often by telling stories (Prusak 2001) and working together to solve problems. Progressively shared insights and skills accumulate over time and contribute to a common community knowledge repository. Through the

Table 16.2   Properties of explicit–tacit knowledge

Explicit knowledge	Tacit knowledge
Can be articulated and codified	Hard to codify and communicate
Generalized knowledge	Hard to formalize
Knowledge of rationality	Individual action and experience
Facts and figures	Subjective, intuitive
Production knowledge (rules)	Knowledge of experience
Procedural knowledge (methods)	Skills and habits
Functional and systemic knowledge	Values and judgements

Source: Adapted from Nonaka and Takeuchi (1995); Wiig (1993).

use of ICT this shared knowledge can be stored and retrieved efficiently. Through the provision of collaborative Internet tools, CoPs achieve richer expression, thus making the discussion of ideas/proposals around specific problems easier.

CoPs, like any social entity, evolve and change over time. Members join, others leave and new areas of interest emerge that change the community character. Typically new communities focus their energy on getting started (e.g. attracting members, specifying policies), whereas established communities are more concerned with domain-related issues. Ideally CoPs develop shared communal resources, such as routines, behaviours, artefacts and vocabulary that help create a sense of community that socially binds members. Policies and norms of behaviour facilitate establishing shared goals and expectations.

Wenger (1998) has identified five stages relating to the evolution of a CoP. According to Wenger (1998), the creation of a CoP is called the potential stage, where the community objectives are defined, membership relationships are still loosely coupled and an Internet presence is initiated. Then the CoP progresses into the second phase, the 'potential' stage, where membership relationships are enhanced with more trust and familiarity and a schedule of events programmed. The third phase is called the maturing and active stage, where a community becomes more focused, a knowledge agenda is developed, a knowledge repository is designed and implemented and community coordinator roles are established. The fourth phase is the stewardship stage, where the community actively seeks to extend the community boundaries through openness and renewal of knowledge contributions from other similar interest communities. The last phase, called the 'transformation' stage, is where the community has fulfilled its learning agenda so that it either disintegrates or expands through merger with another community entity.

The evolution pattern of a CoP is influenced by a dynamic balance of people, process and technology elements (Gongla and Rizzuto 2001). Gongla and Rizzuto also propose five CoP development stages that are very similar in concept to the Wenger (1998) evolution model. The first phase, called the 'potential' stage, is about establishing member connection where individuals find one another and link up. The next phase, called the 'building' stage, signifies the activities the community members engage in to build shared memory and to create a context for collaboration. The engaged stage is primarily about member access and learning. During this stage members build relationships of increased trust and loyalty. The active stage is the mature phase of member collaborations with experts from other communities, with a focus on problem solving and decision making. The final 'adaptive' stage is about changing the community environment to innovate, and generate new products, markets and businesses.

Benefits
Those who have worked with CoPs have long believed that they increase the level and flow of knowledge within an organization (Fontaine and Millen 2002; Brown and Duguid 1991, 2000b; Lesser and Storck 2001). Benefits of CoPs are threefold, viewed from the organizational, community and member perspective.

1. Benefits realized at the organizational level include:
   ● faster problem solving as employees and outside contacts communicate with each other to solve specific problems. Company experts become more accessible;
   ● cross-fertilization of ideas and increased innovation opportunities – most software companies initiate communities to offer an environment for people to communicate, exchange ideas, provide some suggestions about software improvements, etc.;
   ● knowledge retention when employees leave the organization – with the use of a CoP, specific knowledge is not held by a few experts only, which means that should these experts leave the organisation, working projects will not be interrupted.
   ● providing clarity when developing a collaboration strategy (Verna 2000);
   ● developing, recruiting and retaining talent (Verna 2000);
   ● building core capabilities and knowledge competences.
2. Benefits realized at the collaborative community level include:
   ● helping to build common language, methods and models around specific competences;
   ● transfering knowledge and expertise to a larger population.
3. Benefits realized at the individual member level include:
   ● helping people carry out their work more effectively – other community members can provide insights on work-related problems;
   ● fostering a learning-focused sense of identity – people interested in the same project will work together to solve the problem;
   ● helping to develop individual skills and competences (Verna 2000);
   ● helping a knowledge worker to stay current, with access to the latest news and skills about a technology or science subject;
   ● members share control and influence with the formal parts of the organization (Verna 2000).

Table 16.3   Benefits of CoPs

Ability to execute corporate strategy	Job satisfaction
Ability to foresee emerging market, product, technology capabilities and opportunities	Learning and development
Authority, reputation with customers, partners	Learning curve
Collaboration	New business development
Coordination and synergy	New customers
Cost of training	New revenues, new business, product, service or market
Customer loyalty stickiness	Partnering success
Customer responsiveness	Problem-solving ability
Customer satisfaction	Productivity or time savings
Customer service, support and acquisition costs	Professional reputation or identity
Customer turnover	Project success
Employee retention	Quality of advice
Empowerment	Risk management
Higher sales per customer	Supplier relationship costs
Idea creation	Supplier relationships
Identification – access to experts and knowledge	Time to market
Innovation	Trust between employees

Source: Fontaine and Millen (2004).

Fontain and Millen (2004) adopted an organizational viewpoint when identifying the benefits realized by CoPs (Table 16.3).

Performance characteristics of communities

As CoPs usually require an investment in resources, there has been increased pressure, mainly in large organizations, to evaluate the impact of these CoPs on organizational and individual performance. Hildreth and Kimble (2004) conclude that there is growing demand to supplement the evaluation qualitative results with more formal measurements of financial benefits and costs of communities. In fact, measuring CoPs' value is seen as instrumental for communities to gain visibility and influence as well as to educate and guide their own development (Wenger et al. 2002). Table 16.4 is a summary of evaluation studies carried out in the last eight years measuring CoPs' performance.

Table 16.4   Performance studies of CoPs

Analysis methods	Literature sources	Impact on
IT investments vs time	Butler et al. (1997); Clare and Detore (2000); Downes and Mui (1998)	Time savings
ROI based on knowledge workers' perspectives	Davenport and Prusak (1998); US Navy (2001); Lesser and Storck (2001)	• Time/ cost savings • Customer satisfaction • Access to knowledge • Employee performance
Social network analysis	Schenkel et al. (2000)	• Social connectivity • Organization performance • Intellectual capital value
Balanced scorecard	Roberts (2000); Walsh and Bayma (1996)	• Social connectivity • Organization performance • Intellectual capital value
Competitive advantage analysis	Teigland (2000); Liedtka (2000)	• On-time customer performance
Intangible asset valuation	Edvinsson and Malone (1997); Lev (2001); Sveiby (1997)	• Social connectivity • Organization performance • Intellectual capital value
Anecdotes, time use and individual community organizational benefits	Fontaine and Millen (2004)	

Source: Fontaine and Millen (2004).

3.2   Social Collaboration Tools and Web 2.0 Technologies

Collaborative Internet technologies, also known as Web 2.0, are a highly topical subject area. International analyst group Gartner published the respected Gartner (2005) 'Hype Cycle Report', where technologies that enable the development of collaboration, next-generation architecture, and real-world Web are highlighted as being particularly significant. As a brief synopsis, the Gartner (2005) report assesses the maturity, impact and adoption speed of 44 technologies and trends over the coming decade. To compose this comprehensive assessment of technology maturity in the IT industry, more than 300 Gartner analysts evaluated more than 1600 information technologies and trends across more than 60 markets, regions and industries.

The resulting key collaboration technologies designed to improve productivity and transform business practices, as adopted by Gartner (2005), are:

Podcasting	Podcasting is an efficient method for delivering audio and video content to targeted audiences. Podcasting usually involves subscribing to radio programmes and delivering audio content to portable media players and PCs on demand, so that it can be listened to at the user's convenience. Gartner (2005) predict that podcasting will become a widely used corporate communications tool. Also it is predicted that podcast subscriptions will grow as the market for content continues to fragment, which will lead to a change in radio and TV content delivery.
Peer-to-peer (P2P)	Skype is the most widely known and adopted voice-over IP (VoIP) software. New vendor-proprietary P2P VoIP applications are under intense development even though security is a concern. Gartner (2005) predict that the technology will be important for collaborative and multimedia applications as well as low-cost communications for social and business purposes.
Desktop search	This application, also known as personal knowledge search, resides on the desktop and uses local processing power to provide search-and-retrieve functionality for local e-mail, data store, and documents. Google, Microsoft and Yahoo are competing for customer uptake, thus creating market hype.
Really simple syndication (RSS)	RSS is a simple data format that enables Web sites to inform subscribers of new content by bypassing the browser via RSS reader software. RSS is widely used for syndicating weblog content, but its potential is only beginning to be exploited for corporate use such as organizational messaging. Its simplicity makes it easy to implement and add to established software systems. Gartner (2005) predicts that RSS will be useful for content that users find as 'nice to know' rather than 'need to know'.
Wikis	Wikis are widely used as collaborative, authoring tools for online distributed communities, particularly those using open source projects. Wikis are text-based collaborative systems for managing hyperlinked

collections of Web pages, usually enabling users to change pages or comments created by other users. Gartner (2005) predicts that Wikis will affect collaborations and group authoring.

Corporate blogging   This involves the use of online personal journals by corporate employees, either individually or in a group, to further company goals. Its impact will be on projecting corporate marketing messages primarily and secondarily in competitive intelligence, customer support and recruiting.

The significance of collaborative tools (also termed Web 2.0 technologies) is further substantiated by the US *TIME* (2006) article ranking the 50 'coolest' websites in 2006. Most of the websites chosen as the top ones for 2006 are examples of so-called Web 2.0 sites, which give users tools to create and share content online. As *TIME* (2006) states, these social networking sites are 'next-generation sites offering dynamic new ways to inform and entertain, sites with cutting-edge tools to create, consume, share or discuss all manners of media, from blog posts to video clips'. Most prominent examples are MySpace, with 100 million users, and YouTube video-sharing website, recently bought by Google (2006), where 100 million videos are watched per day (BBC 2006). Other examples of the 50 'coolest websites' include:

- Pandora, which offers a radio audio streaming that plays the band selected by the user.
- The 9 videoblog with a daily video-streaming of news.
- Photomuse, a joint project by the International Center of Photography in New York City and the George Eastman House in Rochester (the world's oldest photography museum). Photomuse remains a work in progress.

Other statistical facts substantiating the significance of Web 2.0 technologies are published by the Dow Jones (2006) venture capital service that has published an investment insights report on Web 2.0. In this report it is shown that venture investing in Web 2.0 companies in 2006 were set to double in dollar terms over 2005. Interestingly, the total dollars invested in Web 2.0 companies by venture firms in 2006 were more than the amount in all previous years combined.

General characteristics and examples
While Web 2.0 has initially been embraced in consumer-facing applications, the infrastructure required to build these applications and the scale at

which they are operating suggest that social collaborative technologies will move into the enterprise space.

The best-known definition of Web 2.0 is provided by O'Reilly (2005a):

> Web 2.0 is the network as platform, spanning *all connected devices*; Web 2.0 applications are those that make the most of the intrinsic advantages of that platform: delivering software as a continually updated service that gets better the more people use it, consuming and remixing data from multiple sources, including individual users, while providing their own data and services in a form that allows remixing by others, creating networks through an 'architecture of participation', and going beyond the page metaphor of Web 1.0 to deliver rich user experiences.

In an effort to further formalize definitions and principles of Web 2.0, Musser and O'Reilly (2006) have most recently stated:

> Web 2.0 is a set of economic, social, and technology trends that collectively form the basis for the next generation of the Internet – a more mature, distinctive medium characterized by user participation, openness, and network effects.

In other words, Web 2.0 could be viewed as a computing platform able to serve end users with Web applications created by aggregated content and functionality from websites, thus overcoming isolated information silos. Web 2.0 could also be regarded as an approach of creating and distributing Web content itself, referred to as a social phenomenon characterized by decentralization of authority, freedom to share and reuse, and open communication, rather like using 'the market as a conversation'. Better-developed deep-linking Web architecture making it more categorized and organized in general, is another perception of Web 2.0. Some cynical views consider Web 2.0 as an occurrence of hype and excitement surrounding the possibilities of services and innovative web applications that previously gained a great deal of momentum around mid-2005.

O'Reilly (2005b) summarized the key principles he believed characterized Web 2.0 applications in the Web 2.0 conference such as: the Web as a platform; data as the driving force; network effects created by an 'architecture of participation'; innovation in the assembly of systems and sites composed by pulling together features from distributed, independent developers (a kind of 'open source' development); lightweight business models enabled by content and service syndication; the end of the software adoption cycle ('the perpetual beta'); software above the level of a single device; and leveraging the power of 'the long tail'.

Web 2.0 builds on the view that the Internet moves away from service providers offering static information pages towards facilitating the presence of end users as active, collaborative co-creators with the provision of connectivity tools that will allow an 'architecture of participation'. One other key Web 2.0 characteristic is the notion of 'social software', which facilitates the building and maintenance of virtual communities, self-expression, participation and dialogue through wikis, weblogs and other participatory forms (Wilson 2006). We could observe that the new thinking on the way the Internet is developing follows the trend of open innovation.

Previously, the term Web 2.0 was a synonym for the 'Semantic Web'. FOAF (Friend of a Friend) and XFN (XHTML Friends Network) are examples of social networking systems which, when combined with the development of tag-based folksonomies create a natural basis for a semantic environment when delivered through the medium of blogs and wikis.

So far Web 2.0 has been defined by associating it with products or companies that have embodied its principles. Examples of well-known Web 2.0 entities are Google Maps, Flickr, del.icio.us, digg, last.fm and Technorati.

Use of the Internet is increasingly oriented towards social networks and collaboration. Indeed, Web 2.0 sites act more as points of presence or user-dependent web portals than as traditional websites. Access to consumer-generated content facilitated by Web 2.0 brings the Web closer to Tim Berners-Lee's original concept of the Web as a democratic, personal and DIY medium of communication (Web 2.0 Wikipedia).

The debate on what Web 2.0 is goes on; however, it is generally accepted that Web 2.0 websites exhibit some basic characteristics, such as:

- Information should be easy to enter and retrieve from the system, the concept of a 'walled garden' should not exist with reference to the site.
- Users usually own their data on the site and can modify them at their convenience.
- Web 2.0 applications that can be used almost entirely through a web browser have been found to be some of the most successful and are known commonly by the term 'network as platform'.
- When a user requires data returns, they should not be static, as might have been the case with Web 1.0, but dynamic, and able to change based on variables associated with the user's query such as keywords or location.
- An 'architecture of participation' that allows users to add value to the application as they use it (Web 2.0 Wikipedia).

Figure 16.1 is presented as a clarification aid summarizing Web 2.0 characteristics.

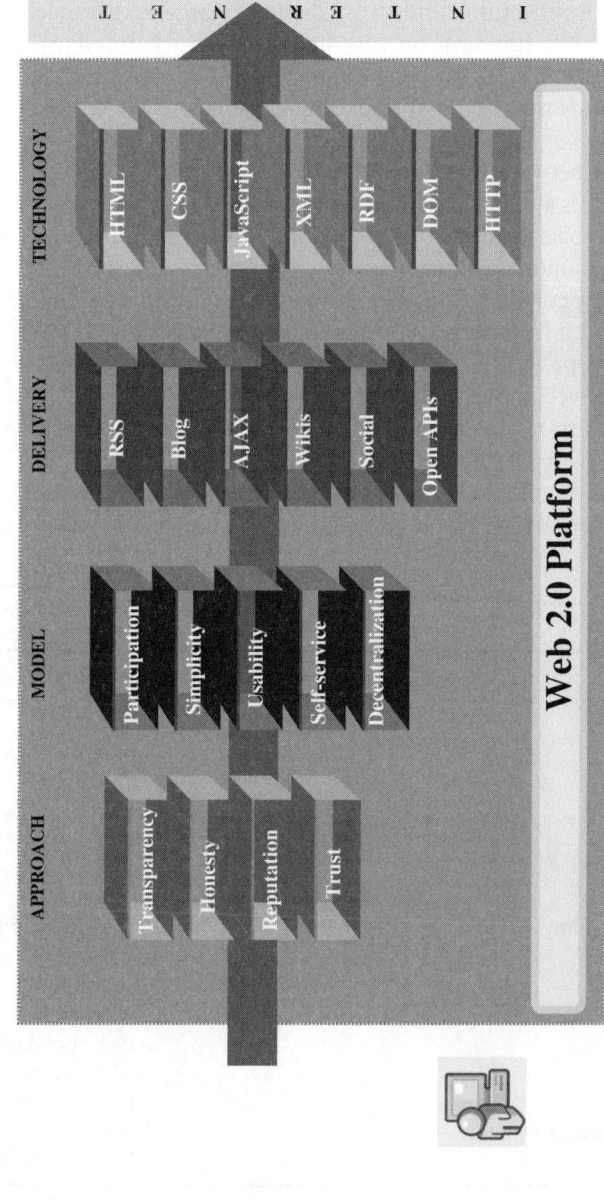

Source: Adapted from O'Reilly (2005a).

Figure 16.1 Common characteristics of the Web 2.0 Platform

375

Turoczy (2005) also argues that there is a visual design element to Web 2.0 that makes it stand apart from Web 1.0. He identified some common design features as employed by Web 2.0 websites such as large colourful icons, often with reflections and drop shadows, large text (especially in comparison with the emphasis on very small text in earlier design), diagonal hatch backgrounds, glossy three-dimensional elements, random highlights and call-outs in text and gradient backgrounds.

Basic differences between Web 2.0 and Web 1.0

Web 2.0 symbolizes a radical departure from a closed, data-rich, application-driven Internet society, to an open, trusting, service-based online society which provides a unique platform for developing new ways of working and, most importantly, collaborating. There are other key features which help to distinguish Web 2.0 from Web 1.0 technologies (Table 16.5).

As mentioned previously, Web 2.0 has been defined by associating the technology with web applications or companies that embody its principles. O'Reilly (2005b) proposed a number of practical examples to clarify the evolution from Web 1.0 to Web 2.0 technologies (Figure 16.2).

One of the most notable applications that demonstrate the Web 1.0 and Web 2.0 evolution are Netscape as opposed to Google. Netscape coined

Table 16.5   Core differences between Web 2.0 and Web 1.0 technologies

Web 1.0	Web 2.0
Static	Dynamic
Companies	Communities
Client-server	Peer to peer
HTML	XML
Home pages	Blogs
Portals	RSS
Stickiness	Syndication
Directories-Taxonomy	Tags-folksonomy
Wires	Wireless
Owning	Sharing
Publishing	Participation
Web forms	Web applications
Screen scraping	APIs, web services
Dial-up	Broadband
Content Management Systems	Wikis
Hardware costs	Bandwidth costs

Source: O'Reilly (2005b).

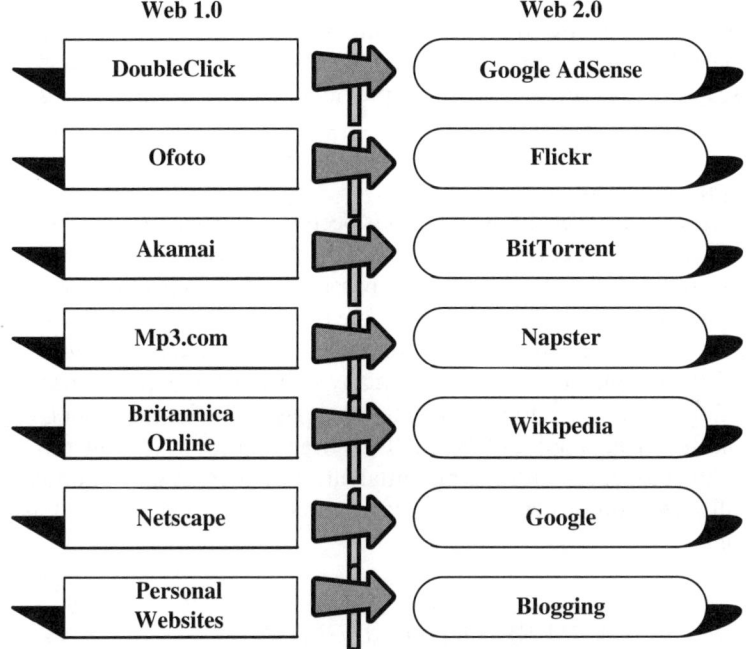

Web 1.0	Web 2.0
DoubleClick	Google AdSense
Ofoto	Flickr
Akamai	BitTorrent
Mp3.com	Napster
Britannica Online	Wikipedia
Netscape	Google
Personal Websites	Blogging

Source: Adapted from O'Reilly (2005b).

Figure 16.2   Web 1.0 and Web 2.0 websites

'the Web as platform' in harmony with the old software paradigm where the Netscape web browser product was a desktop application and the business strategy was to establish a market for high-priced server products. Netscape's 'webtop' would replace the desktop with information updates and applets pushed to the webtop by information providers who would purchase Netscape servers. By having control over standards for displaying content and applications in the browser, Netscape's aim was to attain the kind of market power Microsoft was enjoying in the PC market.

In contrast, Google always positioned itself as a web application free from software releases and licensing that was a free service. Licensing and software releases are irrelevant in Google as the software is never actually distributed but only run. The value of Google as a web application is proportional to the scale and dynamism of the data it helps to manage.

The core competence of Google is that it is a continuously improved, platform-independent application with a large collection of scalable add-ons. As O'Reilly (2006) states, 'Google isn't just a collection of software tools, it's a specialized database. Without the data, the tools are useless; without the software, the data is unmanageable.' Google positions itself between browser, search engine and destination content server, while Netscape aligns itself as a software provider prevalent in the 1980s such as Microsoft, Oracle, SAP.

Critical assessment

One criticism of Web 2.0 is that it has been transformed, reduced even, to a term that can 'mean whatever you want it to mean' and that it has no actual connection to the ideas it was based on. Many websites now declare they are 'Web 2.0' while only using some trivial and basic Web 2.0 features such as blogs or gradient boxes. Regardless of the extent of adoption of Web 2.0 technologies, the fact that many websites proclaim themselves as Web 2.0 indicates the appeal these technologies have for customers.

Another critical point of Web 2.0 has been that there are now too many Web 2.0 companies with the potential effect of market saturation as they are all attempting to create the same product but are lacking in business models.

4. FUTURE RESEARCH AGENDA

Understanding the challenges of linking the innovation agenda with the information systems research requires in turn an understanding of how modern science and industry use globally distributed collaborations involving large datasets. Thus far management research in open innovation and research in information systems have been conducted in separate silos, while in the global economy reality these issues need to be addressed in synergy, holistically. The open innovation theory has provoked many interesting questions about how organizations can benefit from innovations emerging outside the organizational boundaries (Chesbrough 2003). Even though Chesbrough (2003) talks about technologies that spur open innovation, this area is not accentuated in his work. We show how Web 2.0 technologies offer a platform of open innovation in terms of participation and collaboration between community users. In addition, the extensive CoP literature (Brown and Duguid 1991; Wenger and Snyder 2000; Wenger 2001) places its primary focus on learning aspects in intra-organizational communities. This chapter contributes to a more comprehensive thinking about open innovation theory by bringing together research on enabling information tools and knowledge transfer aspects of CoPs.

Future research issues within the area of enabling information infrastructures may include:

- the need to develop information system infrastructures that can be inherently flexible so as to adapt and align with science applications and potential problem solving for the future;
- the need to synthesize the theoretical underpinning of CoPs and social networks with information system development methodologies and Web 2.0 technologies.

REFERENCES

Abecker, A., Bernardi, A. and Sintek, M. (1999), 'Enterprise information infrastructures for active, context-sensitive knowledge delivery', ECIS '99 – The 7th European Conference on Information Systems, Copenhagen, Denmark, June.

BBC (2006), 'YouTube hits 100m videos per day', Online, http://news.bbc.co.uk/1/hi/technology/5186618.stm.

Brown, J.S. and Duguid, P. (1991), 'Organizational learning and communities of practice: toward a unified view of working, learning, and innovation', *Organizational Science*, **2**(1): 40–57.

Brown, J. and Duguid, P. (2000a), 'Balancing act: how to capture knowledge without killing it', *Harvard Business Review*, **78**(3): 73–80.

Brown, J.S. and Duguid, P. (2000b), *The Social Life of Information*, Boston, MA: Harvard Business School Press.

Brown, S.L. and Eisenhardt, K.M. (1995), 'Product development: past research; present findings, and future directions', *Academy of Management Review*, **20**(3), 343–78.

Butler, P., Hall, T., Hanna, A., Mendonca, L., Auguste, B., Manyika, J. and Sahay, A. (1997), 'A revolution in interaction', *The McKinsey Quarterly*, **1**: 423.

Chesbrough, H.W. (2003), *Open Innovation: The New Imperative for Creating and Profiting from Technology*, Boston, MA and Maidenhead, UK: Harvard Business School and McGraw-Hill.

Clabby, J. (2004), 'Clabby Analytics: The Grid Report, 2004 Edition', Grid Literature, IBM, Online, http://www-1.ibm.com/grid/grid_literature.shtml?&ca=Grid&me=W&met=rightnavliteraturelink.

Clare, M. and Detore, A. (2000), *Knowledge Assets: A Professional's Guide to Valuation and Financial Management*, San Diego, CA: Harcourt Professional Publishing.

Cross, R. and Baird, L. (2000), 'Technology is not enough: improving performance by building organizational memory', *Sloan Management Review*, **41**(3): 69–78.

Dahlbom, B. (2000), 'Postface: from infrastructure to networking', in C. Ciborra (ed.), *From Control to Drift*, Oxford: Oxford University Press, pp. 212–26.

Davenport, T.H. and Prusak, L. (1998), *Working Knowledge*, Boston, MA: Harvard Business School Press.

Dodgson, M., Gann, D. and Salter, A. (2005), *Think, Play, Do*, Oxford: Oxford University Press.

Dow Jones (2006), Report on 'Web 2.0 Investments', Online, http://www.ventureone.com/ii/YTD3Q06_Web2.0Release.xls.

Downes, L. and Mui, C. (1998), *Unleashing the Killer App*, Boston, MA: Harvard Business School Press.

Edvinsson, L. and Malone, M.S. (1997), *Intellectual Capital*, New York: HarperCollins.

e-science (2001), Research framework: 'About the UK e-Science Programme', Online, http://www.rcuk.ac.uk/escience/default.htm.

Evans, P. and Wurster, T. (1999), *Blown to Bits: How the New Economics of Information Transforms Strategy*, Boston, MA: Harvard Business School Press.

Fontaine, M.A. and Millen, D.R. (2002), 'Understanding the value of communities: a look at both sides of the cost benefit equation', *Knowledge Management Review*, **5**(3): 24–7.

Fontaine, M.A. and Millen, D.R. (2004), 'Understanding the benefits and impact of communities of practice', in Paul M. Hildreth and Chris Kimble (eds), *Networks: Innovation Through Communities of Practice*, Hershey, PA: Idea Group Inc., pp. 1–13.

Gartner (2005), 'Hype Cycle Report: Gartner highlights key emerging technologies in 2005 Hype Cycle', Online, http://www.gartner.com/press_releases/asset_134460_11.html.

Gongla, P. and Rizzuto, C.R. (2001), 'Evolving communities of practice: IBM Global Services experience', *IBM System Journal*, **40**(4): 842–62.

Google (2006), Online, http://www.google.com/press/pressrel/google_youtube.html.

Grid Research Integration Deployment and Support Centre (2006), Grid Projects Database, Online, http://www.grids-center.org/news/news_deployment.asp.

Harrison, T.M. and Zappen, J.P. (2005), 'Building sustainable community information systems: lessons from a digital government project', ACM International Conference Proceedings Series, vol. 89, *Proceedings of the 2005 National Conference on Digital Government Research*, pp. 145–50.

Hildreth, Paul M. and Kimble, C. (eds) (2004), *Knowledge Networks: Innovation Through Communities of Practice*, Hershey, PA: Idea Group Inc.

Joseph, J. and Fellenstein, C. (2004), *Grid Computing*, Upper Saddle River, NJ: Prentice-Hall PTR.

Kessler, E.H. (2003), 'Leveraging e-R&D processes: a knowledge-based view', *Technovation*, **23**, 905–15.

Kourpas, E. (2006), Report: 'Grid computing: Past, Present and Future: An Innovative Perspective', Online, www-1.ibm.com/grid/pdf/innovperspective.pdf.

Kubicek, H. and Wagner, R.M. (2002), 'Community networks in a generational perspective: the change of an electronic medium within three decades', *Information, Communication and Society*, **5**: 291–319.

Lesser, E. and Storck, J. (2001), 'Communities of practice and organizational performance', *IBM Systems Journal*, **40**(4): 831–41.

Lev, B. (2001), *Intangible: Management, Measurement, and Reporting*, Washington, DC: Brookings Institution Press.

Liedtka, J. (2000), 'Linking competitive advantage with communities of practice', in E.L. Lesser, M.A. Fontaine and J.A. Slusher (eds), *Knowledge and Communities*, Boston, MA and Oxford: Butterworth-Heinemann, pp. 133–50.

Luftman, J. and Brier, T. (1999), 'Achieving and sustaining business–IT alignment', *California Management Review*, **42**(1): 109–22.

McDermott, R. (1999), 'Why information technology inspired but cannot deliver knowledge management', *California Management Review*, **41**(4): 103–17.

McGough, A.S., Cohen, J., Darlington, J., Katsiri, E., Lee, W., Panagiotidi, S. and Patel, Y. (2006), 'An end-to-end workflow pipeline for large-scale grid computing', *Journal of Grid Computing*, **3**: 259–81.

Musser, J. and O'Reilly, T. (2006), Report: 'Web 2.0 Principles and Best Practices', Online, http://www.oreilly.com/catalog/web2report/chapter/web20_report_excerpt.pdf.

Nonaka, I. and Takeuchi, H. (1995), *The Knowledge-Creating Company*, Oxford: Oxford University Press.

O'Reilly, T. (2005a), 'Web 2.0: Compact Definition?', *O'Reilly Radar* blog, 1 October, URL (accessed April 2006): http://radar.oreilly.com/archives/2005/10/web_20_compact_definition.html.

O'Reilly, T. (2005b) , 'What Is Web 2.0? Design Patterns and Business Models for the Next Generation of Software', Online, last accessed October 2006, http://www.oreillynet.com/pub/a/oreilly/tim/news/2005/09/30/what-is-web-20.html.

Prusak, L. (2001), Online, http://www.creatingthe21stcentury.org/Larry-I-overview.html.

Qureshi, S. (2000), 'Organisational change through collaborative learning in a network form', *Group Decision and Negotiation*, **9**, 129–47.

Rad, P.F. and Levin, G. (2003), *Achieving Project Management Success Using Virtual Teams*, Boca Raton, FL: J. Ross Publishing.

Roberts, B. (2000), 'A balanced approach', *Knowledge Management* (September), 26–33.

Rothwell, R. (1994), 'Towards the fifth-generation innovation process', *International Marketing Review*, **11**(1): 7–31.

Schenkel, A., Teigland, R. and Borgatti, S.P. (2000), 'Theorizing communities of practice: a social network approach', paper submitted to the Academy of Management Conference, Organization and Management Theory Division. ID No. 11307.

Science and Innovation framework (2006), Report: Science and innovation investment framework 2004–2014: next steps report, Online, www.hm-treasury.gov.uk/media/1E1/5E/bud06_science_332.pdf.

Shapiro, C. and Varian, H.R. (1998), *Information Rules: A Strategic Guide to the Network Economy*, Boston, MA: Harvard Business School Press.

Sproull, L. and Kiesler, S. (1991), *Connections. New Ways of Working in the Networked Organization*, Cambridge, MA: MIT Press.

Sveiby, K.E. (1997), *The New Organizational Wealth: Managing and Measuring Knowledge-based Assets*, San Francisco: Berrett Koehler Press.

Teece, D.J. (1998), 'Capturing value from knowledge assets', *California Management Review*, **40**(3): 55–78.

Teigland, R. (2000), 'Communities of practice at an internet firm: netovation vs. on-time performance', in E.L. Lesser, M.A. Fontaine and J.A. Slusher (eds), *Knowledge and Communities*, Boston, MA and Oxford: Butterworth-Heinemann pp. 151–78.

Tidd, J., Bessant, J. and Pavitt, K. (1997), *Managing Innovation: Integrating Technological, Market and Organisational Change*, Chichester, UK: Wiley.

TIME (2006), '50 Coolest Websites', Online, http://www.time.com/time/2006/50coolest/index.html.

Turoczy, R. (2005), Report: 'Web 2.0 interface design checklist', *hypocritical.* Online, Retrieved 2006.

US Navy (2001), *Metrics Guide for Knowledge Management Initiatives*, United States Department of Navy, August.

Verna, A. (2000), 'Knowledge networks and communities of practice', *OD Practitioner*, Fall/Winter, **32**(4), available at http://www.odnetwork.org/odponline/vol32n4/knowledgenets.html.

Walsh, J.P. and Bayma, T. (1996), 'Computer networks and scientific work', *Social Studies of Science*, **25**: 661–703.

Wenger, E. (1998), *Communities of Practice: Learning, meaning, and identity*, Cambridge: Cambridge University Press.

Wenger, E. (2001), 'Supporting communities of practice: a survey of community-oriented technologies', version 1.3, March, available at http://www.ewenger.com.

Wenger, E. and Snyder, W.M. (2000), 'Communities of practice: the organizational frontier', *Harvard Business Review*, **78**(1): 139–45.

Wenger, E., McDermott, R.M. and Snyder, W.M. (2002), *Cultivating Communities of Practice*, Boston, MA: Harvard Business School Press.

West, J. and Gallagher, S. (2006), 'Patterns of open innovation in open source software', in H. Chesbrough, W. Vanhaverbeke and J. West (eds), *Open Innovation: Researching a New Paradigm*, Oxford: Oxford University Press, pp. 82–106.

Wiig, K. (1993), *Knowledge Management Foundations – Thinking about Thinking – How People and Organizations Create, Represent and Use Knowledge*, Arlington, VA: Schema Press.

Wilson, J. (2006), '3G to Web 2.0? Can mobile telephony become an architecture of participation?', *Convergence: The International Journal of Research into New Media Technologies*, **12**(2): 229–42.

Zhang, M.J. and Lado, A.A. (2001), 'Information systems and competitive advantage: a competency based view', *Technovation*, **21**: 147–56.

17. Conclusion

John Bessant and Tim Venables

The shift towards knowledge-based economic development is playing out as we write. Moving from local-level exchange of goods and services to 'world trade' emerged when trading nations were able not only to exploit natural advantages such as availability of raw materials and energy supplies, but also deploy technologies to minimize the cost of labour, transport and distribution. In the twenty-first century the volume of such trade has grown massively and includes not only physical goods and services but also a high proportion of intangible services. But significantly the balance of trade – especially in terms of manufactured goods – has moved hugely in favour of low wage cost and rapidly growing economies like India and China.

The arguments for knowledge economies have been well made but they risk missing a key part of the puzzle. If we are moving to an era of knowledge-based competition, then we need to understand far more about how this model works – and how to build and run a knowledge-based economy. It is too simplistic to see this as a matter of increasing the volume of resources on the input side, hoping that an increase in the absolute volume of knowledge produced will magically convert into competitive goods and services. Quite apart from the direct costs of doing so on the necessary scale, there is the difficulty that this investment will increasingly be diluted by the parallel investments of a growing number of players. Where the world is spending close to a trillion US$ and the number of serious players in the game is rising to include not only giants like India and China but niche players like Singapore and Ireland, the spoils are decreasingly likely to come simply to those who spend most.

So the case for spending more and more becomes hard to sustain and we move instead to looking at the *efficiency* with which this knowledge production converts to knowledge application – i.e. innovation. How does this process work and how could it be improved? As we have seen, this old issue becomes more complex as a question in a context where both supply-side and demand-side factors are shifting and interacting. Meeting the needs of customers is (relatively) easy when the demand exceeds supply (as Henry Ford found when developing the principles of mass production) – but what does a firm do when faced with fragmenting demand and increasing pressure for individually customized products and services? And how can firms understand markets when they are shifting in their composition – and where a growing element is inhabiting virtual as well as physical space (the online community My Space – a key driving force in markets of music and other media – had an estimated 106 million members in September 2006, equivalent to being the eleventh largest country in the world by population (between Japan and Mexico).

Another dramatic shift in the context has been increasing emphasis on 'open innovation' (Chesbrough 2003). Although – as Linus Dahlander and David Gann argue in Chapter 3 of this book – the concept is not new, the extent to which firms are moving towards exchanging and sharing knowledge previously held tightly in closed innovation systems is changing the innovation game in dramatic fashion. When there is recognition among even the largest firms (such as Procter and Gamble with a \$3 bn R&D spend) that 'not all the smart people work for us', then understanding the dynamics of knowledge flow and knowledge trading becomes a critical issue in building and sustaining a knowledge-based economy.

In part this reflects a shift in perspective triggered by the rising costs of doing R&D and the difficulties of tracking across increasingly wide technological, market and geographical frontiers. But it also reflects the huge expansion of the enabling technological infrastructure which facilitates innovation – a process called 'innovation intensification' by Gann, Salter and Dodgson in the book *Think, Play, Do* (Dodgson et al. 2005). Examples include the growth of knowledge networking sites such as www.innocentive.com, which acts as an 'eBay for ideas'. Here 'seekers' – including many global giants like P&G or Eli Lilly (founders of the site) – pose questions that 'solvers' – currently around 80 000 of them, drawn from the ranks of professors, students, retired scientists, professional consulting firms and even moonlighting employees of other organizations – try to answer. The model seems to work – Eli Lilly recently reported that over 50 per cent of its questions received worthwhile and implementable answers – giving them, in the process, massive leverage on their own in-house R&D capability.

WEALTH FROM KNOWLEDGE IN AN OPEN INNOVATION CONTEXT

This is the message from the first part of the book – that the innovation game itself is changing and we may need to let go of old myths like linear models or R&D investment and look instead at new ones based on knowledge trading as opposed to knowledge production and ownership. It's a global activity and different players need to manage their trading patterns in different ways – for example the Australian question is different to that of the UK, but both imply a much more active management of an inter-connected open innovation game. And policies to help make this happen move from intervention around picking winners or R&D leadership to more complex instruments that facilitate knowledge flow.

Chapter 2 by Pablo D'Este and Andy Neely shows that, as world knowledge production increases and becomes much more globally distributed, so the issue of flow and trading rises in prominence. In turn this suggests taking a lesson from the firm-level book and looking for regional and national innovation policy to think along 'open innovation' lines. But doing this successfully implies a need to develop many links and networks, and places emphasis on broking and bridging skills and mechanisms. And it does not remove the need to carry out R&D in order both to have something to trade with and also to develop the 'absorptive capacity' needed to understand and select what we wish to acquire.

This issue comes sharply into focus in the chapter by Mark Dodgson and John Steen, who look at the case of Australia. Australia's economy is based at present on successful resource exports, but there is an underlying need to develop a broader base of trade – and an even greater need for it to pay its dues as a member of the international R&D club. But despite its huge physical scale, Australia is a relatively small country, and committing R&D investments on the scale needed to keep up with major players like the USA will be difficult. The alternative strategy of open innovation, involving specialization in some key fields and intelligently trading and using knowledge in others, seems a more viable strategy in order to became a strong player in the global knowledge economy.

Much of the challenge comes down to *how* we manage this networking at different levels – and this questions the role of different actors in the innovation system. Even if those roles were effectively played in previous decades, the argument is increasingly that they need to change and explore additional and complementary ones. There is also scope for reviewing and challenging some of the myths that have grown up about how the innovation game is played. One such challenge is to the concept of universities as contributors to wealth creation by the process of spinning off technology-based businesses, rather as stars form from intense gas clouds. In Chapter 4, Alan Hughes reviews the evidence for this widely held view and suggests that there is little support for it. Even in the USA, the country where the model is widely believed to be cost-effective, the evidence shows that most productivity growth has come not from dramatic high-tech spin-offs but rather from much more mundane incremental changes, with the lead coming from the service sector. As Hughes puts it, 'US productivity growth owes more to

Sam Walton than to Bill Gates!' This does not mean that universities are irrelevant in the innovation system – far from it. But their role is much more valuable in creating 'public spaces' around which a variety of knowledge flow activities can take place – a view that echoes the findings of the UK's Lambert Review, which saw the value in diversity across universities rather than trying to create a set of institutions with the same research-intensive character as Oxford or Cambridge (Lambert 2003).

Tim Minshall's review of innovation support policies (Chapter 6) suggests that while they are getting sharper, there is still much to improve. Emphasis has moved from simply throwing money at the problem or trying to pick winners to more varied and targeted policies for innovation – but the underlying assumption is still somewhat linear, with assumptions made about the ways in which innovation happens. In particular, two stumbling blocks have not been overcome – assumptions about homogeneity among firms and a continuing albeit implicit belief in a linear model of the innovation process. The first gives rise to somewhat narrow views about where the market fails and how to target policy support, whereas the reality is of very different segments with differing needs and therefore requiring different support. And the second militates against the development of policies which support knowledge *flow* along widely spread different routes rather than clearly defined 'traditional' channels. On the positive side there is growing recognition of what the knowledge economy really means, especially the need to look at such knowledge flows – and there are encouraging signs of useful experimentation around this theme, such as the creation of knowledge transfer networks.

LEARNING TO MANAGE OPEN INNOVATION SYSTEMS

A core theme in the emerging discussion of the 'new' innovation context is how to enable effective knowledge *flows* as opposed to simply production or ownership. Different knowledge communities – users and producers – are increasingly being connected via different physical and virtual bridges to expand innovation, but at the same time we are better aware of how knowledge has always been combined – and recombined – in the innovation process. Andrew Hargadon's work, for example, highlights the ways in which this has been happening since the days of Edison and Ford, and how important such network models become in the twenty-first century (Hargadon 2003). And the seminal ideas of Tom Allen at MIT help us to understand the key role that agents such as 'gatekeepers' play in this process (Allen 1977). As we use tools to map knowledge flows – like social network analysis – so we realize that innovation has always been a connecting networked activity, a 'spaghetti' model where the skills in weaving the strands together are what matters.

This plays out across the economic life cycle of firms. Our understanding of what makes a knowledge-based firm form and grow has come a long way from the simple image of the 'Eureka!' moment or the light bulb flashing on above a cartoon character's head. We now realize – through studies such as those of Erik Stam and Elizabeth Garnsey – that there is a complex ecosystem that enables the flow of different knowledge types as firms move from start-ups to becoming serious economic players (Phelps et al. 2007).

Stam and Garnsey show that new knowledge in science and technology is an important and localized source of entrepreneurial opportunities, but public and corporate sector players do not necessarily commercialize this knowledge because they lack the vision or incentives. New firms arise that seek to do so, but the recognition of emerging opportunities and the mobilization of the necessary resources in order to create new economic value is achieved by only a few – but these high-growth start-ups are of more importance for economic growth than new firms in general. Corporate spin-offs are more likely to turn into these high-growth firms than university spin-offs. The international variation in realized firm growth is far greater than the variation in ambitious entrepreneurship, suggesting that entrepreneurs in certain countries face severe constraints preventing their firms from realizing intended growth.

Stam and Garnsey's empirical studies on new firm growth show that high levels of human, social and financial capital are enabling endowments, facilitating the growth of new business. Despite or because of the many problems facing new firms, among the successful innovations that are achieved by dedicated entrepreneurial teams there are some that have a very major impact on their firm and industry. The fact that these firms more often originate in the USA than any other country can partly be traced back to the huge (direct and indirect) government support to new technology-based firms in the USA.

In particular, Stam and Garnsey's studies stress the importance of seeing such new firm growth as a complex process but one in which the role of social capital is significant and where it is deployed in a highly networked fashion. Successful innovation is certainly not based on simple linear models of spin-offs of ideas.

Another myth that we need to challenge is that innovation-led growth is only the province of high-tech, high-growth firms. Even if we were able to accelerate their formation and increase their survival and growth chances, they remain only part of the national economic engine. A key question surrounds *how* the knowledge flow argument plays out among established businesses seeking to renew their product/service and process offerings. In the book we have looked at two extremes and find similar conclusions.

In the Adams and Bessant chapter (Chapter 10) the focus is on slow or non-adoption of knowledge proven elsewhere to have value. Is this simply stubbornness and stupidity among certain firms – and if it is, should it not be left to the market to weed out such ineffective players? Or does it require a more nuanced understanding of different user needs and concerns in the adoption process – with the implication that we need to build new types of connections along which knowledge flow can be enabled? Reviewing the evidence around diffusion of manufacturing innovations, much seems to depend on looking at different types of users and connections and needs to be seen in terms of a complex communication path – essentially a knowledge flow model. Improved understanding of this flow and the partners can help point to new and more effective policy for different intervention actors.

At the other extreme lies the challenge of breakthrough innovation – a frontier that depends on the firm acquiring new knowledge from unlikely and not normally explored places. Discontinuous shifts take place in organizational environments – new technologies or markets emerge, the regulatory framework moves, new business models redefine the rules of the game – and only by picking up on these early and acting can incumbents maintain their position. Tushman and Anderson gave us a window on this in their studies of 'competence-enhancing' and 'competence-destroying' technological change – essentially firms can ride out the waves of discontinuity but only if they have early warning systems that identify signals and then link to capacity (theirs or via alliances) to deal with the new picture (Tushman and Anderson 1987).

Other writers draw similar conclusions – that discontinuous change can pose a threat and an opportunity to established incumbents (Christensen 1997; Foster and Kaplan 2002; Philips et al. 2006). Importantly, it is not always new entrant firms that benefit under these conditions, but the ability of established players to survive – to ride the waves of discontinuous change – requires them to build a different organizational capability for finding and using new knowledge. The reality is that such new knowledge doesn't flow along existing channels – and even if the firm develops new early warning systems, the risk is that the signals it brings in may conflict with the established selection and resource allocation approaches and mindset (Henderson and Clark 1990; Tripsas and Gavetti 2000).

In their chapter Simone Ferriani, Elizabeth Garnsey and David Probert suggest that two fundamental enabling conditions can be identified. For an established firm wishing to engage in breakthrough innovation, first it must create an environment conducive to idea generation; and second it must have the fortitude and risk tolerance to persevere and allow the most promising ideas to have a fair chance to succeed. The first is the upstream creative challenge of developing the ability to 'see differently'. Since radical concepts often spring from the imagination of individuals or teams, the challenge is to create an organizational context where creativity may flourish. The second is the downstream implementation challenge of successfully applying and marketing the unique concept, which requires the ability to implement the concept by matching it to the actual market needs. Without the ability to see differently, the firm is unable to change the rules of the game, and without the ability to implement it, the firm will join the ranks of companies that failed to capitalize on their pioneering inventions such as Xerox with personal computers and EMI with scanner technology.

But *how* do they do this? The chapter by John Bessant and Bettina von Stamm reports on work with large incumbent firms seeking to develop modified or completely new routines in order to manage the search process. They identify 12 types of approach with considerable experimentation around each of them – underlining the need to develop new ways of managing knowledge flow in an era of increasingly open innovation.

In similar fashion the strategic selection and resource allocation issue requires new approaches. Ferriani, Garnsey and Probert identify three dominant 'selection regimes' – individual driven; user driven; application domain driven. They suggest that all of these approaches, in their various declinations, may prove viable in choosing among promising alternatives and thereby supporting radical innovations in large companies.

Whether a new firm start-up, a growing small business or an established incumbent seeking to renew its products, processes and services through infusion of new knowledge, one issue remains of critical importance. We often make assumptions about the organizational learning process, using concepts such as absorptive capacity as if they applied to the institutions themselves. But – as Hedberg and others remind us – it is not organizations that learn but rather the people within them (Hedberg 1981). So we need to explore more closely the ways in which such learning takes place – and what can block it. This theme is explored in the chapter by Sue Morton and Neil Burns, who review some of the literature on innovation adoption with particular reference to the psychological aspects. They conclude that in a climate where long-term employment can no longer be relied upon, the psychological contract between employer and employee plays a key role in innovation and organizational learning.

CLOSER CONNECTIONS . . .

Critical in all of this is the bridge building between sources of knowledge – universities and research and technology institutes (RTIs) – and the ways they connect. Again we see simplistic linear models moving to more complex understanding and new mechanisms. Importantly, we have known for some time that it is people who represent the best channels to move knowledge around – and so universities are most valued for their export of trained human resources. But we are also now seeing a more subtle and complex set of structural mechanisms that help extend the interconnections between universities and their potential users.

Understanding the richness and complexity of university–industry linkages is the focus of the chapter by Markus Perkmann and Kathryn Walsh. They argue that effective interaction depends on building long-term relationships rather than managing spot knowledge transactions – and that we need to understand the many different mechanisms through which such bridges and connections can be made. They suggest a continuum along which different approaches can be mapped and highlight the importance of some mechanisms – such as academic consultancy – which play an often underestimated role in developing and sustaining such relationships.

In their second chapter Pablo D'Este and Andy Neely (Chapter 13) show that the literature on knowledge transfer activities between university and business remains fragmented and inconclusive in two crucial respects. First, there is a lack of understanding both about *who* in academia interacts with industry, and *why* they interact. Second, there is the contested question of the impact of knowledge transfer activities on both the quality and quantity of academic research and innovation.

These two issues are particularly important for those responsible for formulating policy in the area of business and university collaboration. On the one hand, a better understanding of the mechanisms that shape the inclination of researchers to interact with industry, and the most pervasive channels through which such interactions occur, should contribute to the formulation of more nuanced and effective policies. On the other hand, by understanding the implications that university–business interactions have not only for business innovation, but also for the quality and quantity of academic research, we would be better positioned to infer some normative implications regarding which policies are most beneficial to encourage university–business interactions, and which policies could have unintended, counterproductive effects.

D'Este and Neely highlight three core implications for further research. First, while a large proportion of the existing literature is based on data provided by patent records, there is comparatively little evidence on the behaviour of academic researchers in connection with 'softer' forms of interaction with industry, such as joint research collaborations or consultancy work. This is particularly important in the light of some preliminary evidence highlighting that: (a) the proportion of academic researchers involved in patenting is small; and (b) compared to other channels of interaction, patenting is relatively infrequent.

Second, an important area of research that remains wide open concerns the factors that shape both the inclination of academic researchers to interact with business, and also the huge heterogeneity among academics in terms of their degree of engagement in knowledge transfer activities. Why do some academic researchers engage so heavily, while others, the large majority, interact very little or not at all? D'Este and Neely's review highlights the influence exerted by the institutions where academics conduct their research activities: departments, universities and the scientific communities to which researchers belong. While a large volume of literature has addressed the role played by technology transfer offices in facilitating technology transfer, we still know very little about what are the institutional settings most conducive to knowledge transfer activities. This is particularly important given that a large volume of knowledge transfer appears to occur outside the remit of the technology transfer offices.

But D'Este and Neely's review also stresses the importance of individual characteristics of researchers as important factors behind their propensity to, and degree of, engagement in knowledge transfer activities. We still need to better understand the motivations that drive researchers to interact, and we still know very little about the factors that lead to the development of the individual skills necessary to effectively integrate the worlds of scientific research and application.

D'Este and Neely's third set of unsolved issues relates to the impact of knowledge transfer activities on both the innovative activities conducted by business and the nature of academic research. While their review mainly discusses the latter, they also note that on the former there is much more we need to know on: (a) whether channels of interaction significantly differ by industry; and (b) what type of universities or departments are more likely to attract the attention of business.

They point out that there is little agreement among scholars on whether an increase in the engagement in knowledge transfer activities is causing a shift in the direction of research towards more applied (rather than fundamental research), or a lower quality and/or quantity of research. It is reasonable to conjecture that such relationships are likely to be contingent on a number of environmental factors. Further research should disentangle the circumstances that are more conducive to complementarities between knowledge transfer activities and academic research, and those that are more conducive to substitution effects.

There is some evidence that universities are moving beyond simplistic models of technology 'transfer' and ideas of spin-offs as the main route for linkage with industry. One area in which this is clear is the changing role of the specialized agencies established by universities to carry this mission forward. 'Technology transfer offices' (TTOs) have been growing in number but also – and enabled in part by building a professional community of practice around their operation and management – in the range and quality of their relationship building. In their chapter reviewing the UK experience, Hossein Sharifi and colleagues argue that developing the role of TTOs requires a better understanding of the underlying theories around knowledge flows and open innovation, and a flexibility to match the role of different universities within the user communities in which they operate. There are echoes here again of the Lambert Review argument for a 'mixed economy' of knowledge with different kinds of university providing different – and complementary – knowledge links.

Developing these relationships depends on universities redefining a role for themselves within a variety if different communities – essentially the 'public space' argument outlined in the early chapter by Alan Hughes. Doing so raises some interesting questions about the importance – or otherwise – of geographical proximity, especially in an increasingly networked and connected world. Are universities becoming virtual knowledge centres at the hub of innovation networks? Or is there still an important physical component to their activities and to defining the networks in which they operate? These themes are explored in the chapter by Kate Bishop, Toke Reichstein and Ammon Salter, who conclude that despite technological advances, proximity does still matter, especially for certain types of interaction. They support the view that the research activities of universities may shape the behaviour of local firms through various channels: the movement of skilled labour, access to recent research, or face-to-face contacts.

The extent to which technology opens up the possibilities of different or complementary forms of collaboration is explored further by Roula Michaelides and Dennis Kehoe in their chapter which looks particularly at information and communications technologies. There is little doubt that such advances enable closer working and also different ways of working collaboratively, whether between researchers or across different forms of collaboration with industry. These range from simple aids such as teleconferencing and collaborative work environments – essentially substituting current face-to-face models with their virtual equivalents – to exotic new tools and techniques for visualization, prototyping and experimentation. With rising interest – and investment – in more advanced systems, and projects such as the UK government's 'e-science' initiative, there is an urgent need to review the ways in which such technologies can be effectively used. Moves to more 'open' innovation systems can undoubtedly make use of such technological aids, but their use needs to be informed by understanding how knowledge flows and how communities of practice emerge and operate.

Their main conclusion is that making effective use of such technological possibilities requires us to pay close attention to the design and implementation issues – not a new thought but one with considerable relevance in making sure the new technologies support and enable effective collaboration. The risk is that they will be adopted as expensive 'fashion' accessories and under-utilized because the core relationships and interactions which they should support are not fully understood or reflected in their design.

MEETING THE INNOVATION CHALLENGE

Where next? Essentially this book offers a stocktaking and mapping exercise, trying to identify some of the emerging challenges in managing innovation in a rapidly shifting global context. We have only scratched the surface of the many issues being raised as we move to an era of 'open innovation' in which the management of knowledge flows becomes critical. But the evidence presented in the book suggests the emergence of a new game within which we need to upgrade and develop our innovation solutions – the fifth-generation story. We've drawn mainly from literature and secondary reviews here, but have highlighted a rich and exciting empirical research agenda. A key challenge for researchers in academe, policy and practice worlds is to pick up the 'open innovation' challenge – not just in terms of the content of their work, but also in the ways they themselves study and develop new approaches for effective innovation management.

REFERENCES

Allen, T. (1977), *Managing the Flow of Technology*, Cambridge, MA: MIT Press.

Chesbrough, H. (2003), *Open Innovation: The New Imperative for Creating and Profiting from Technology*, Boston, MA: Harvard Business School Press.

Christensen, C. (1997), *The Innovator's Dilemma*, Cambridge, MA: Harvard Business School Press.

Dodgson, M., Gann, D. and Salter, A. (2005), *Think, Play, Do: Technology and Organization in the Emerging Innovation Process*, Oxford: Oxford University Press.

Foster, R. and Kaplan, S. (2002), *Creative Destruction*, Cambridge, MA: Harvard University Press.

Hargadon, A. (2003), *How Breakthroughs Happen*, Boston, MA: Harvard Business School Press.

Hedberg, B. (1981), 'How organisations learn and unlearn', in H. Nystrom and W. Starbuck (eds), *Handbook of Organisation Design*, Oxford: Oxford University Press, pp. 3–27.

Henderson, R. and Clark, K. (1990), 'Architectural innovation: the reconfiguration of existing product technologies and the failure of established firms', *Administrative Science Quarterly*, **35**: 9–30.

Lambert, R. (2003), *Lambert Review of Business–University Collaboration*, London: HM Treasury.

Phelps, R., Adams, R.J. and Bessant, J. (2007), 'Models of organizational growth: a review with implications for knowledge and learning', *International Journal of Management Reviews*, **9**(1): 53–80.

Philips, W., Noke, H., Bessant, J. and Lamming, R. (2006), 'Beyond the steady state: managing discontinuous product and process innovation', *International Journal of Innovation Management*, **10**(2): 175–96.

Tripsas, M. and Gavetti, G. (2000), 'Capabilities, cognition and inertia: evidence from digital imaging', *Strategic Management Journal*, **21**: 1147–61.

Tushman, M. and P. Anderson (1987), 'Technological discontinuities and organizational environments', *Administrative Science Quarterly*, **31**(3): 439–65.

Index

Titles of publications are in *italics*.